COMPUTATIONAL NEUROSCIENCE

REALISTIC MODELING FOR EXPERIMENTALISTS

METHODS & NEW FRONTIERS IN NEUROSCIENCE

Series Editors
Sidney A. Simon, Ph.D.
Miguel A.L. Nicolelis, M.D., Ph.D.

Published Titles

Apoptosis in Neurobiology
Yusuf A. Hannun, M.D. and Rose-Mary Boustany, M.D.

Methods of Behavior Analysis in Neuroscience
Jerry J. Buccafusco, Ph.D.

Methods for Neural Ensemble Recordings
Miguel A.L. Nicolelis, M.D., Ph.D.

Neural Prostheses for Resoration of Sensory and Motor Function
John K. Chapin and Karen A. Moxon

COMPUTATIONAL NEUROSCIENCE

REALISTIC MODELING FOR EXPERIMENTALISTS

Edited by
Erik De Schutter, M.D., Ph.D.
Born Bunge Foundation
University of Antwerp
Antwerp, Belgium

CD-ROM by Robert C. Cannon, Ph.D.
Institut de Neurobiologie de la Méditerranée
INSERM, Unité 29
France

CRC Press
Boca Raton London New York Washington, D.C.

Library of Congress Cataloging-in-Publication Data

Computational neuroscience : realistic modeling for experimentalists / edited by Erik de Schutter.
p. ; cm. — (Methods & new frontiers in neuroscience series)
Includes bibliographical references and index.
ISBN 0-8493-2068-2 (alk. paper)
1. Computational neuroscience. I. Schutter, Erik de. II. Methods and new frontiers in neuroscience.
[DNLM: 1. Models, Neurological. 2. Computational Biology. WI 20 C738 2000]
OP357.5 .C636 2000
573.8'01'5118—dc211 00-060806
CIP

Direct all inquiries to CRC Press LLC, 2000 N.W. Corporate Blvd., Boca Raton, Florida 33431.

International Standard Book Number 0-8493-2068-2
Library of Congress Card Number 00-060806
Printed in the United States of America 1 2 3 4 5 6 7 8 9 0
Printed on acid-free paper

Series Preface

Our goal in creating the *Methods & New Frontiers in Neuroscience Series* is to present the insights of experts on emerging experimental techniques and theoretical concepts that are, or will be, at the vanguard of neuroscience. Books in the series will cover topics ranging from methods to investigate apoptosis, to modern techniques for neural ensemble recordings in behaving animals. The series will also cover new and exciting multidisciplinary areas of brain research, such as computational neuroscience and neuroengineering, and will describe breakthroughs in classical fields like behavioral neuroscience. We want these books to be what every neuroscientist will use in order to get acquainted with new methodologies in brain research. These books can be given to graduate students and postdoctoral fellows when they are looking for guidance to start a new line of research.

The series will consist of case-bound books of approximately 250 pages. Each book will be edited by an expert and will consist of chapters written by the leaders in a particular field. The books will be richly illustrated and contain comprehensive bibliographies. Each chapter will provide substantial background material relevant to the particular subject. Hence, these are not going to be only "methods books." They will contain detailed "tricks of the trade" and information as to where these methods can be safely applied. In addition, they will include information about where to buy equipment, Web sites that will be helpful in solving both practical and theoretical problems, and special boxes in each chapter that will highlight topics that need to be emphasized along with relevant references.

We are working with these goals in mind and hope that as the volumes become available, the effort put in by us, the publisher, the book editors, and individual authors will contribute to the further development of brain research. The extent to which we achieve this goal will be determined by the utility of these books.

Sidney A. Simon, Ph.D.
Miguel A. L. Nicolelis, M.D., Ph.D.
Duke University
Series Editors

Foreword

In theory, theoretical and practical neuroscience should be the same. In practice, they are not. Everybody knows that to bring about their convergence, theoretical neuroscientists should become familiar with the experimental literature and the practical aspects of neuroscience research. Lately it has become increasingly clear that experimental neuroscientists must also become more familiar with theoretical tools and techniques. Experimentalists have the advantage of knowing that they work on the real nervous system. However, only half of the experiments can claim to be the real nervous system. The other half is the working model on which the experiment is based. The informal models employed by experimentalists in the design of experiments and interpretation of results are becoming too complex to remain informal.

Experimental neuroscientists, even those studying subcellular events such as synaptic transmission and intracellular signaling, find themselves with working models consisting of tens to thousands of interacting components, each with its own dynamics. The components are nearly always engaged in interactions forming feedback loops, and they quickly evade attempts to predict their behavior using the experimentalist's traditional intuitive approach. Luckily for experimentalists wishing to escape from this predicament, the tools of formal model building have become much more accessible. Skill at cracking the tough integrals is still valued, but not truly essential for the experimentalist trying to refine his working hypothesis into a formal model. It has become possible (at least in theory) for every trained scientist to begin an experimental project with a simulation of the experiment. This formal model can be manipulated to determine if the ideas behind a complex project are consistent and if all the possible experimental outcomes are interpretable. After the experiment, the model makes a quantitative description to which the results may be compared. The model becomes a curve to place over the data points. In addition to the advances in availability of fast computers and modeling software, experimentalists wishing to generate formal models look for examples of successful work by others. Of course, there are the traditional examples of experimental model making, for example, the series of papers on the action potential by Hodgkin and Huxley. While these papers are still outstanding examples of how to proceed in general, the computational techniques employed there are increasingly out of date. Experimentalists may additionally choose to study some or all of the examples in this book. They are concerned with problems of current interest, spanning the range from molecular models of single ion channels and interacting members of intracellular signaling pathways to complex models of active dendrites and networks of realistic neurons.

Practical modeling involves some methodological concerns such as accuracy and efficiency of computation, fitting multidimensional experimental data to theoretical outcomes, and selection and correct use of simulation software. The chapters cover a theoretical issue within the context of a practical experimental problem, and they include methodological information on modeling related to the individual problem. Each contains a lesson in generation and evaluation of a formal model, but also presents contemporary findings within the context of a problem of current interest. They will be valuable to both experimentalist and theoretician, but especially to those seeking to escape this classification.

Charles Wilson, Ph.D.
University of Texas
San Antonio, Texas

Introduction

Over the last decade computational neuroscience has entered the mainstream. Some of us still remember the early 1980s when one had to explain to the average neuroscientist what modeling was about (not to be confused with the fashion industry). Nowadays the Society of Neuroscience has poster sessions on modeling of specific brain systems and computational neuroscience meetings attract hundreds of participants. Several specialized institutes, graduate courses, and summer schools have been created and targeted funding programs exist in most industrial nations.

Nevertheless, modeling too often remains the prerogative of specialized research groups or of theoreticians leading an isolated life at the fringe of an experimental lab. This segregation of experimentalists and theoreticians in the field of neuroscience impedes scientific progress. Therefore, the contributors to this book strongly believe that experimentalists should become part-time modelers, as well. If training programs give neuroscience students the opportunity to play with passive and active computer models of neurons (the neuroscience equivalent of Lego™ and great fun, too), this practical experience may lead to application of models during later research projects.

While self-evident to physicists, not every biologist realizes that the use of quantitative models is a fundamental ingredient of all exact sciences. The more complex the system under investigation, the higher the probability of counterintuitive findings and the higher the need for quantitative models. Given that the brain is one of the most complex systems being studied, modeling is particularly important in neuroscience. Surprisingly some neuroscientists still believe that they can do without modeling, forgetting that they are already using a qualitative mental model, such as, for example, "this calcium current is important in bursting" or "this synapse is uniquely placed to influence the cells' firing pattern," etc. Quantifying such a mental model and making it explicit in mathematical expressions is as important toward proving or falsifying a hypothesis as demonstrating that blocking the calcium current stops bursting. The model may, for example, show that the system is not completely understood ("what terminates the bursting?") or that intuition about it was wrong ("low input impedances at the site of contact make the synapse ineffective despite its unique place.").

We hope that this book will help many experimentalists to take their first steps along this exciting, though sometimes arduous, path. It is conceived to help both the beginning modeler and the more advanced one, but it assumes that you are already a neuroscientist. Different from many other computational neuroscience books, we do not explain the basic neuroscience needed to understand the examples, though we provide references to the basic literature in case you need to refresh your memory. Conversely, not many assumptions are made about your mathematical skills. We have tried to keep the mathematics to an introductory level, though this was not entirely possible in Chapters 3, 4, and 12. To help you get started, the first chapter introduces the basic mathematical skills required to understand the most important tools of the trade, differential equations and parameter optimization.

The rest of the book is organized in an increasing scale of description: from molecular reactions, to synapses and cells, then to networks, and finally to simulations of musculoskeletal systems. While every chapter can be read separately, they are extensively cross-referenced. We have minimized overlap between the chapters and the later ones build on the earlier ones. A novice in the field should read them consecutively; more experienced modelers may read back and forth. The experienced modeler will also enjoy the extensive coverage of topics which have not been included in other computational neuroscience methods books: molecular reactions, stochastic modeling of synapses, neuronal morphology, including its development, and musculoskeletal systems. In between, one finds the more classical topics of diffusion, Hodgkin–Huxley equations, passive and active compartmental models, small circuits and large networks. But in each of those chapters, the authors have provided unique insights or novel approaches which have not been covered elsewhere.

All chapters have in common the fact that they are primarily geared toward realistic models, i.e., models which represent the simulated system in detail and use many parameters to do so. This does not imply that we believe such models are intrinsically better than more abstract models with only a few parameters, in fact several chapters (3, 7, 10, and 11) explicitly describe more abstract models too. But for the experimentalist, realistic models are more accessible because their parameters often correspond to directly measurable quantities and the mathematical tools are highly standardized. Therefore we believe that this class of models is most likely to be useful to the readers of this book and that it is worthwhile to devote an entire book to this approach.

And then there is, of course, the CD-ROM. Have you inserted it in your drive yet? On almost every computer platform you should be able to start playing immediately. The contributors have worked very hard to make their chapters come alive on the CD-ROM. The software provided allows you to get an intuitive feeling for the equations and models described and even allows you to start your own modeling project. We have also set up a Web site *www.compneuro.org* which contains updates to the software and additional examples and models.

If we succeed in our mission, soon you will be the proud parent of a realistic model. You may then discover that not everyone approves of your modeling effort. Criticisms can be divided into two categories: those by die-hard experimentalists and those by true theoreticians. The first will typically ask, "How can you be sure your model is correct?" The reasoning developed above should already provide a good motivation why modeling is useful and hopefully your own experience will add to the story. But over time I have become slightly more aggressive in my response to this recurring question. I now ask, "How can you be sure your experiment is correct?" and then point to barium currents being used to characterize calcium channels, slice experiments done at room temperature to predict circuit properties *in vivo,* etc. Almost all experiments are done on reduced preparations that are considered to be physical models of the real nervous system. This considered, the difference between a mathematical model and experiment becomes less extreme.

A more challenging criticism to respond to is "realistic models are useless because they do not provide insights in the underlying principles; analytical models are the only way to understand how the brain computes." Of course, there is a basic truth in saying that a realistic model may be as complex as the real preparation and therefore does not provide a simple, succinct description. However, the practice of using realistic models demonstrates that they provide a better understanding of the often counterintuitive dynamics of nervous systems. And while it would be ideal to translate these insights into more abstract model descriptions (e.g., Chapter 11), it may be too early to do so in the many instances where the system is not entirely understood. Because, after all, how can we be sure the analytical model describes the relevant properties? Just imagine using a rate coding model in a system where spike timing is important. Realistic models suffer less from this problem because they are based on more comprehensive descriptions of the biophysics and biochemistry of the system.

Finally, I wish to thank the many people who made this book possible. This includes, of course, all the contributing authors who did a great job and patiently accepted a "rookie" editor. Sadly, one of the invited authors, Joel Keizer, passed away before the writing started, but fortunately, Greg Smith carried the load by himself. Robert Cannon did much of the hard work, including writing the simulator code and creating most of the tutorials on the CD-ROM. Without Robert or the series editors, Drs. Sidney Simon and Miguel Nicolelis, who were so kind to ask me to be an editor, you would not be reading this book now. Last but not least, I thank Barbara Norwitz, Publisher at CRC Press, who was not only a great help, but was also so kind to accept several delays in manuscript submissions.

Erik De Schutter, M.D., Ph.D.
University of Antwerp
Antwerp, Belgium

Material Available on the CD-ROM

The CD-ROM contains images, data files, simulation scripts, and software for the models described in the text. It is organized as a self-contained Web site with one section for each chapter in the book. To view the *Contents Page*, direct your Web browser at the file index.html which can be found in the top directory of the CD-ROM.

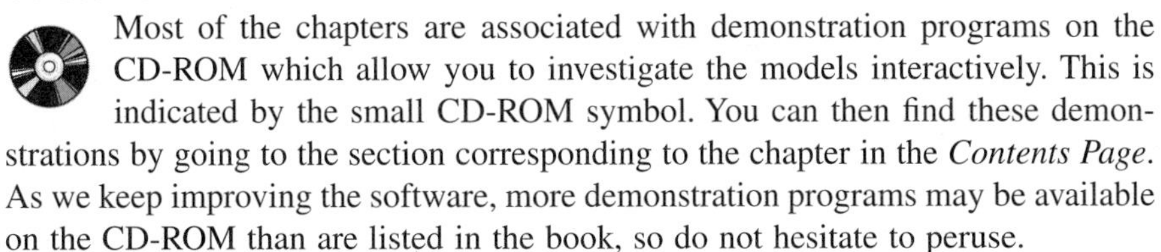

Most of the chapters are associated with demonstration programs on the CD-ROM which allow you to investigate the models interactively. This is indicated by the small CD-ROM symbol. You can then find these demonstrations by going to the section corresponding to the chapter in the *Contents Page*. As we keep improving the software, more demonstration programs may be available on the CD-ROM than are listed in the book, so do not hesitate to peruse.

Most of the demonstration programs are written in *Java*, which allows you to run them on any computer platform. The easiest way to use them is through a Java-enabled Web browser provided it is compatible with Java version 1.1.7. This includes Netscape 4.5 or later for Windows and UNIX platforms and Internet Explorer 4 for Windows platforms. See the *Software* section on the *Contents Page* for further details about other operating systems.

The demonstration programs are *Java applets* derived from a larger interactive modeling package called *Catacomb*, also included on the CD-ROM in the *Software* section. Whereas the applets present only one view of a model and access to some of the parameters, *Catacomb* affords free exploration of its properties and even allows such models to be created from scratch through the graphical user interface. Java engines in Web browsers have program size and security restrictions and are often slower (ten times or more on some platforms) than stand-alone Java engines. It may therefore be more convenient to run the demonstrations directly from the *Catacomb* environment. This requires a platform specific Java engine which is available free for most systems. Full details of how to obtain and install Java, including the necessary files where permitted by copyright, can be found in the *Catacomb* installation instructions on the *Contents Page*. When running the stand-alone version of *Catacomb*, exactly the same demonstration models as in the Web version can be found under the "models" menu on the main window. Notice that you still need to go to the *Contents Page* for non-Java software and additional text or illustrations.

Several chapters also contain simulation scripts for the GENESIS (*http://www.bbb.caltech.edu/GENESIS/genesis.html*) or NEURON (*http://www.neuron.yale.edu/*) simulation environments. For convenience, parts of these Web sites are mirrored on the CD-ROM, including all material necessary to install the software and run the various simulation scripts.

Finally, we keep improving the software and the additional material available on the CD-ROM. You can find instructions on how to update your software in the *Software* section on the *Contents Page* or by going to our Web site at *http://www. compneuro. org*.

Robert C. Cannon, Ph.D.
Institut de Neurobiologie de la Méditerannée
INSERM, Unité 29
France

About the Editor

Erik De Schutter, M.D., Ph.D., was born in Antwerp, Belgium, in 1959. He earned his medical degree from the University of Antwerp in 1984. He went on to do a clinical residency in neuropsychiatry and a Ph.D. in medicine at the University of Antwerp. From 1990 to 1994, Dr. De Schutter was a postdoctoral fellow in the Computation and Neural Systems program of the California Institute of Technology. In 1994 he returned to the University of Antwerp, where he is currently a professor, to start the Theoretical Neurobiology Group.

Dr. De Schutter's research focuses on the function and operations of the cerebellar cortex. He is a computational neuroscientist who has contributed to several popular neural simulation programs. His published work includes single cell and network modeling studies and physiological and brain mapping studies of the cerebellum. He played a seminal role in starting and directing a series of European summer schools: the Crete and Trieste courses in computational neuroscience.

Contributors

Thomas M. Bartol, Ph.D.
Computational Neurobiology Laboratory
The Salk Institute
LaJolla, California

Upinder S. Bhalla, Ph.D.
National Centre for Biological Studies
Bangalore, India

Ronald L. Calabrese, Ph.D.
Department of Biology
Emory University
Atlanta, Georgia

Robert C. Cannon, Ph.D.
Institut de Neurobiologie de la Méditerannée
INSERM, Unité 29
France

Erik De Schutter, M.D., Ph.D.
Born–Bunge Foundation
University of Antwerp
Antwerp, Belgium

Alain Destexhe, Ph.D.
Unité de Neurosciences Integratives et Computationnelles
CNRS
Gif-sur-Yvette, France

Örjan Ekeberg, Ph.D.
Department of Numerical Analysis and Computing Science
Royal Institute of Technology
Stockholm, Sweden

Michael E. Hasselmo, Ph.D.
Department of Psychology
Boston University
Boston, Massachusetts

Andrew A.V. Hill, Ph.D.
Department of Biology
Emory University
Atlanta, Georgia

John Huguenard, Ph.D.
Department of Neurology and Neurological Sciences
Stanford University Medical Center
Stanford, California

Dieter Jaeger, Ph.D.
Department of Biology
Emory University
Atlanta, Georgia

Ajay Kapur, Ph.D.
Department of Psychology
Boston University
Boston, Massachusetts

Gwendal LeMasson, M.D., Ph.D.
Laboratoire de Physiopathologie des Réseaux Médullaires
Institut de Neurosciences François Magendie
Bordeaux, France

Reinoud Maex, M.D., Ph.D.
Born–Bunge Foundation
University of Antwerp
Antwerp, Belgium

Guy Major, Ph.D.
University Laboratory of Physiology
Oxford, United Kingdom

Gregory D. Smith, Ph.D.
Department of Mathematics
Arizona State University
Tempe, Arizona

Volker Steuber, Ph.D.
Born–Bunge Foundation
University of Antwerp
Antwerp, Belgium

Joel R. Stiles, Ph.D.
Biomedical Applications Group
Pittsburgh Supercomputing Center
Pittsburgh, Pennsylvania

Harry B.M. Uylings, Ph.D.
Netherlands Institute for Brain Research
Amsterdam, The Netherlands

Stephen D. van Hooser
Department of Biology
Emory University
Atlanta, Georgia

Arjen van Ooyen, Ph.D.
Netherlands Institute for Brain Research
Amsterdam, The Netherlands

Jaap van Pelt, Ph.D.
Netherlands Institute for Brain Research
Amsterdam, The Netherlands

Charles Wilson, Ph.D.
Division Life Sciences
University of Texas at San Antonio
San Antonio, Texas

Table of Contents

1 Introduction to Equation Solving and Parameter Fitting

Gwendal LeMasson and Reinoud Maex

CONTENTS

0-8493-2068-2/01/$0.00+$.50

1.1 INTRODUCTION

Modeling is casting systems into formula.[1] In this chapter we briefly introduce the mathematical tools that are essential to comprehend and apply the techniques described in later chapters. In particular, we focus on two aspects: differential equations, which describe the dynamics of a system (Sections 1.2–1.3), and optimization algorithms that can be used to obtain the parameters to such equations (Sections 1.4–1.5).

A system in a *steady state* can generally be described with *algebraic equations*. This means that the value of an unknown variable is completely defined by the applied input and by the values observed, or calculated, for the other variables. For example, the equation $V = I R_{in}$ expresses the electrical potential V over an ohmic membrane resistance R_{in} when applying a constant current I.

Biological systems are rarely in a steady state. Because of their intrinsic dynamics they reach the steady state imposed by a constant input only asymptotically with time. As a consequence, the state of the system depends on the time elapsed since the input was applied. When the inputs are fluctuating continually, furthermore, the values observed for a variable will keep changing with time, unless sophisticated equipment is used such as a voltage clamp circuit.

This time-dependent behavior of biological systems is described with *differential equations*. Differential equations express the *rate of change* of a variable as a function of the current status of the system, i.e., in terms of the values of the state variables and of the input (see e.g., Equation 1.1).

Differential equations are also used to describe the variation of a quantity along the spatial dimensions of a system. Examples in this book are particle diffusion (Chapter 3) and electrotonic propagation (Chapter 8). Particle or current injection at a single position can maintain a gradient of concentration or voltage even when a steady state is reached, i.e., when the concentration (Section 3.3) or voltage (Section 8.2) at each single position no longer varies with time.

1.2 HOW TO READ DIFFERENTIAL EQUATIONS

1.2.1 ORDINARY DIFFERENTIAL EQUATIONS (ODEs)

The simplest type of differential equation describes the evolution of a single dependent variable, say V, relative to a single independent variable, say t. Consider a system obeying

$$\frac{dV(t)}{dt} = \frac{-V(t) + I(t)R_{in}}{\tau}, \tag{1.1}$$

with the *ordinary derivative* of V with respect to t denoted and defined by

$$\frac{dV(t)}{dt} = \frac{dV}{dt} = \dot{V} = V' = V^{(1)} = \lim_{\Delta t \to 0} \frac{V(t+\Delta t) - V(t)}{\Delta t}. \tag{1.2}$$

Equation 1.1 specifies how the value of V decreases at a rate proportional to V itself, and how V increases proportionally to the applied input. Many biological systems can be described with this kind of differential equation, at least so long as their state variables remain bounded within certain intervals. Examples are a passive membrane around its resting potential (Chapter 8) and Ca^{2+} buffering in the excess buffer approximation (Section 3.2.1.2). For the sake of clarity, we consider here Equation 1.1 as a minimal model of an isopotential membrane compartment with V, t, and I denoting the transmembrane potential, time and the applied current, respectively. The parameters τ and R_{in} are the membrane's time-constant (in seconds) and resistance (in Ohm). From Equation 1.1 it follows that V changes at a rate inversely proportional to τ. We mention only that in reaction kinetics it is customary to use the inverse of τ, i.e., rate constants; time then appears in the denominator of the parameter's units (Chapters 2 and 3).

Mathematically spoken Equation 1.1 is an *ordinary differential* equation (ODE) because V is ordinarily differentiated with repect to a single independent variable t. The ODE is *first-order* because $V^{(1)}$ is the highest-order derivative, and *linear* because V and its derivative appear only in linear expressions, i.e., they are not raised to a power or multiplied with other dependent variables. Rearranging Equation 1.1 gives

$$\tau \frac{dV(t)}{dt} + V(t) = I(t)R_{in}. \tag{1.3}$$

The left- and right-hand side now describe the intrinsic dynamics of the system and the input applied to it, respectively. If the input is zero, the ODE is called *homogeneous*.

At steady state, V no longer varies with time, by definition. The steady state value V_∞ to which the variable V evolves during the application of a constant current I can therefore be calculated by putting to zero the time-derivative in Equation 1.1, giving

$$V_\infty = I\,R_{in}\,, \tag{1.4}$$

where V_∞ denotes the value of V after infinite time t. Hence, the algebraic equation at the beginning of this chapter can be regarded a special instantiation of the ODE Equation 1.1. From Equations 1.1 and 1.4, it is also clear that V always changes in the direction of its steady state: if V is larger than the steady-state value, its rate of change becomes negative (this is the right-hand side of Equation 1.1) and its value decreases.

As a modeler, we want to calculate the value of V at any time t. For the linear, first-order ODE Equation 1.1 with constant input I

$$\tau \frac{dV}{dt} = V_\infty - V(t), \tag{1.5}$$

the analytical solution is

$$V(t) = V_\infty - (V_\infty - V_0)e^{-\frac{t-t_0}{\tau}}, \tag{1.6}$$

where V_0 is the steady-state value of V before time t_0, when the constant input I is being applied. Figure 1.1 shows a graph of this solution, and a geometrical interpretation of how this solution satisfies Equation 1.5.

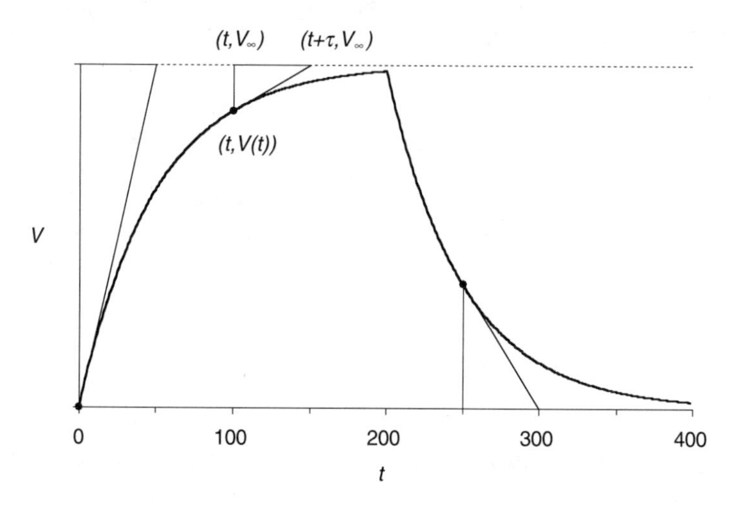

FIGURE 1.1 Graphical interpretation of ODE Equation 1.1. A constant input is applied from $t = 0$ to $t = 200$, during which V rises from its zero resting level V_0 to the new steady-state V_∞ (horizontal broken line). After the input is withdrawn, V returns to the old steady-state $V_\infty = V_0$. The tangent at each point $(t, V(t))$ of the graph has a slope equal to the derivative dV/dt, and intersects the steady-state curve $V = V_\infty$ at $t + \tau$. At three points rectangular triangles are drawn. The vertical side of each triangle is the difference between the actual value of V and the steady-state value V_∞, and hence represents the right-hand side of Equation 1.5. Elementary geometry shows that this vertical side has indeed a length equal to $|\tau dV/dt|$.

Of course, in a system described with multiple variables, a single variable V does not change in isolation. The change of each other variable relative to dt or dV must therefore be cast into differential equations too, resulting in a *system* of ODEs describing the entire model. For example, in an active membrane model, a small change in potential can change the conductance of the voltage-gated channels, which in turn affects the membrane potential (Section 5.2.1). Note that an nth order ODE can also be rewritten as a system of n first-order ODEs by a substitution of variables. For example, a second-order ODE $a\ddot{y} + b\dot{y} + cy = d$ is equivalent to a system of two first-order ODEs: $\dot{y} = z$ and $a\dot{z} + bz + cy = d$.

1.2.2 Partial Differential Equations (PDEs)

If Equation 1.1 describes the membrane voltage in a single, isopotential compartment, then extending this point neuron to a one-dimensional cable yields the following cable equation[3]

$$\tau\frac{\partial V(x,t)}{\partial t} = -V(x,t) + I(x,t)R_{in} + \lambda^2\frac{\partial^2 V(x,t)}{\partial x^2}, \tag{1.7}$$

with a *partial derivative* denoted and defined as

$$\frac{\partial V(x,t)}{\partial t} = \frac{\partial V}{\partial t} = V_t = \lim_{\Delta t \to 0}\frac{V(x,t+\Delta t) - V(x,t)}{\Delta t}. \tag{1.8}$$

Equation 1.7 is called a *partial differential* equation (PDE) because V is partially differentiated with respect to two independent variables t and x. The variable x represents the position along the cable and has the dimension of a length; the parameter λ has consequently also the dimension of a length.

Equation 1.7 describes how the local membrane potential V at a particular position x changes with time t through the effect of three processes, listed at the right-hand side (see also Section 8.2). The first two terms at the right-hand side of Equation 1.7 represent, as in Equation 1.1, the voltage decay due to leakage of current through the membrane and the local input, experimentally applied or intrinsically generated by ionic channels, respectively. The last term describes electrotonic propagation along the cable.

Extending the model to three spatial dimensions requires that the partial derivative of V with respect to x be replaced by

$$\left[\frac{\partial^2 V(x,y,z,t)}{\partial x^2} + \frac{\partial^2 V(x,y,z,t)}{\partial y^2} + \frac{\partial^2 V(x,y,z,t)}{\partial z^2}\right],$$

which is also denoted as

$$\left[\frac{\partial^2}{\partial x^2} + \frac{\partial^2}{\partial y^2} + \frac{\partial^2}{\partial z^2}\right]V(x,y,z,t),$$

where

$$\left[\frac{\partial^2}{\partial x^2} + \frac{\partial^2}{\partial y^2} + \frac{\partial^2}{\partial z^2}\right] = \nabla^2$$

is called the *Laplacian operator* in (orthogonal) cartesian coordinates[2] (see Section 3.3.1.1 for the Laplacian in spherical coordinates).

Note that the steady-state solution of the PDE Equation 1.7 is a second-order ODE

$$\lambda^2 \frac{d^2V(x)}{dx^2} = V(x) - I(x)\,R_{in}.$$

1.3 HOW TO SOLVE DIFFERENTIAL EQUATIONS

1.3.1 Ordinary Differential Equations (ODEs)

Equation 1.9 is the canonical form of a first-order ODE. The right-hand side of Equation 1.1 is abbreviated as $f(V,t)$, i.e., a function of the dependent variable V and the independent variable t

$$\frac{dV}{dt} = f(V,t). \tag{1.9}$$

Using Equation 1.9, it is possible to calculate the rate of change (dV/dt) of variable V for all possible values of V and t, but not the value of V itself. Indeed, the solution $V(t)$ remains ambiguous unless an "anchor" value of V at a particular time is given. This value is called, by convenience, the *initial condition*

$$V(t = 0) = V_0. \tag{1.10}$$

The problem described jointly by Equations 1.9 and 1.10 is called an initial-value problem. A solution gives the value of V at any time t.

In exceptional cases, such as for Equation 1.1 with a constant input IR_{in}, *analytical* integration rules yield for the dependent variable V an expression in terms of elementary functions of the independent variable t, like Equation 1.6. When stimuli are used that cannot be represented as integrable functions, or when the models become more realistic,[4] the differential equations need to be solved numerically. A numerical solution is a table of pairs $(t_i, V(t_i))$ for discrete time instants t_i.

In *numerical integration*, the limit constraint in the definition of a derivative (Equation 1.2) is relaxed. More particularly, the derivative is replaced with a *finite difference,* reducing the ODE Equation 1.9 to an algebraic equation of which $V(t + \Delta t)$ is the unknown variable. For example,

$$\frac{dV(t)}{dt} \approx \frac{V(t+\Delta t) - V(t)}{\Delta t} = f(V(t),t)$$

gives the forward-Euler rule

$$\begin{aligned} V(t+\Delta t) &= V(t) + \Delta t\, f(V,t) \\ &= V(t) + \Delta t V'(t) \end{aligned} \tag{1.11}$$

Equation 1.11 calculates V at $t + \Delta t$ through a linear extrapolation of V at t (see Figure 1.2, upper panel). At $t = 0$, Equation 1.11 reads

$$V(\Delta t) = V(0) + \Delta t\ f(V(0),0) = V_0 + \Delta t\ f(V_0,0)$$

using Equations 1.9 and 1.10. At $t = \Delta t$, $V(2\Delta t)$ is calculated starting from the value of $V(\Delta t)$. Repeating this procedure $k = t_i/\Delta t$ times yields the value of V at any discrete time t_i.

It is important to realize that relaxing the limit constraint of a derivative induces an *error*. The order of magnitude of this error can be evaluated indirectly by reference to the error induced by truncating a Taylor series. To understand this, remember that, in a small interval around t, a smooth function y can always be approximated to any desired accuracy by a polynomial so that $y(t + h) = a + bh + ch^2 + dh^3 + \ldots$. The value of a can be determined by evaluating the equation for $h = 0$, and hence equals $y(t)$. The value of b is determined by evaluating for $h = 0$ the equation that results when both sides are differentiated with respect to h. The coefficient of each higher-power term of h is found in succession by repeated differentiation and evaluation of the polynomial equation, yielding the well-known Taylor-series expansion

$$y(t+h) = y(t) + hy^{(1)}(t) + \frac{1}{2!}h^2 y^{(2)}(t) + \frac{1}{3!}h^3 y^{(3)}(t)$$

$$+ \ldots + \frac{1}{n!}h^n y^{(n)}(t) + \frac{1}{(n+1)!}h^{n+1} y^{(n+1)}(t) + \ldots$$

(Please note that superscripts to h indicate powers, whereas superscripts to the function y indicate orders of differentiation.)

If this series converges, then truncating it after the term containing the nth derivative $y^{(n)}$ is said to induce an error of order h^{n+1} (also denoted $O(h^{n+1})$). Because the principal truncated term is proportional to h^{n+1}, its value will decrease z^{n+1} times when h is taken z times smaller.

It is now possible to describe formally the error which solving Equation 1.11 induces at each integration step, i.e., the error on $V(t + \Delta t)$ assuming that there is no error on $V(t)$. To this end, the value of $V(t + \Delta t)$ calculated formally with the integration rule Equation 1.11 is compared to the theoretically correct value of $V(t + \Delta t)$ obtained with the Taylor-series formalism (Equation 1.12)

$$V(t+\Delta t) = V(t) + \Delta t V^{(1)}(t) + \frac{1}{2!}(\Delta t)^2 V^{(2)}(t) + \frac{1}{3!}(\Delta t)^3 V^{(3)}(t) + \ldots$$
$$\ldots + \frac{1}{n!}(\Delta t)^n V^{(n)}(t) + \frac{1}{(n+1)!}(\Delta t)^{n+1} V^{(n+1)}(t) + \ldots \tag{1.12}$$

From this comparison, Equation 1.11 is equivalent to the Taylor-series expansion of $V(t + \Delta t)$ truncated after the first derivative $V^{(1)}$. Relaxing the limit constraint in Equation 1.11, therefore, induces a *local truncation error* with a principal term

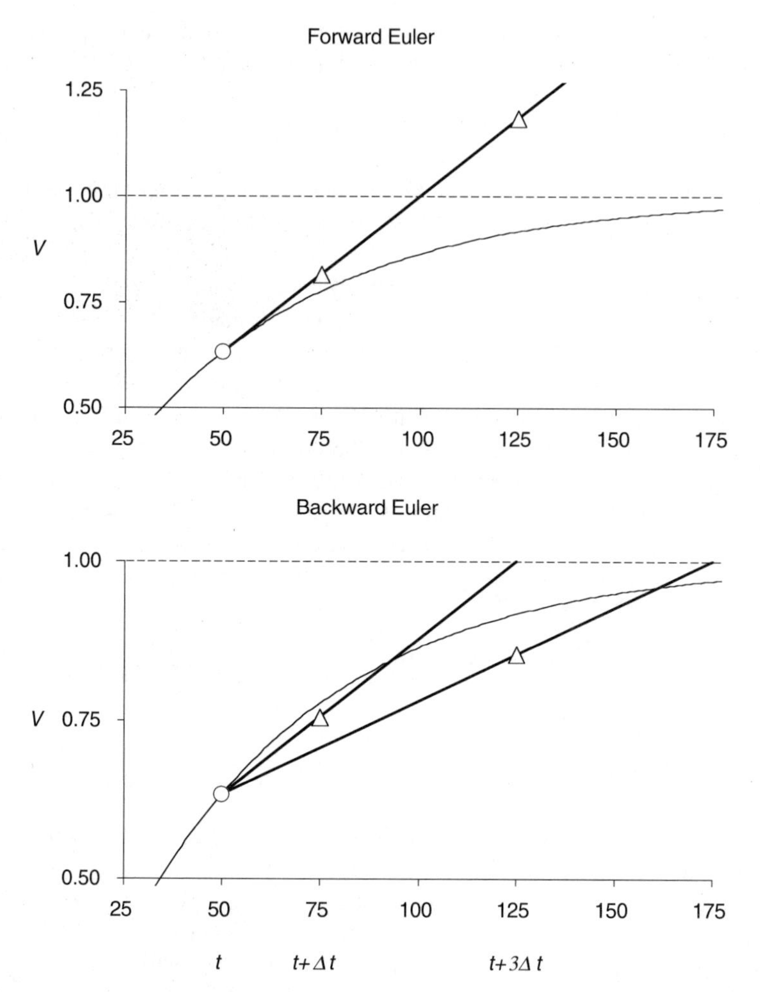

FIGURE 1.2 Numerical integration of ODE Equation 1.1 using Euler's rules (Equations 1.11 and 1.14). A single integration step calculates $V(t + \Delta t)$ from $V(t = 50)$ using step sizes $\Delta t = 25$ and $\Delta t = 75$. In forward-Euler (Equation 1.11; upper panel), the values $V(t + \Delta t)$ are found through linear extrapolation along the tangent through $(t, V(t))$. If $\Delta t > \tau$ (with $\tau = 50$ in this particular example), $V(t + \Delta t)$ overshoots the steady-state value V_∞ (horizontal dashed line). This does not only yield a very inaccurate value of $V(t + \Delta t)$, but causes also instability. In backward-Euler (Equation 1.14; lower panel), the new point $(t + \Delta t, V(t + \Delta t))$ lies on the straight line connecting $(t, V(t))$ with $(t + \Delta t + \tau, V_\infty)$. The value $V(t + \Delta t)$ never overshoots the steady-state V_∞, irrespective of the size of Δt.

proportional to $(\Delta t)^2$. In general, when the principal term of the difference between the calculated and the theoretical value of $V(t + \Delta t)$ is proportional to $(\Delta t)^{n+1}$, the rule is said to have a local truncation error of order $(\Delta t)^{n+1}$ (denoted $O((\Delta t)^{n+1})$), to be accurate through terms of order $(\Delta t)^n$ ($O((\Delta t)^n)$), or in brief to be *a method of order n*. The forward-Euler integration rule Equation 1.11 is therefore first-order.

As mentioned above, the local truncation error induced by applying Equation 1.11 is the error on $V(t + \Delta t)$ provided that the value of $V(t)$ is correct. However, only the value $V(t)$ at $t = 0$, given by Equation 1.10, is correct so that only $V(\Delta t)$, calculated in a single step from V_0, has an error equal to the local truncation error. This error inevitably propagates, and accumulates with the local error induced at each further integration step, so that all values of $V(t + \Delta t)$ at $t > 0$ are calculated from increasingly erroneous values of $V(t)$. The actual resulting error on $V(t + \Delta t)$, called the *global* truncation error, is therefore larger than the local truncation error. It has an order of magnitude that depends primarily on two factors: the order of magnitude of the local truncation error, which is generated at each integration step and which is $O((\Delta t)^2)$ for Equation 1.11, and the number of applied integration steps, i.e., the number of intermediate values of V calculated since $t = 0$. This number equals $t/\Delta t$ and is thus inversely proportional to Δt. From both dependencies together, the actual error on $V(t + \Delta t)$, or the global truncation error of Equation 1.11, appears to be of order Δt instead of order $(\Delta t)^2$. *In general, a numerical integration method of order n, using an integration step size Δt, produces values of V with an error approximately proportional to $(\Delta t)^n$.*[5–8]

Although Equation 1.12 might suggest that higher-order accuracy can be obtained only by evaluating higher-order derivatives, which are not explicitly given in the problem statement Equations 1.9 and 1.10, this is not strictly the case.[1] Higher-order accuracy can be achieved by iteratively evaluating the first derivative, i.e., the function f in Equation 1.9. In general, a method of order-n evaluates then n times the function f at every integration step. A popular method of this kind is the fourth-order Runge–Kutta rule

$$V(t+\Delta t) = V(t) + \Delta t \frac{(F_1 + 2F_2 + 2F_3 + F_4)}{6} \tag{1.13}$$

with

$$F_1 = f(V, t)$$

$$F_2 = f\left(V + \frac{\Delta t}{2} F_1, t + \frac{\Delta t}{2}\right)$$

$$F_3 = f\left(V + \frac{\Delta t}{2} F_2, t + \frac{\Delta t}{2}\right)$$

$$F_4 = f(V + \Delta t\, F_3, t + \Delta t)$$

The function f, which is the first derivative and hence the slope of V (see Equation 1.9), is now evaluated four times in the interval $[t, t + \Delta t]$: once at t (F_1) and $t + \Delta t$ (F_4), twice at $t + \Delta t/2$ (F_2 and F_3) (see numbered symbols in Figure 1.3). The resulting values F_1, F_2, F_3 and F_4 are then averaged to calculate $V(t + \Delta t)$. It can be shown[9] that this weighted sum in Equation 1.13 reduces the error on $V(t + \Delta t)$ to order five; hence the rule, due to Runge and Kutta, is of order four.

The above forward-Euler and Runge–Kutta rules are *explicit* rules. Indeed, in Equations 1.11 and 1.13, the unknown variable $V(t + \Delta t)$ appears only at the left-hand side of the equation, so that its solution is explicitly given. In *implicit* rules, $V(t + \Delta t)$ appears on both sides of the equation. If, for example, the following finite-difference approximation is used for the derivative

$$\frac{dV(t)}{dt} \approx \frac{V(t) - V(t - \Delta t)}{\Delta t} = f\big(V(t), t\big),$$

the first-order rule becomes

$$V(t + \Delta t) = V(t) + \Delta t\, f(V(t + \Delta t), t + \Delta t), \tag{1.14}$$

which is called the *backward-Euler* rule (Figure 1.2, lower panel). A hybrid of forward- and backward-Euler is the *trapezoidal* rule

$$V(t + \Delta t) = V(t) + \frac{\Delta t}{2}\Big(f(V, t) + f\big(V(t + \Delta t), t + \Delta t\big)\Big), \tag{1.15}$$

which is of order two, i.e., its local and global truncation error are $O(\Delta t)^3$ and $O(\Delta t)^2$, respectively.

In order to retrieve the unknown variable $V(t + \Delta t)$ from an implicit rule, an equation needs to be solved (e.g., Equations 1.14 or 1.15). When a *system* of ODEs is to be integrated, a system of equations has to be solved. In matrix notation, this requires an inversion of the matrix of coefficients, which is computationally hard and impractical, except when the matrix is sparse, i.e., when most of its coefficients are zero. In that case, efficient substitution rules exist to solve the system of equations. An example is the tri-diagonal matrix that results when the cable equation (Equation (1.7)) is discretized.[10,11,12]

As an illustration, we apply the above integration schemes to the ODE Equation 1.1, giving:

$$V(t + \Delta t) = \left(1 - \frac{\Delta t}{\tau}\right) V(t) + \frac{\Delta t}{\tau} I(t)\, R_{in}, \qquad \text{forward-Euler} \tag{1.16}$$

$$V(t + \Delta t) = \left[\frac{1}{1 + \dfrac{\Delta t}{\tau}}\right] V(t) + \left[\frac{1}{1 + \dfrac{\tau}{\Delta t}}\right] I(t + \Delta t)\, R_{in}, \qquad \text{backward-Euler} \tag{1.17}$$

$$V(t + \Delta t) = \left[\frac{1 - \dfrac{\Delta t}{2\tau}}{1 + \dfrac{\Delta t}{2\tau}}\right] V(t) + \left[\frac{1}{1 + \dfrac{2\tau}{\Delta t}}\right] \big(I(t) + I(t + \Delta t)\big)\, R_{in}, \qquad \text{trapezoidal} \tag{1.18}$$

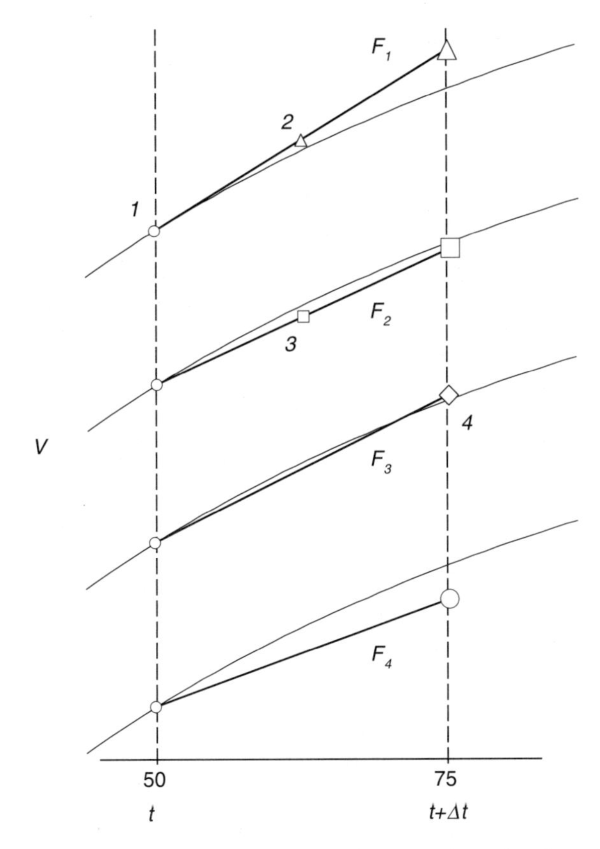

FIGURE 1.3 Numerical integration of ODE Equation 1.1 using a fourth-order Runge–Kutta rule (Equation 1.13). A single integration step from time t to $t + \Delta t$ comprises four linear extrapolations from the same starting point $(t, V(t))$ (open circle at t) but along lines with different slopes, labeled F_1 through F_4. The four extrapolations are drawn separately, but shifted vertically to avoid overlap. Each extrapolation yields a different presumptive value of $V(t + \Delta t)$, shown as a large open symbol at $t + \Delta t$. Eventually, the rule returns a value $V(t + \Delta t)$ that is a weighted mean of these four presumptive values, according to Equation 1.13. This mean point is not drawn, but lies exactly on the curved line. The values F_1, F_2, F_3, and F_4 are the slopes of the "tangents" to the curve at the points labeled *1*, *2*, *3*, and *4*, respectively. The slope of the tangent at a particular point can be calculated by evaluating the right-hand side of Equation 1.1. For this example, the tangents can also be determined geometrically following the same procedure as that used in Figures 1.1 and 1.2 (not shown).

Finally, a simple method for solving first-order ODEs was introduced by MacGregor,[13] and is commonly called exponential Euler (Chapter 2). Applied to our problem Equation 1.1, the rule reads

$$V(t+\Delta t) = e^{-\frac{\Delta t}{\tau}} V(t) + \left(1 - e^{-\frac{\Delta t}{\tau}}\right) I(t)\, R_{in}\,. \tag{1.19}$$

This explicit, first-order method is more accurate and stable than forward Euler, to which it reduces when the exponentials in Equation 1.19 are replaced by the first two terms of their series expansion,

$$e^{-\frac{\Delta t}{\tau}} \approx 1 - \frac{\Delta t}{\tau}.$$

By substituting $t + \Delta t$ for t in Equation 1.6, it can be seen that Equation 1.19 calculates the exact solution of Equation 1.1, irrespective of the size of Δt provided the input I is constant.

1.3.2 Partial Differential Equations (PDEs)

Integration of the second-order PDE Equation 1.7 introduces two more constants of integration. Two more constraints are therefore needed to produce a unique solution. These constraints are called *boundary conditions*. In the present case they can be given as the steady-state values $V_\infty(x)$ at two points x_1 and x_2 (usually the end-points of the cable), as the steady-state value $V_\infty(x)$ and its first derivative $dV_\infty(x)/dx$ at a single point x_1, or as combinations of these. The initial condition $V(x, t = 0)$ must now also be specified at each position x. Numerical integration of Equation 1.7 requires in addition a finite-difference approximation of the second-order partial derivative, for example

$$\frac{\partial^2 V(x,t)}{\partial x^2} \approx \frac{V(x+\Delta x,t) - 2V(x,t) + V(x-\Delta x,t)}{(\Delta x)^2}. \tag{1.20}$$

By summating the Taylor–series equations for $V(x + \Delta x)$ and $V(x - \Delta x)$, this finite-difference approximation can be shown to have a truncation error of $O((\Delta x)^2)$.[14] Equation 1.20 can then be incorporated into one of the temporal integration schemes from above. Combining Equation 1.20 with the trapezoidal rule Equation 1.15 yields the method first proposed by Crank and Nicolson.[14] This Crank–Nicolson method is second-order in both space and time, with a local truncation error of $O((\Delta t)^3 + (\Delta t)(\Delta x)^2)$ and a global truncation error of $O((\Delta t)^2 + (\Delta x)^2)$.[15] Note that when the cable is discretized into N–1 intervals of width Δx, a system of N equations with N unknowns ($V(x_i, t + \Delta t)$, $i = 1...N$) results.

1.3.3 Accuracy and Stability

As explained above, a solution obtained through numerical integration is expected to be more accurate if a higher-order rule and a smaller integration step are used. It should be kept in mind, however, that the order of a rule only states how the error scales with the integration step size (linear for first-order rules, quadratic for second-order rules, etc.). The actual magnitude of the error cannot be predicted, and depends on the actual problem to be solved.[7,16]

Moreover, the expected higher accuracy of higher-order methods cannot always be exploited to use larger integration steps. Indeed, *instability* puts an upper limit on the integration step size that can be used with any explicit rule. In the examples above (Equations 1.16 and 1.18), the coefficient of $V(t)$ becomes negative when the integration step is too large ($\Delta t > \tau$ in the forward-Euler rule, Equation 1.16, and $\Delta t > 2\tau$ in the trapezoidal rule, Equation 1.18). As a result, the value of V can change sign at every integration step, and grow unbounded (Equation 1.16) or show damped oscillations around the solution (Equation 1.18). When a system contains processes with different time constants, the integration step size must be small enough to ensure stability of the fastest process (i.e., with the smallest τ), even if this process produces such a negligible contribution that igoring it would hardly affect the accuracy of the solution. For such systems, called *stiff* systems, implicit integration methods are preferred, which do not suffer from instability (Figure 1.2). Finally, when PDEs like Equation 1.7 are solved with explicit rules, the size of the temporal integration step that is critical for stability (Δt_c) does not only depend on the system's parameters (λ and τ in Equation 1.7), but also on the size of the spatial integration step Δx. For the PDE Equation 1.7, Δt_c is proportional to $(\Delta x)^2$.[6,8]

1.4 FITNESS MEASUREMENTS

1.4.1 The Problem to Solve

In modeling equations, which describe changes in variables over time, often contain unknown parameters. For example, if we assume the probability p for a channel to be in the open state to be a sigmoid function of the membrane potential V, the mathematical form is (Section 5.2.2)

$$p(V) = \left(1 + e^{\frac{a-V}{b}}\right)^{-1}. \tag{1.21}$$

The problem is to determine the values of the parameters a and b for the specific channel we are interested in to model (a representing the membrane potential for half-open probability and b determining the slope of p in a). In this case, a and b are not experimentally measurable quantities and their estimation may require complex experimental protocols such as voltage clamp. Such experiments will measure some discrete points of the sigmoid relationship of Equation 1.21, but for modeling purposes we need to determine a and b in order to be able to evaluate p for any value of V.

This type of problem is a *fitting problem*, where one likes to describe a finite number of experimentally obtained data points by a continuous function. To address this question we decide which mathematical form to use (e.g., Equation 1.21) and identify the number and type (integer or float) of the parameters to be determined. We then make a first guess about the values of these parameters. Using these values we calculate a large number of points of the chosen function. On these points, a so-called *fitness function* is applied to numerically estimate "how good" the fit is compared to the experimental data. The idea is to use this fitness measurement to correct the initial guess of the parameters so that the next one will give a better fit. Thus, using an iterative scheme, we hope to converge to an acceptable solution for a and b.

One can easily understand that in this general procedure, two steps are of fundamental importance.

1. The fitness function must ideally return an *error value* equal to 0 if the fit is perfect and to 1 (for a normalized function) if the fit is extremely poor.
2. The *optimization procedure*, i.e., the correction scheme that guesses the parameter values from iteration to iteration, must ensure that the fitness coefficient will converge to a minimum value.

Fitness measurement and optimization protocols are not necessarily coupled and can be viewed as separate problems, although the performance of the optimization scheme in converging rapidly to a "good" solution clearly depends on the quality of the fitness function.

We first describe the general problem of fitness measurement, the objective of which is to use both the experimental data available and the model under construction to estimate how good a given model is at reproducing the data. Next, we give a short overview of some efficient optimization algorithms.

1.4.2 The Fitness Coefficient

Suppose that you can experimentally measure the relationship between an independent variable x and a dependent variable y. Standard numerical acquisition procedures give you a set of N data points (x_i, y_i), and the repetition of experiments will lead to points that are averaged responses associated with a standard deviation. This relationship could be a current-voltage (I/V) curve, a current-frequency (I/f) curve or a voltage-dependent activation function (V/p). Suppose that you have a model (some equation) suitable to reproduce this relationship with adjustable parameters. The problem is then to find good values for the parameters, so that the equation will give a good estimation of y for any value of x. A fitness measure, or fitness coefficient quantifies the goodness of this estimation. We first present some standard fitness measurements and then expose new directions in fitness measurement for the case of time-dependent functions.

1.4.3 Least Squares and Chi-Square Error Coefficients

One of the most commonly used fitness coefficients is known as *least squares*. Given the experimental relationship $y_{real}(x)$ and the model representing it $y_{model}(x)$, the least-square error is given by

$$E = \sum_{i=1}^{N} \left(y_{real}(x_i) - y_{model}(x_i) \right)^2 . \tag{1.22}$$

This measure is a *maximum likelihood* that quantifies the probability of getting the experimental data with the model for a given set of parameters. Several important statistical assumptions are made in this equation. First, it is assumed that the experimental measurement errors are independent for each experimental point y_i, and second that these measurements are distributed around the true model following a Gaussian with the same standard deviation σ at each measurement point. In the case of non-equal standard deviations, one should use an extended form of Equation 1.22, called the *chi-square estimator*

$$\chi^2 = \sum_{i=1}^{N} \left(\frac{y_{real}(x_i) - y_{model}(x_i)}{\sigma_i} \right)^2. \quad (1.23)$$

Several statistical methods exist[11,17] to deal with cases where the error measurement of y is not normally distributed (so-called *robust* statistical estimators, and Monte Carlo simulation for a known distribution reference).

If least-squares and chi-square coefficients are by far the most commonly used fitness functions for non-time-dependent functions, they are not very satisfactory for time-dependent functions. The direct fitting of time-dependent data series, such as intracellular recording samples, may lead to specific problems. The most crucial problem is to obtain the correct time phase between the data points and the model. This point is illustrated in Figure 1.4 where the least-square coefficient is calculated between two finite samples of intracellular recordings. Although the two sets of raw data are strictly identical, the coefficient strongly depends on the time phasing between them. In practice, using this coefficient for optimization purposes requires a procedure to phase-lock the data, which can be difficult for complex signals such as bursting, pseudo-periodic spiking or even a non-periodic signal. We can see from Figure 1.4 that for zero phase-locking, the coefficient is low, as expected for identical models, but that for anti-phase-locking the coefficient becomes much larger. Therefore, for time-dependent signals a different fitness measurement is needed.

1.4.4 A New Method Based on Phase Plane Trajectories

Let us take the case of a neuron for which the only experimental data available are finite time samples (seconds or minutes) of intracellular membrane potentials obtained during different current clamp protocols. It is always possible to build a generic model of the neuron we are studying (for instance based on the Hodgkin–Huxley formalism, see Section 5.2.1). This usually requires some hypothesis about the number of compartments required (Chapter 9) and about the type of the different voltage-dependent channels involved. Of course, many of the model parameters will be unknown. One can apply the general framework of any fitting and optimization scheme, but in this case, the fitness function will have to assign a fitness coefficient between two sets of membrane potentials sampled over time (one coming from experimental data, the other from the model's behavior).

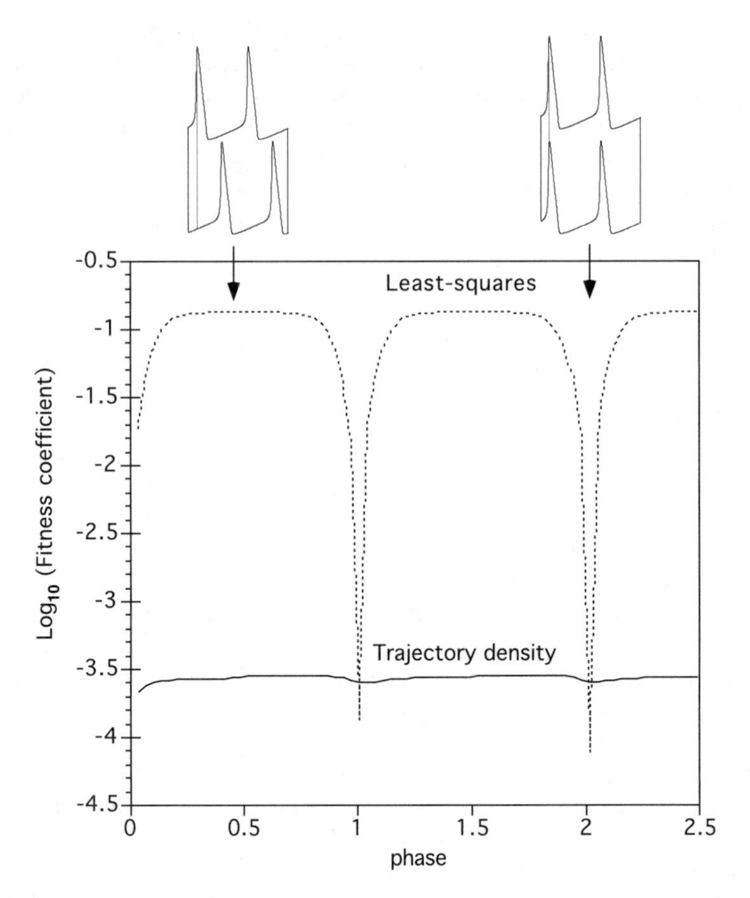

FIGURE 1.4 Phase dependence of the least-square estimator. The least-square estimator is computed between two identical current clamp time series produced by a spiking Hodgkin–Huxley model neuron (see Section 5.2.1). Depending on the phase relationship between the spikes, the least-square fitness value changes. For zero phase locking (phase = 0,1,2) the least-square value is close to 0 but for non-zero phase locking the coefficient increases rapidly. In solid line the same computation is made using a different estimator based on a trajectory density measurement (see text for details).

We recently developed a new method based on the concept of trajectory density in a phase plane to compute a robust fitness coefficient. The principle of a phase plane is to exclude the time parameter by plotting one time-dependent variable against another time-dependent variable. In the case of an intracellular sample $V(t)$, one possibility is to use as the other variable the first time-derivative $\dot{V}(t)$, so that the coordinates of points in this plane will be $(V, \dot{V})$. This derivative can be estimated using

$$\frac{dV}{dt} \approx \frac{V(t+\Delta t)-V(t)}{\Delta t}. \tag{1.24}$$

Color Figure 1.1* illustrates such a representation for a simple Hodgkin–Huxley spiking neuron (A1, A2). We see that the classical way of plotting the membrane potential is replaced by trajectories in the $(V, \dot{V})$ plane, where specific features of the signal, such as periodicity, become a closed loop that is easily recognizable and can be analyzed geometrically. Note that each spike describes one loop and that if several spikes are present their successive loops superimpose. It is then possible, for each point of the plane to count how many times it has been hit during the entire recording. This value (coded by a color scale in Color Figure 1.1, A3) is called a *trajectory density*. It can also be viewed as the average time being spent at a given membrane potential and with a given slope (derivative) during the whole recording. To compute this density numerically, we define a *flag function* $\eta(x,y)$ (a function returning 1 if x and y are equal). This function between a given point $x(V_x, \dot{V}_x)$ of the phase plane and a given point $y(V_{y_t}, \dot{V}_{y_t})$ of the raw data can be defined as

$$\eta(x, y_t) = \begin{cases} = 1 \; if \; x = y_t \\ = 0 \; if \; x \neq y_t \end{cases} \tag{1.25}$$

Then for each point $x(V_x, \dot{V}_x)$ of the phase plane we define its normalized trajectory density as

$$\delta(x, y) = \frac{1}{N} \sum_{t=0}^{T} \eta(x, y_t). \tag{1.26}$$

T is the time duration of the recording and N the number of discrete points in each set. In practice we also have to digitize the phase plane in small squares $\Delta V \times \Delta \dot{V}$, and normalize $\delta(x,y)$ by this product. We can define more formally this density in the continuous case as

$$\delta(x, y) = \int_0^T dt \; \Delta(x - y(t)), \tag{1.27}$$

where $\Delta()$ is the delta function. This delta function (also called Dirac function) is an ideal impulse with infinite amplitude but infinitely small width, which implies that its value is zero except for zero arguments.

We are now ready to define the fitness coefficient based on trajectory density functions. Given two time series (a recorded membrane potential and a model's activity), we sum the squared density differences of the two series over the entire plane, so we write:

$$E\left(V_{real}(t), V_{model}(t)\right) = \sum_{V} \sum_{\dot{V}} \left[\delta_{real}(V, \dot{V}) - \delta_{model}(V, \dot{V})\right]^2. \tag{1.28}$$

* Color Figure 1.1 follows page 140.

This equation can be viewed as the least-squares estimator between the two trajectory densities.

To validate our fitness function, we tested it in a single-parameter search mode using a Hodgkin–Huxley model spiking neuron (Section 5.2.1). We first run the model with a fixed value of the maximal conductance for potassium ($\bar{g}_K = 20$ mS/cm^2) (Chapter 9), and then use the model's activity produced with this value as a "fake" experimental recording. Figure 1.5 shows the fitness coefficient obtained when the reference was compared with the model using a linear increase of $\bar{g}_K$ between 0 and 40 mS/cm^2. We can clearly see that a minimum of the fitness function appears for a $\bar{g}_K$ value that is equal to the $\bar{g}_K$ used as reference (20 mS/cm^2). No pre-processing of the signal was done prior to fitness measurement (both reference and model voltage were computed for 0.5 s simulated time and using a 0.05 ms integration step). We have done several other one-dimensional systematic parameter scans, with different durations and resolutions and with different parameters (maximal conductance, half-activation potential of voltage-dependent gating variable, passive properties, etc.) and we were always able to retrieve with good accuracy the value used in the reference recording.

1.5 PARAMETER SEARCH ALGORITHMS

1.5.1 SEARCHING PARAMETER SPACE

Given the experimental data and a model, we can now measure how well the model reproduces the experiment with a particular parameter set. Our next goal is to find the parameter set that gives the lowest fitness coefficient. This problem is a classical optimization problem, where we like to find the minimum of a function, in our case the fitness function.[11] The basic constraints are (1) we do not know *a priori* the shape of the fitness function *f*, and (2) we do not want to calculate this function for all the possible values of the model's parameters. We need a clever way to travel in parameter space and retrieve the best sets. How can we design a computer scheme that will always converge to a minimum point, with as little computer time as possible? Note that even if there is no guarantee that the minimum found is global (truly the lowest possible point) or local (the lowest within a neighborhood), we want to be reasonably sure that there are no better solutions around.

Different kinds of methods can be used[18,19] depending on two main factors: (1) Do we have more information than just the value of *f*? More specifically, can we sense the slope in the nearest neighborhood (can we compute the partial derivatives), so as to guess where the best downward parameter direction can be found? That direction gives information about how we should change the model's parameters to get a better fit (to decrease the fitness coefficient and eventually reach a minimum). (2) During our scheme, can we accept transient increases of the fitness coefficient? This allows the search algorithm to escape from small local minima in the parameter space and to converge to a better solution (to reach a lower minimum and eventually the global one). A common metaphor to sketch this problem is to imagine a ball rolling down in a basin. We can imagine that small irregularities in the landscape (small bumps) will stop the ball very far from the trough of the basin. A way to

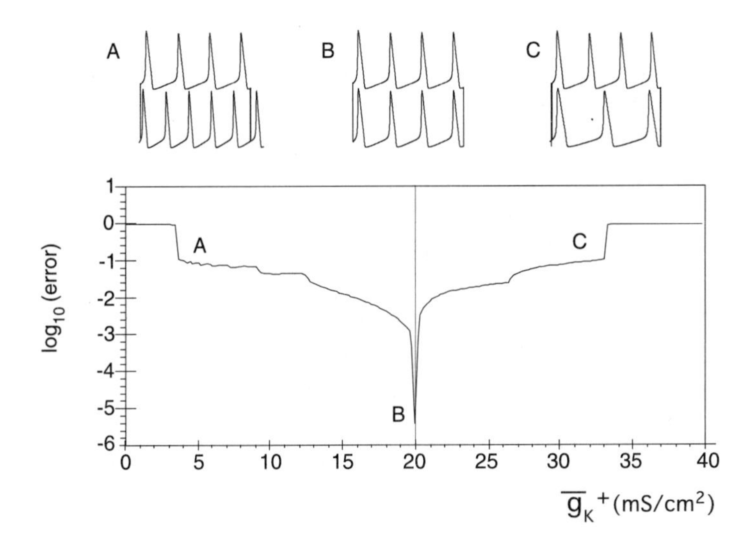

FIGURE 1.5 Validation of the trajectory density fitness function. Scan of maximal potassium conductance ($\overline{g}_K$) and measure of the fitness coefficient between a Hodgkin–Huxley spiking model using this $\overline{g}_K$ value and the same model using a fixed reference value ($\overline{g}_K$ set to 20 mS/cm^2). A minimum value for the coefficient is found (B) when $\overline{g}_K$ reaches the value used in the reference. For lower (A) or higher (C) values of $\overline{g}_K$, the fitness coefficient is higher. Any optimization scheme that will find this fitness minimum will give a correct estimate of the $\overline{g}_K$ used in the reference. Note that no phase locking is required in this example.

avoid this problem is to shake the ball with some random force to give it a chance to escape from this little trap. In optimization techniques, it is often very useful to introduce some noise (randomness) in the procedure, either in the choice of the parameters or in the fitness coefficient itself.[20,21] This strategy is efficient for bumpy parameter spaces, which is frequently the case when dealing with real experimental data.

1.5.2 Parameter Search Methods without Randomness

1.5.2.1 Downhill Simplex

This general method does not require a computation of the derivative of the function that needs to be minimized.[11] It is usually efficient and fast. Starting from any random vector of N parameters (the first guess), the idea is to construct a geometrical figure called a *simplex* made of $N + 1$ points in the N-dimensional parameter space. Usually, the different points of this simplex figure are chosen along the N unit vectors, but the distance between the points can be random. The optimization process per se consists of the evaluation of the fitness coefficient at each point of the simplex, and then a move of the point of worst fit in the direction of the points of better fit. Different algorithms exist to change the distance between points but they all apply successive contractions and expansions to the simplex figure in one or more dimensions. These allow an adaptive search in the parameter space, thus reducing as much as possible the number of iterations needed to converge. When the algorithm comes close to a solution, the simplex contracts to approach the minimum. One often compares this process to the "crawling" of an amoeba-like geometrical figure.

1.5.2.2 Conjugate Gradient Method

This is probably the most popular optimization scheme, but it requires the computation of the function and its partial derivatives (used to compute *the gradient*). One can understand that being able to sense the slope of the function in the different parameter directions (the gradient) gives important information about where to go to find the minimum of the function. Thus, following the steepest gradient is intuitively and mathematically a reliable way of finding the route to the minimum value of the function. Of course computing the gradient requires computing the derivative and this can be a time consuming process, but one usually worth doing. The gradient descent method is often coupled with the concept of *conjugate directions*. This is based on the notion that to explore any multi-parametric function, we need to search in several directions (because there are several parameters in the model). Conjugate directions are non-interfering directions along which one can optimize without taking too much risk of redundancy with respect to the other directions already searched. This avoids cycling problems and inefficient computing.

Practically, the algorithm goes as follows. Start from an initial point P in N-dimensional parameter space and then try to minimize along the line from P in the direction of the steepest downhill gradient. Each time you move in one direction to its minimum, you change direction for a conjugate one in respect to the previous one and if possible to all previous traversed directions. Several efficient algorithms exist. They are particularly good for smooth (no irregularities) functions that have a quadratic form (a function where variations are governed by a square law). To minimize a more complex fitness function with a lot of local minima, you might be interested in other methods making use of noise and randomness.

1.5.3 Parameter Search Methods with Randomness

1.5.3.1 Random Walk

This method is quite straightforward. After choosing an initial center of exploration (first guess), this point is considered as the mean of a Gaussian distribution with a given variance. The next point is then randomly picked up from this distribution. If the fitness coefficient at this point is the best so far, it is set as the next center for further exploration. If not, another point is again randomly chosen. During this process, the variance is slowly decreased until it reaches a minimum value, then eventually re-expanded. Together with the random selection of the model's parameters, successive cycles of shrinkage and extension of the variance give a chance to escape from local minima.[22]

1.5.3.2 Genetic Algorithms

These methods are inspired by evolutionary processes.[23–25] First a random population of vectors each describing a parameter set is chosen (the first guesses) and then the model is evaluated for each of them. The fitness coefficient is computed for each of these vectors. During the optimization scheme, a parameter set may be conserved into the *next generation* (next iteration) with a probability that depends on its fitness value. In addition, at each generation some degree of *crossover* (mixing of parameters between sets) together with some random variation (analogous to a mutation process) occurs. After several generations, the best parameter sets (corresponding to the best fitness coefficients and therefore to a higher probability of transmission to the next generation) start to dominate the population. This strategy is extremely powerful although it requires a lot of model evaluations and consequently computation time. It may be the most effective method for large neuronal models.[18,23]

1.5.3.3 Simulated Annealing

The experimental evidence behind this method comes from thermodynamics and the control of cooling of liquid matters in order to obtain a crystalline structure (low energy states) as perfect as possible which avoid amorphous states (higher energy states). If one slows down the cooling and controls the liquid to solid transition, one gives a better chance to the atoms to find their lowest energy arrangement. The core of the optimization procedure is often based on a classic gradient descent or simplex method, but with a *temperature* parameter (or annealing schedule) that adds some noise to the fitness value. As the algorithm converges, the temperature parameter is slowly decreased. Here also, the noise allows escaping from local minima. Different schemes are described using slight variations of the same general principles. A recent comparison of optimization methods for conductance-based neuron models found simulated annealing to be the most effective method.[18]

1.6 CONCLUSION

The most commonly used fitting procedure by both modelers and experimentalists is still the trial-and-error method of a *fit-by-eyes* fitness evaluation. It would be unfair to say that no successful work has been done with this principle, which is based on the impressive capacity of the brain to estimate the goodness-to-match between experimental data and a model. Nevertheless, it is clear that automated parameter search methods become necessary due to the large amount of data available and the increasing complexity of models made possible by these data. It is unrealistic to suppose that one could process all these data comparisons by manual procedures. More exciting is the real perspective of having automated methods to build complete models out of detailed sets of current clamp or voltage clamp data. Furthermore, the use of fast computing techniques (parallelism, analog computing through analog VLSI models,[26] etc.), broaden these perspectives to very fast or even real time optimization and model construction.

ACKNOWLEDGMENTS

R. M. thanks Mike Wijnants for his computer assistance and Volker Steuber for his corrections and suggestions on the first part of this chapter.

REFERENCES

1. Cheney, W. and Kincaid, D., *Numerical Mathematics and Computing*, Brooks/Cole Publishing Company, Monterey, California, 1985.
2. Arfken, G., *Mathematical Methods for Physicists*, Academic Press, Orlando, Florida, 1985.
3. Jack, J. J. B., Noble, D., and Tsien, R.W., *Electrical Current Flow in Excitable Cells*, Clarendon Press, Oxford, 1975.
4. Wilson, M. A. and Bower, J. M., The simulation of large-scale neural networks, in *Methods in Neuronal Modeling: From Synapses to Networks*, Koch, C. and Segev, I., Eds., 1st ed., MIT Press, Cambridge, 1989, 291.
5. Hornbeck, R. W., *Numerical Methods*, Quantum Publishers, New York, 1975.
6. Dahlquist, G. and Björck, Å, *Numerical Methods*, Prentice-Hall, Englewood Cliffs, NJ, 1974.
7. Kuo, S. S., *Computer Application of Numerical Methods*, Addison-Wesley, Reading, MA, 1972.
8. Stoer, J. and Bulirsch, R., *Introduction to Numerical Analysis*, Springer-Verlag, New York, 1980.
9. Romanelli, M. J., Runge–Kutta methods for the solution of ordinary differential equations, in *Mathematical Models for Digital Computers*, Ralston, A. and Wilf, H. S., Eds., John Wiley & Sons, New York, 1960, 110.
10. Hines, M., Efficient computation of branched nerve equations, *Int. J. Bio-Med. Comput.* 15, 69, 1984.
11. Press, W. H., Teukolsky, S. A., Vetterling, W. T., and Flannery, B. P., *Numerical Recipes in C*, Cambridge University Press, London, 1992.
12. Mascagni, M. V. and Sherman, A. S., Numerical methods for neuronal modeling, in *Methods in Neuronal Modeling: From Ions to Networks*, Koch, C. and Segev, I., Eds., 2nd ed., MIT Press, Cambridge, 1998, 569.
13. MacGregor, R. J., *Neural and Brain Modeling*, Academic Press, San Diego, California, 1987.
14. Crank, J., *The Mathematics of Diffusion*, Clarendon Press, Oxford, 1975.
15. Mitchell, A. R., *Computational Methods in Partial Differential Equations*, John Wiley & Sons, London, 1969.
16. Hansel, D., Mato, G., Meunier, C., and Neltner, L., On numerical simulations of integrate-and-fire neural networks, *Neural Computation,* 10, 467, 1998.
17. Tawfik, B. and Durand, D. M., Parameter estimation by reduced-order linear associative memory (ROLAM), *IEEE Trans. Biomed. Eng.*, 44, 297, 1997.
18. Vanier, M. C. and Bower J. M., A comparative survey of automated paramet-search methods for compartmental models, *J. Comput. Neurosci.,* 7, 149, 1999.
19. Baldi, P., Vanier M. C., and Bower, J. M., On the use of Bayesian methods for evaluating compartmental models, *J. Comput. Neurosci.*, 5, 285, 1998.
20. Tabak, J. and Moore, L. E., Simulation and parameter estimation study of a simple neuronal model of rhythm generation: role of NMDA and non-NMDA receptors, *J. Comput. Neurosci.,* 5, 209, 1998.

21. Wright, W. N., Bardakjian, B. L., Valiante, T. A., Perez-Velazquez, J. L., and Carlen, P. L., White noise approach for estimating the passive electrical properties of neurons, *J. Neurophysiol.,* 76, 3442, 1996.
22. Foster, W. R., Ungar, L. H., and Schaber, J. S., Significance of conductances in Hodgkin-Huxley models, *J. Neurophysiol.*, 54, 782, 1993.
23. Eichler West, R. M., De Schutter, E., and Wilcox G. L., Evolutionary algorithms to search for control parameters in a nonlinear partial differential equation, in *Evolutionary Algorithms*, Vol. 111 of the IMA Volumes in Mathematics and its Applications, Davis, L. D., De Jong, K., Vose, M. D., and Whitley, L. D., Eds., Springer-Verlag, New York, 33, 1999.
24. Goldberg, D., Genetic Algorithms in Search, Optimization and Machine Learning, Addison-Wesley, New York, 1989.
25. Booth, V., A generic algorithm study on the influence of dendritic plateau potentials on bistable spiking in motoneuron, *Neurocomputing*, 26, 69, 1999.
26. LeMasson, S., Laflaquière, A., Bal, T., and LeMasson, G., Analog circuits for modeling biological neural networks: design and applications, *IEEE Trans. Biomed. Eng.,* 46, 638, 1999.

2 Modeling Networks of Signaling Pathways

Upinder S. Bhalla

CONTENTS

2.1 INTRODUCTION

The inside of a cell is a remarkable environment in which to find computation. At first glance, it is as if a thick soup somehow manages not only to keep track of many unique signals but also to have them interact in specific ways to give rise to computation. This feat is accomplished through a mapping of signal identity onto molecular identity. In this mapping lies the power of the cellular computer. Every chemical reaction is a molecular transformation, and computation is born through this manipulation of chemical symbols.

In a chemical plant, or for that matter, in the conventional "Harvard" computer architecture, the instructions are neatly separated from the raw materials or data respectively. In the cell, however, computation merges seamlessly with all aspects of cellular chemistry. The cellular machinery of the cytoskeleton and protein trafficking are as much a part of the signaling apparatus as they are substrates for the

0-8493-2068-2/01/$0.00+$.50

signaling to act upon. The all encompassing nature of cellular signaling means that it touches upon most disciplines of biology. It also makes it hard to separate the wood from the trees, that is, understand the overall biological function in terms of the individual enzymatic reactions. A central theme of this chapter is how simulations may help to bridge this gap. As is done in this book, we can identify three main levels of analysis of biological signaling: "well stirred" (i.e., point or non-diffusive) systems, spatially structured reaction-diffusion systems (Chapter 3), and stochastic models dealing with probabilities of individual molecular events (Chapter 4). One could think of these as beginner, intermediate and advanced topics in the subject. The well-stirred analysis is reasonably well-characterized experimentally, and is comparatively straightforward to specify and solve. It is also necessary to first specify the basic reactions before introducing additional details such as space and probability. The emphasis in this chapter is to develop methods for building empirically accurate models of signaling at the level of well-stirred cells. Test-tube biochemistry is relatively simple to simulate, and programs for modeling enzyme kinetics have existed since the days of punch-cards. The difficulties in scaling up to tens of signaling pathways and thousands of reactions are not computational, but have to do with interface design and most of all with converting experimental data into kinetic parameters in a model. The general approach outlined in this chapter is to modularize the problem in terms of individual signaling pathways, usually involving just one key signaling enzyme. The grand picture is built up by drawing upon a library of models of individual pathways. This fits well with standard experimental approaches, and also lends itself to plug-and-play upgrades of complex models as more complete individual enzyme models are developed.

2.2 METHODS

2.2.1 EQUATIONS

Signaling pathways are fundamentally based on chemistry. A description of signaling in terms of a series of chemical reactions is therefore about as general as one can get in the domain of the well-stirred cell. Fortunately, the numerical methods for handling such reaction systems are simple and fast enough to embrace this generality without having to make compromises. Furthermore, the reaction formalism appears to work well even for empirical and approximate models, as we shall see below. As experimental data are frequently incomplete, this is important in adopting this formalism for biochemical models. Our starting point for such models is the basic reaction:

$$A + B \ldots \underset{k_b}{\overset{k_f}{\rightleftharpoons}} X + Y + \ldots \tag{2.1}$$

This can generalize to any number of terms on either side, and any stoichiometry. Reactions of order greater than two are uncommon, and when they are encountered they are often a reflection of incomplete mechanistic data. At the molecular level, higher-order reactions would require a simultaneous collision between multiple molecules. The probability of this is extremely low. Instead, apparent high-order reactions typically proceed through fast intermediate reaction steps.

Two rules govern Equation 2.1:

$$d[A]/dt = -kf \cdot [A] \cdot [B]... + kb \cdot [X] \cdot [Y]... \tag{2.2}$$

and

$$d[A]/dt = d[B]/dt = -d[X]/\mathrm{dt} = -d[Y]/dt... \tag{2.3}$$

or equivalently,

$$[A0] - [A] = [B0] - [B] = [X] - [X0] = [Y] - [Y0]... \tag{2.3a}$$

$$[A] + [B] + \ldots + [X] + [Y] + \ldots = \text{constant} \tag{2.3b}$$

Equation 2.2 is the differential equivalent of the rate equation. The various forms of Equation 2.3 express the stoichiometry of the reaction and are equivalent to conservation of mass.

For small systems of equations, it is useful to use the mass conservation relationship directly to reduce the number of differential equations:

$$[B] = [B0] - ([A0] - [A]). \tag{2.4}$$

For large, extensively coupled systems, however, keeping track of the conservation relationships among many diverging reaction pathways is cumbersome and does not save much computation. In this case it may be simpler to write out the full differential equation for each reaction component and solve each independently, secure in the knowledge that the differential equations embody the stoichiometry.

$$d[B]/dt = -kf \cdot [A][B]\ldots + kb \cdot [X] \cdot [Y]\ldots \tag{2.5a}$$

$$d[C]/dt = +kf \cdot [A][B]\ldots - kb \cdot [X] \cdot [Y]\ldots \tag{2.5b}$$

$$d[D]/dt = +kf \cdot [A][B]\ldots - kb \cdot [X] \cdot [Y]\ldots. \tag{2.5c}$$

An added advantage of this approach is that it lends itself to modular and object-oriented simulation schemes. Suppose we have an additional reaction

$$A + P + \ldots \underset{k_{b2}}{\overset{k_{f2}}{\rightleftharpoons}} Q + R + \ldots \tag{2.6}$$

The equation for A alone would need to change to incorporate the additional terms:

$$d[A]/dt = -k_f[A][B]\ldots - k_{f2}[A][P] + k_b \cdot [X] \cdot [Y]\ldots + k_{b2} \cdot [Q] \cdot [R] \quad (2.7)$$

If one relies on the differential equations to maintain stoichiometry, it is possible for numerical errors to introduce embarrassing problems with stoichiometry and mass conservation. This can be turned to our advantage, by using conservation as a test for numerical accuracy and as a quick empirical way of deciding if the time-steps need to be shorter.

The prototypical reaction scheme for a Michaelis–Menten enzyme[1,2] crops up frequently in signaling interactions:

$$E + S \underset{k_2}{\overset{k_1}{\rightleftharpoons}} E.S \xrightarrow{k_3} E + P \quad (2.8)$$

This is readily represented by two reactions in sequence. The final reaction is conventionally assumed to be unidirectional. This is a good approximation for most systems with low product concentrations, though it is also reversible. The Michaelis–Menten scheme is so common that the modeler may wish to represent it as a self-contained module (Box 2.1).

Box 2.1 Useful Equations

1. Michaelis–Menten Equation and Relationship to Rate Constants

Standard formulation:

$$E + S \underset{k_2}{\overset{k_1}{\rightleftharpoons}} E.S \xrightarrow{k_3} E + P \quad \text{(1) (From Equation 2.8)}$$

Where S is substrate, E is enzyme, $E.S$ is enzyme-substrate complex, and P is product.

$$V_{max} = \text{maximum velocity of enzyme} = k_3. \quad (2)$$

Derivation: Substrate is saturating, so all of E is in $E.S$ form. So

$$V_{max} \cdot [E_{tot}] = [E.S] \cdot k_3 = [E_{tot}] \cdot k_3 \quad (3)$$

The units of V_{max} come in a variety of forms, due care is necessary in conversion as described below.

(continued)

Box 2.1 (continued)

$$K_m = (k_3 + k_2)/k_1 \text{ (by definition)} \tag{4}$$

In most kinetic experiments, only V_{max} and K_m are obtained for a given enzyme. This means that we are short one parameter for constraining k_1, k_2, and k_3. An assumption which seems to work well is

$$k_2 = k_3 * 4 \tag{5}$$

This is not entirely arbitrary. In many enzymes $k_2 >> k_3$. If k_2 is small compared to k_3 then a large proportion of the enzyme will be in the complex form, assuming K_m is fixed. In such cases it may be better to explicitly model the enzyme reaction as a series of conventional reaction steps, rather than hide the enzyme complex. The factor of four keeps the amount of enzyme complex fairly small while avoiding extreme rate constants which might cause numerical difficulties. In previous studies, this factor has been varied over a range from 0.4 to 40, with very little effect on the simulation results.[11]

2. *Standard Bimolecular Reaction*

$$A + B \underset{k_b}{\overset{k_f}{\rightleftharpoons}} C \tag{6}$$

Where A and B are reactants and C is the product. Note that this is completely reversible. The equilibrium dissociation constant is

$$K_d = k_b/k_f \quad \text{(by definition)} \tag{7}$$

So, if B is limiting, and half of B is bound, then at equilibrium:

$$[A][B_{half}] \cdot k_f = [C_{half}] \cdot k_b = [B_{half}] \cdot k_b \tag{8}$$

Since the other half of B has been converted to C. So we get the standard relationship for K_d:

$$[A_{Bhalf}] = k_b/k_f = K_d \tag{9}$$

or, K_d is that conc of A at which half of B is in the bound form. Obviously the association or binding constant is

$$K_a = k_f/k_b = 1/K_d \tag{10}$$

2.2.2 Integration Methods

Three characteristics of biological signaling reactions help select between integration methods:

1. Reaction loops
2. Rates spanning at least three orders of magnitude
3. Very large numbers of reactions

Reaction loops are a very common motif: even enzyme reactions contain such a loop (Figure 2.1). Other situations include interconversion between states of a system, and when an enzyme can be reversibly bound to multiple activators.

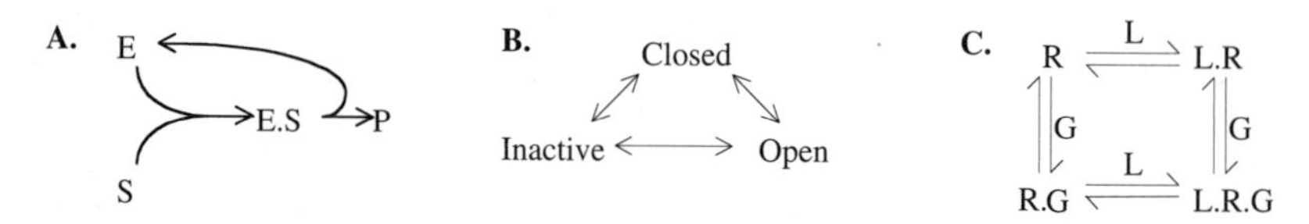

FIGURE 2.1 Reaction loops. (A) Enzyme reaction; E, Enzyme; S, Substrate; P, Product. (B) States of a voltage-gated ion channel. (C) Activation scheme for a G-protein coupled receptor: R, Receptor; L, Ligand; G, G-protein.

These loops make it difficult to apply implicit integration methods (see Section 1.3) efficiently to reaction systems. As described above, reaction loops can be eliminated through conservation relationships but this may give rise to other computational problems. For non-looping systems, it is sometimes possible to order the implicit solution matrix in a manner that can be solved using backward elimination in order (N) operations. This was described for branching neuronal models by Hines[3] and is utilized in Chapter 9. In the presence of loops, however, one needs to solve the matrix of reactions, which takes order (N^3) operations in the general case.[4] Iterative methods and clever ordering of the matrices can reduce this requirement somewhat, but implicit integration remains computationally expensive for large reaction systems. This is unfortunate, as the wide range of time-constants (point 2 above) means that one needs to use time-steps of the order of the fastest rate in the system unless one is using implicit methods.[5] Higher-order integration methods, especially with variable timestep, may be useful in solving reaction systems under many conditions. Runge–Kutta and Bulirsch–Stoer are among the techniques that have been examined.[4] However, both methods are non-local, that is, they are difficult to decompose in a modular manner. They may also suffer from instability or drastic speed reduction when confronted with abrupt stimuli typical of experimental manipulations. The wide range of reaction rates also slows down such methods (see Equation 1.19). The exponential Euler method[6] is a simple integration scheme predicated on the assumption that variables will follow exponential decay time-courses. This happens to fit well with most reaction systems, and the simplicity of the method makes it easy to implement in a modular fashion. It is also extremely robust and can handle abrupt stimuli gracefully. It is, of course, not as efficient as the higher-order methods operating at their best, and does not have built-in variable time-step calculations. These can sometimes be imposed from above if the modeler knows when a particular simulation is expected to be in a slowly changing state.

Box 2.2 Useful Conversions

1. Biological Ratios (for Mammalian Cells)

Proportion of protein to cell weight: 18% (Reference 17, pp. 92) (11)

Proportion of lipid to cell weight: 3% (Reference 17, pp. 92). (12)

Approx. 50% of membrane weight is protein (up to 75% in some organelles) (Reference 17, pp. 264). (13)

So maybe 3% of cell mass is membrane protein. (14)

2. Concentration and Other Units

When working within a single compartment, it is simplest to use standard concentration units such as μM. However, transfers between compartments or to and from the membrane are best handled using the number of molecules. Indeed, the concentration of a molecule in the membrane is a somewhat tricky quantity.

To illustrate, we will consider conversion of standard enzymatic units to number units in a cubical model cell of 10 μM sides.

$$\text{Vol of cell} = \text{1e-6 } \mu\text{l} \tag{15}$$

$$1\ \mu\text{M in cell will have 1e-18 moles} \sim \text{6e5 molecules/cell} \tag{16}$$

$$V_{max} \text{ of } 1\ \mu\text{mol/min/mg will convert to Mwt/6e4 \#/sec/\#} \tag{17}$$

where Mwt is the molecular weight of the enzyme and numbers/cell are represented by the # symbol. The units can be interpreted as "# of molecules of product formed per second per # of enzyme molecules."

Similarly, rates for binding reactions will need to be scaled to the appropriate units. For reaction 6 above:

$$A + B \underset{k_b}{\overset{k_f}{\rightleftharpoons}} C \tag{18}$$

(continued)

Box 2.2 (continued)

Take

$$k_b \sim 1/\tau \quad \text{sec}^{-1} \tag{19}$$

Then

$$k_f = k_b/K_d \sim 1/(\tau * K_d * 6e5)\ \text{sec}^{-1}\#^{-1} \tag{20}$$

So the units of K_d would be #.

These equations give us the rates k_f and k_b in terms of K_d and τ, and would be a good place to start refining a model to fit an experimental concentration-effect curve.

2.2.3 Software

Given that chemical rate equations entail the simplest kind of differential equations, it is not surprising that there is a plethora of useful software available for carrying out the numerical modeling (e.g., Matlab, Mathematica, others). Assuming that all packages do a reasonable job with the numerical integration, the choice of method should be governed more by availability and familiarity than by small differences in efficiency. For most models and modern computers, simulation times are likely to be much faster than real time, and one has the luxury of using whatever method is most convenient. Therefore, the real challenge in specifying complex networks of signaling pathways is user-interface and data management related. There are rather fewer packages designed specifically for handling large numbers of chemical reactions in a friendlier format than simply entering rates and differential equations (XPP, Vcell, GENESIS/kinetikit). Given the interest in the field, such packages are likely to evolve rapidly.

Box 2.3 Recurring Biochemical Motifs in Signaling

There are a few specific motifs that recur in models of biological signaling, and are worth examining more closely.

1. The Two-State System

Several enzymes, especially the protein kinases, can be reasonably well modeled as two-state systems: either completely on or completely off. The level of activity of the system is determined simply by the proportion of the enzyme in the "on" state. For example, protein kinases typically have an autoinhibitory domain, which binds to and blocks the active site. Activators of the enzyme, to first approximation, simply release this block.[11] As modeled for PKC, one plausible mechanism for such an enzyme is for the activators to bind to the inactive enzyme, and convert it to the active state. Similar "switches" seem to take place in many enzymes by phosphorylation. (Figure B2.1A).

(continued)

Box 2.3 (continued)

2. The Empirical Enzyme

A remarkably good empirical description for an enzyme about which little is known is simply that each activator "turns on" a distinct level of activity (Figure B2.1B). This mechanism is relatively easy to relate to experimental data, but it does lead to a proliferation of activity states and enzyme sites with different rates. PLA2 is an enzyme which has been modeled in this manner.[11]

3. The Unregulated Enzyme

If a pathway has an enzyme whose activity is unlikely to be saturated, and which does not undergo any known regulation, it is safe to treat it as a simple unidirectional reaction with a fixed rate constant (Figure 2.1C). This situation frequently arises in degradative pathways of second messengers (e.g., DAG and AA).

4. The Reaction Loop

This is a common trap for the unwary. Consider the quite plausible scheme for G-protein coupled receptor activation as illustrated in Figure 2.1C. There is a strong constraint on the rates: the product of the $K_d s$ around the loop must always be one. The proof of this is left as an exercise for the reader. (Hint: Consider free energy.) There is no problem with reaction loops where there is an energy-releasing step, such as phosphorylation or ATP hydrolysis.

5. Buffers

Almost all test-tube experiments rely on buffers (see also Section 3.2.1). For example, the standard way to obtain a concentration-effect curve, either experimentally or in a simulation, is to buffer the control molecule at a series of concentrations and monitor the resulting effect on the target. A good chemical buffer is actually quite unpleasant to model. By definition, a good buffer must a) be able to compensate for the largest chemical concentration changes that may happen in the system, and b) do so faster than other reactions. Ca^{2+} buffering systems, for example, have large amounts of avid chelators like EGTA with balancing amounts of Ca^{2+}. Cooperative binding makes such buffers even more effective. Accurate models of such buffered systems therefore require very small timesteps. If one is willing to settle for a perfect rather than a realistic buffer, it is almost embarrassingly easy to do so numerically: Simply fix the concentration of the molecule at the target level. One drawback of such perfection is that any one-way reaction fed by such buffers will quite happily build up infinite levels of product.

(continued)

Box 2.3 (continued)

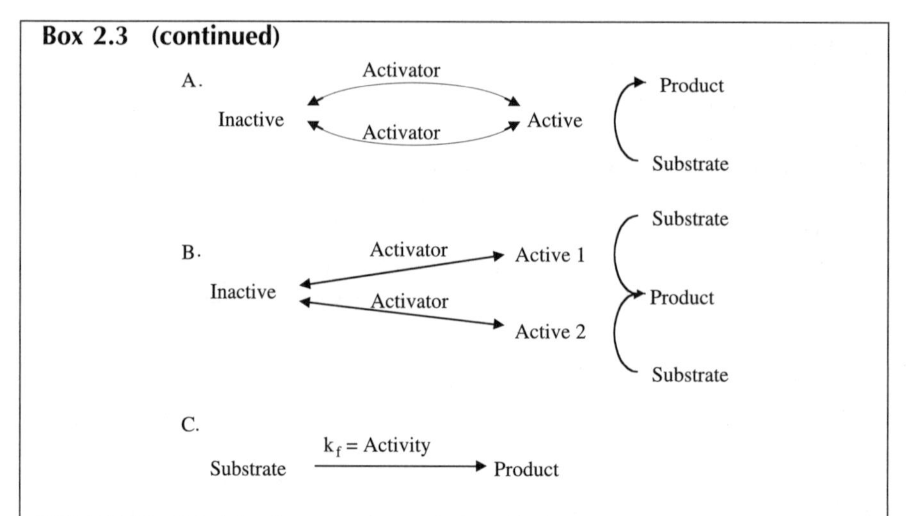

FIGURE B2.1 Models of enzymatic regulation. (A) Two-state enzyme where either of two activators form the same active state. (B) Empirical enzyme. Each activator produces a distinct activity level. (C) Unregulated enzyme. A simple unidirectional reaction suffices.

2.3 PARAMETERIZATION

Parameterization is by far the most difficult aspect of modeling signaling pathways (and most other biological systems as well). In developing a model one has to draw a balance between known biological detail, and the applicability of such detail to the problem at hand. This is constrained further by the availability of data, and the sheer scale of the problem of modeling extremely complex systems.

2.3.1 From Test-Tube Data to Models

Cellular signaling is complex both in terms of the number of players, and the properties that emerge from their interactions. A systematic reductionist approach to cataloging and characterizing the major pathways may seem daunting. Nevertheless, the admirably quantitative traditions of biochemistry have led to the accumulation of a remarkably complete base of data for many signaling molecules. Within the constraints of test-tube chemistry, it is possible to derive faithful models of many individual pathways. For example, the major protein kinases PKC,[7] PKA,[8] CaMKII,[9] and MAPK[10] have each been well-characterized. Such data are essential for developing internally consistent models of pathways. It is significant, however, that the data are complete enough that *independently* developed models of the pathways are also consistent.[10,11] Further validation of these models can be done by examining their predictive capacity in composite models. For example, the MAPK response to EGFR stimulation involves the receptor and its immediate signaling cascade, the Ras pathway as well as the MAPK cascade (see Equation 2.10 below). Replication of such complex signaling sequences suggests that one can reasonably scale up the reductionist approach to larger problems.

In order to scale up from a collection of rate constants to a framework which can tackle interesting biological problems, a necessary first step is the development of a library of models of individual pathways[11] (*http://www.ncbs.res.in/~bhalla/ltploop/* or *http://piris.pharm.mssm.edu/urilab/*). The next stage is to merge relevant pathway models into a composite simulation of the system of interest, inserting system-specific interactions and parameters where necessary. This is a multi-step process. Where experimental results involving combinations of a few models are available, these can provide excellent constraints on model parameters. Finally, the composite model is explored with a variety of stimuli, emergent properties are examined, and explanations for known phenomena are sought.

2.3.2 Data Sources

Primary data for model building come from three principal kinds of experimenal publications. The first category of paper tends to be very qualitative, and is frequently of the form of "Gene A expressed in cell type B activates pathway C." Nevertheless, such data is critical for establishing the mechanisms of the pathway and the causal sequence of signaling events. It is instructive to follow the early literature on the MAP Kinase cascade (Reviewed in Reference 12). Initial studies indicated co-activation of various stages of the cascade. Intermediates were suspected, but not identified.

$$EGF \rightarrow EGFR \rightarrow ? \rightarrow Activator\ (phosphorylated\ ?) \rightarrow\ MAPK \qquad (2.9)$$

It is a measure of the rapid pace of the field that by 1992, some clear experiments were done which established the current version of our understanding of the cascade.[13]

$$EGF \rightarrow EGFR \rightarrow\ GRB2 \rightarrow\ p21ras \rightarrow\ Raf-1 \rightarrow\ MAPKK \rightarrow\ MAPK \qquad (2.10)$$

In the laboratory, as well as in model development, such studies are usually followed by quantitative analyses which put some numbers onto the mechanistic arrows. A typical paper in this category would seek to purify a newly identified signaling molecule, verify its mechanism of activation in the test tube, and obtain rate constants. For enzymes this purification usually involves a sequence of chromatographic separations leading to ever higher specific activities and lower yields. Table 2.1 illustrates such a purification sequence for MAPK (adapted from Reference 14).

Such papers provide not only activity parameters, but also an estimate of concentration of the signaling molecule in the preparation. These numbers are, of course, interdependent. The good news is that the degree of purification affects the activity and concentration in inverse proportion, so one does have a good estimate of total cellular enzyme activity even if concentration values are inaccurate. The bad news is that most regulatory interactions require a good estimate of concentration. Continuing with our MAPK example, it turns out that the concentrations for MAPK in the hippocampus are some ten times that in most other neural tissues.[15] Such system-specific parameters greatly complicate quantitative modeling. As discussed below,

TABLE 2.1
Purification Series for MAPK

Purification Stage	Volume (ml)	Total Protein (mg)	Total Activity $nmol.min^{-1}$	Specific Activity $nmol.min^{-1}.mg^{-1}$	% Recovery	Purification
Cytosol	366	11488	1241	0.108	100	1
DEAE-cellulose	432	5490	898	0.16	72	1.5
Phenyl-Sepharose	70	28	378	13.5	30	125
Mono-Q	10	0.05	46.9	938	3.8	8685

there are still more fundamental difficulties in extracting system-specific parameters using most current experimental approaches.

The third category of paper, in which detailed reaction mechanisms and kinetics are analyzed, is much to be treasured. Such papers typically contain numerous concentration-effect curves for enzyme activation and binding of molecules, often with estimates of time-course.[8,16] These curves tightly constrain parameters and reaction mechanisms.

2.3.3 Iterative Parameterization

An overview of the process of parameterization is provided in the flowchart in Figure 2.2. The key aspect of this flowchart is that it is iterative. We start with a simple, plausible mechanism for a signaling molecule and results for a specific interaction involving it. Once this is parameterized to satisfaction, another interaction may be considered. This elaboration of the reaction mechanism will usually mess up the first stage of parameterization, which must be repeated, until now both interactions behave well. In systematic experimental papers, interactions are often examined in a progressive sequence which fits very well with such an iterative approach. The process of parameterization is illustrated in a worked example for developing a model of PKC (Box 2.4). There is a great deal of similarity between this process and the Towers of Hanoi: each additional interaction may mean that the modeler must go through the whole process again. It is well to apply Occam's razor ruthlessly to prune elaborate mechanisms: do not assume any reactions unless the data demand it. The reader is invited to work out the number of iterations required to match, for example, five graphs of experimental results, assuming that each new graph introduces reaction changes that require adjustment to all previously fixed parameters. This iterative process of model development typically goes through 50 or more versions for each new pathway modeled. A similar process may be needed to refine models in the light of new data or to incorporate additional mechanistic details such as cytoskeletal interactions.[17]

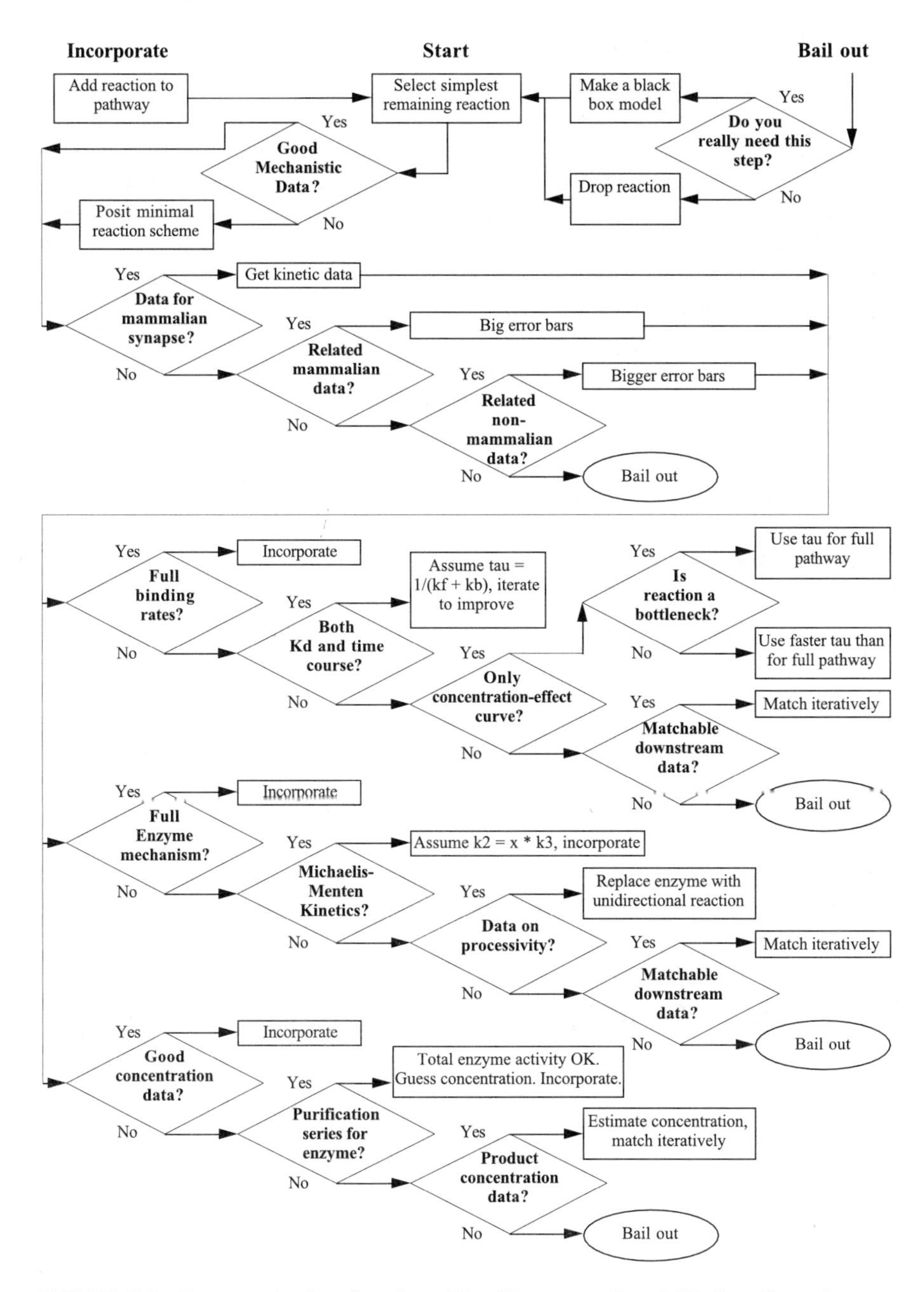

FIGURE 2.2 Parameterization flowchart. The "Incorporate" and "Bail out" routines are indicated on top.

2.3.4 Parameter Searching

Several of the stages in the flowchart require parameter estimation based on fits between simulated and experimental data. This is an aspect of modeling that is especially suited to judicious automation. Parameter searching can be done in three main ways: user-guided, brute force, and automated. The user-guided approach takes advantage of human insight into the system to select the best of a small number of versions of the model. Rather complex judgments, for example, about the shape of a stimulus response curve, typically enter into such searches. It is obviously a labor intensive and somewhat subjective method. Nevertheless, it is often the best approach especially in the initial stages of refining a simulation. The brute force and automated searching methods both require a measure of the "goodness" of a given set of parameters and this can be difficult to define. Obvious methods include mean-squared error of the simulation results as compared to experimental data. The key requirement for such a measure of "goodness" of fit is that it should improve monotonically and smoothly as the simulation improves. Often the mean-squared term (Equation 1.22) behaves well near the best fit, but does not help much if the parameters are significantly off. More sophisticated measures may behave better but are correspondingly harder to set up. Brute force methods are effective if there are only a few parameters to examine, and if each simulation runs rather quickly. In this approach each parameter is scaled up and down in small increments which are expected to span the range of interest. If there are N parameters each run for M scales, then the number of required simulations = M^N. A practical upper limit for N is typically four, using an M of 3 or 4. A variety of automated methods exist, such as simplex, gradient descent, and conjugate gradient.[4] These methods as well as genetic algorithms have been compared for neuronal models.[18] It is difficult to provide a general prescription here; each specific problem has its own unique features. The only common rule is not to let the searches proceed blindly. The modeler usually has a pretty good feel for reasonable parameters, and can catch the search algorithm when it threatens to launch off into the realm of the absurd.

Box 2.4 PKC Parameterization

PKC is a "mature" enzyme: it has been studied in sufficient depth that most of the relevant data are available. Here we use it as an example of the parameterization process. We start with the following mechanistic data:[7]

- PKC is activated by Ca^{2+}, AA, and DAG. We will examine only the first two for this exercise.
- PKC, like many serine/threonine kinases, is kept in the inactive form by an inactivation domain on the kinase itself which contains a pseudo-substrate. (Figure B2.2A) Normally the protein folds over onto itself so that the pseudo-substrate domain covers the catalytic site and prevents access to other substrates. Activators unfold the protein and expose the catalytic site. This suggests that we could model

(continued)

Box 2.4 (continued)

it as a two-state enzyme, where the activity is always the same when open, and different activators open different fractions of the total enzyme (or open it for different times).

- Membrane translocation is known to be involved in PKC activation.
- The time-course of activation of PKC is under 10 sec. Ca^{2+} activation is under 1 sec.[7,26]

In order to model the kinase, we go sequentially through the data plots in Figure B2.3. The solid lines represent experimental data.[26]

1. Basal Activation (Figure B2.3A). Inspecting the plot, we first note that there is a basal level of activation even at very low Ca^{2+}. This amounts to about 5% of maximal activity. Ignoring the 1 nM Ca^{2+}, we can set up this activity by postulating a basal reaction where 5% of the cytosolic enzyme goes to the active state. So,

$$k_{f1}/k_{b1} \approx 1/20 \tag{1}$$

Setting this to a 1 sec time course,

$$1 \text{ sec} \approx 1/(k_{b1} + k_{f1}) \tag{2}$$

Solving:

$$k_{b1} \sim 1/\text{sec},\ k_{f1} \sim 0.05/\text{sec} \tag{3}$$

2. Ca^{2+} Activation. The curve in Figure B2.3A appears to have a half-max of about 1 μM Ca^{2+}. Unfortunately this produces nothing at all like the desired curve. The maximal activity with Ca^{2+} stimulation alone is only about 30% of peak PKC, whereas a simple Ca-binding reaction will give us 100% activity. This is a situation where the known mechanistic information about membrane translocation gives us a useful hint. Let us keep the half-maximal binding where it was, at about 1 μM Ca^{2+} (Reaction 2), and assume that the membrane translocation step (Reaction 3) is what allows only 1/3 of the kinase to reach the membrane. This additional degree of freedom lets us match the curves nicely. The final parameters are pretty close to our initial guesses:

$$k_{f2} = 0.6/(\mu\text{M.sec}), \quad k_{b2} = 0.5/\text{sec} \tag{4}$$

$$k_{f3} = 1.27/\text{sec}, \quad k_{b3} = 3.5/\text{sec} \tag{5}$$

(continued)

Box 2.4 (continued)

It is important to keep in mind that this is still just an empirical model for the true reaction mechanisms. Although we can justify the additional reaction step by invoking membrane translocation, strictly speaking we do not have any direct experimental basis for this. Indeed, a blind parameter search will quite effectively utilize the two additional parameters to give us the desired curve, without any insight into mechanistic details.

3. Matching the AA activation in the absence of Ca^{2+}. Examination of the curves (Figure B2.3B) indicates that the activation is almost linear in this concentration range. Therefore, almost any large K_d should do as a starting point. After a little exploration, values of

$$k_{f4} = 1.2\text{e-}4\ /(\mu\text{M.sec}),\quad k_{b4} = 0.1/\text{sec} \tag{6}$$

seem reasonable. We use a slower 10-sec time-course for the AA binding steps.

4. Matching AA activation with 1 μM Ca^{2+}. A quick run with the previous version of the model shows that it is not able to account for the synergistic activation of PKC by AA and Ca^{2+} combined (Crosses in Figure B2.3C). We therefore must introduce this synergy in the form of combined binding of Ca^{2+} and AA to the kinase (Reaction 5). We already have Ca^{2+}- and AA-bound forms of the kinase. Which should we use as the starting point? Given that Ca^{2+} binding is faster, it is reasonable to use the Ca^{2+}. PKC form as our first intermediate, and subsequently bind AA to this to give our synergistically active form. Furthermore, this combined form should account for the bulk of the kinase when both Ca^{2+} and AA are present, so its reaction should have a much tighter affinity for AA than does the PKC-AA reaction. Working through the matches, we find that a tenfold increase in affinity produces a good match:

$$k_{f5} = 1.2\text{e-}3/(\mu\text{M.sec}),\quad k_{b5} = 0.1/\text{sec} \tag{7}$$

5. Matching Ca^{2+} activation with fixed 50 μM AA. This is the test of the completeness of our model. In principle we have already accounted for all combinations of Ca^{2+} and AA activation. As it turns out, our model does indeed do a good job of replicating this curve, without adding any further reactions (Figure B2.3D). Not only is our model empirically able to match the previous figures, it has been able to generalize to a new situation. This improves our confidence in the model being applicable to other signaling situations as well. Our final model is shown in Figure B2.2B. As an exercise, the reader is invited to try to match further interactions involving DAG.

(continued)

Box 2.4 (continued)

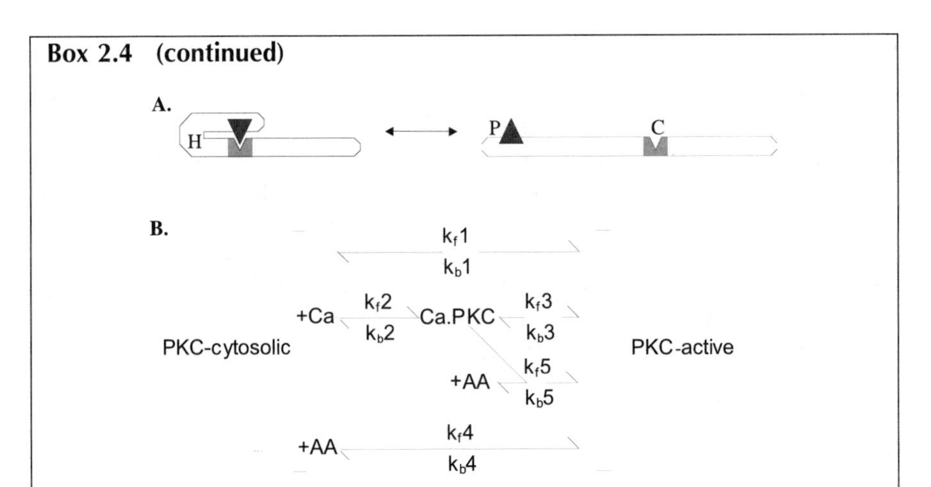

FIGURE B2.2 Activation of PKC. (A) Activation of PKC. In the inactive form it is folded over at the hinge region H so that the catalytic site C is covered by the pseudosubstrate P. When activated, the catalytic site is exposed. (B) Reaction scheme for PKC activation model. The cytosolic form can undergo any of reactions 1, 2 or 4. The active form of PKC is formed by each of the reactions 1, 3, 4, or 5.

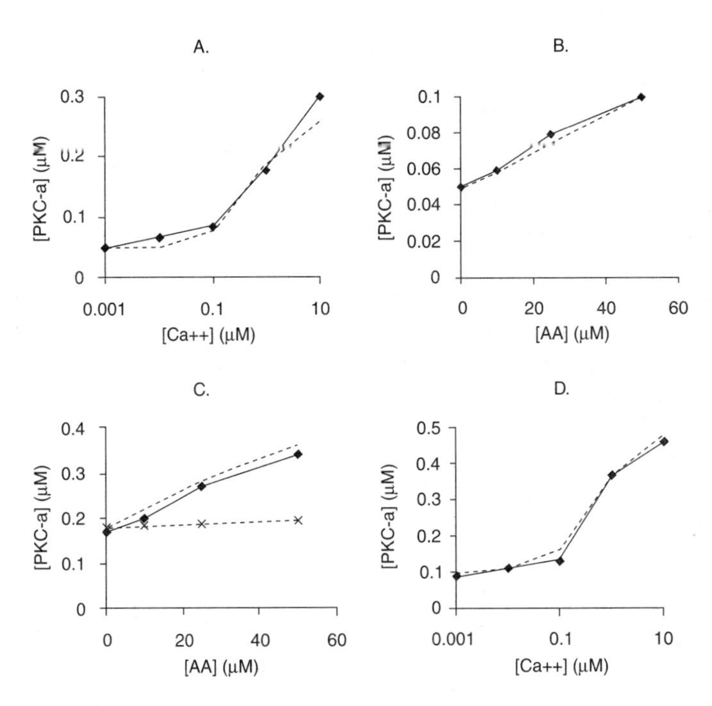

FIGURE B2.3 PKC Regulation: experimental (solid lines) and simulated (dashed lines). (A) Basal and Ca-stimulated activity. (B) AA-stimulated activity without Ca. (C) AA-stimulated activity in the presence of 1 μM Ca. Crosses indicate curve obtained without using a separate reaction to represent synergy. (D) Ca-stimulated activity in the presence of 50 μM AA. The simulated curve was predicted from the model based on panels A–C.

2.4 MODEL INTEGRATION

A well-designed library of pathway models would ideally be plug-and-play: one should be able to take individual pathway modules and connect them together as if one were drawing a block diagram. For this to work, the interactions between pathways should be incorporated into the pathway models themselves, rather than pasted on after the composite simulation has been assembled. This principle makes the parameterization process relatively independent of subsequent use of the pathways. It is also cleaner in the sense that emergent properties of the system are not susceptible to being built in by tweaking all parameters together. Three common motifs for inter-pathway interactions are second messengers, multimer formation, and covalent modification.

Interactions between pathways via second messengers are trivial to implement: cAMP produced by any of the 15-odd adenylyl cyclases (or applied through a pipette) will activate PKA identically.

Multimer formation (such as binding of Ca_4.CaM to CaMKII) or dissociation of multimers (as in G-proteins) is somewhat more involved. One complication is that all components of the multimer may undergo modification of activity in a substrate-specific manner. For example, the catalytic activity of the G-protein alpha subunit is different in isolation, and when bound to target enzymes.[19] Furthermore, the GTPase activity depends on which target enzyme it is bound to. Another difficulty is that unlike second messengers, every participating protein in a multimer may be present in multiple isoforms. Such detailed data may be difficult to obtain, and incorporation of all isoform permutations into a library can be difficult.

Communication between pathways through covalent modification, such as by phosphorylation, is still more problematic. First, enzyme-substrate interactions are often highly specific. The isoform permutation problem mentioned above is more problematic. In addition to specificity of rates for enzyme action, there is also isoform specificity of signaling responses following covalent modification. Determination of the signaling responses due to a specific enzyme-substrate pair often boils down to educated guess work: Can one assume that the phosphorylation rates for PKC-α on ACII are the same as for PKC-β on ACIII? Second, covalent interactions are often modulatory rather than switch-like. Unfortunately, the modulatory effects are typically applicable for a specific activity state of the enzyme. An example of this situation might be the experimental result that MAPK phosphorylates PLA_2 to increase its activity some threefold.[20] Closer examination of the methods reveals that the activity was measured in 5 mM Calcium. It would now be useful to know what happens over a more physiological range of Ca^{2+}. Yet more detailed models would need to incorporate increase in activity as compared to other activators of PLA_2 as well. Practically speaking, pathway interactions may involve even more parameters than the pathways themselves.

Model integration, that is, converting a set of models of individual pathways into a network, is still a long way from the ideal plug-and-play situation. Current models may have to fall back on the following compromises between the modular library approach and completely uninhibited parameter tweaking.

- Depending on data sources, it may be simpler to develop two or more tightly coupled pathways as a single unit. The MAPK pathway is a sequence of three kinases, with interspersed inhibitory phosphatases. These reactions are so tightly coupled that in experiments as well as in modeling, it is simplest to treat them as a single entity.
- Data may be more readily available for a series of pathways than for the individual elements. The EGFR pathway and the MAPK cascade in Equation 2.10 illustrate a situation where the direct outputs of the EGFR are difficult to measure, but a good deal of data are available for MAPK activation by EGF (e.g., Reference 21. Given the input (EGF) and the output (MAPK activity) and known parts of the pathway (Ras, MAPK), one can solve for the unknown (EGF pathway). In other words, the MAPK pathway just acts as an output "black box" in our parameter selection process for the EGF pathway.
- In several cases (e.g., MAPK $\rightarrow$ PLA_2), the individual models are well defined and the composite effect is known but the details of the interaction are not. Here the interaction strength is the unknown which can be parameterized while the structure of the individual models are kept fixed.

The final step of model integration, when several individual pathway models are merged into a grand simulation, is precisely the point at which models scale beyond the grasp of most unaided human minds. Well-designed interfaces play a key role at this stage. First, they can provide sanity checks for trivial problems like duplication of enzymes in different pathway models, or incompatible units. Second, they allow quick checks for convergence of molecule concentrations to reasonable steady state values. Most importantly, the interface should make it easy for the modeler to match simulations against experiments which provide data for an entire network of pathways. This is, after all, the purpose of the entire exercise.

2.4.1 Model Evaluation

Model evaluation involves at least two questions: How good is the data; and, how good is the model? The "garbage in, garbage out" rule means that the first sets fundamental limits to accuracy. This is reflected in the flowchart above. Error bars, while they inspire confidence in a given experiment, may not be the best way of evaluating data accuracy. It is common for different groups to have widely divergent results for the same enzyme, each internally accurate to around 10% (Table 2.2).

Subtle differences in the methods may lead to severalfold differences in rates. One useful rule of thumb is that enzyme rates *in vivo* are likely to be higher than test-tube rates: all else being equal, choose the highest reported rate. Multiple data sources for the same parameter are extremely valuable as they provide a feel for the range of values and often a range of conditions and supplementary results. The preferred technique for model evaluation is to test its predictive capacity. Within a paper, some plots may contain mutually redundant data, though it would be all but impossible to derive one from the other without explicitly making a model. As part

TABLE 2.2
Phosphorylation of Neurogranin by PKC

Source	K_m (μM)	V_{max} (min^{-1})
Paudel et al., JBC 268, 6207, 1993	16.6 ± 4.7	643 ± 85
Huang et al., ABB 305, 570, 1993 (Using PKC β)	26.4 ± 1.3	285 ± 11

Note: Both sets of experiments were run at 30°C; however, Huang et al. separately measured rates for the α, β, and γ isoforms of PKC (β isoform reported here). The methods used in each paper appear tightly controlled and reliable.

of the iterative process described earlier, one might imagine approaching each new graph in a paper hoping that the existing interactions and model are already capable of predicting it (Figure B2.3D). The disparity between results from different papers usually makes it difficult to accurately predict a figure in one paper completely from another. One should, however, be able to adapt a model to a specific paper with relatively little effort if the model is fundamentally sound. It is worth making a distinction here between the fundamental mechanisms and rates of a model (which should be the same for a given enzyme) and the actual concentrations in a given experiment. In test-tube experiments in particular, concentrations are entirely up to the experimenter. This lays a particular onus on the modeler to accurately model experimental conditions. This seemingly trite recommendation is nontrivial. For example, consider a two-state enzyme whose activity is measured using a test substrate and a concentration-effect curve for levels of product.

An obvious shortcut in simulating such an experiment would be to simply measure (in the simulation) the levels of the active form of the enzyme. There are at least two pitfalls here. First, experiments to measure enzyme activity often utilize saturating substrate levels. If the enzyme acts on some other regulatory component the substrate will competitively inhibit this other enzyme activity (Figure 2.3A). Unless the substrate assay is modeled explicitly, this competitive interaction will be missed. Second, the mechanism for regulation of the enzyme usually assumes that the enzyme-substrate complex is a transient, and is present at very low levels. This assumption breaks down at saturating substrate levels. Here we would have to explicitly include the enzyme-substrate complex as a participant in other interactions. (Figure 2.3B).

Steady state data often account for most of the parameters in a simulation of signaling pathways. Such experiments can often be replicated by more than one possible reaction mechanism. Models based on such data are clearly poorly constrained. Experimental designs which explicitly incorporate time-series measurements[22] provide much tighter constraints on model parameters.

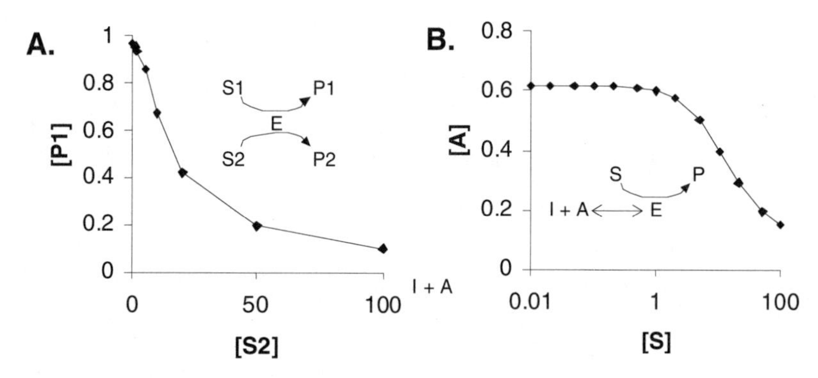

FIGURE 2.3 Effects of high substrate concentrations. (A) Competitive inhibition of formation of P_1 by high levels of S_2. The reaction runs for 100 sec, with 1 μM of S_1 and 1 μM of enzyme. (B) Shifted activation curve depending on substrate levels. The inactive enzyme I and activator A start out at 1 μM. When substrate levels approach saturation the amount of free enzyme declines, leading to the shift in the curve. Any other reaction depending on the activator would be affected. In both cases all enzyme rates are set to 0.1, so $V_{max} = 0.1$ and $K_m = 2$.

2.5 COMPLICATIONS AND FUTURE DIRECTIONS

A cynic would identify two serious problems with current attempts to model signaling pathways. The first is that there is not enough detail in the models. The second is that there is too much. The first criticism refers to the fact that biological systems are far more intricate and computationally rich[23] than crude test-tube chemistry allows. As described much more comprehensively in subsequent chapters, our simple differential equations are only a first approximation to the three-dimensional, compartmentalized, stochastic events in a real cell. Even at the very limited level of well-stirred biochemistry, current models barely begin to describe the known menagerie of signaling molecules. The second problem relates to the limitations of experimental data. It is perhaps presumptuous to attempt to devise models involving dozens of signaling pathways given the current state of our knowledge. Thus far, no more than two or three signaling molecules have ever been simultaneously monitored in real time *in vivo*.[24] There is also a growing recognition that macroscopic, test-tube experiments are seriously flawed as models of cellular signaling. Cellular chemistry is not a point process, or even simply diffusive. Complex sequences of reactions are known to occur in an assembly-line fashion on cytoskeletal scaffolds.[25] Such reactions are not necessarily well-described by bulk concentrations, rates, or even mechanisms.

From a practical viewpoint, both problems are addressed by the simple observation that many models do indeed replicate and predict experimental data. These models are therefore, at least, empirically useful. To the extent that such models remain faithful to biology, they promise to provide the theoretical counterpart to the biology of signaling. It is increasingly evident that such a theoretical understanding will be critical to overcoming two of the major difficulties in the field: complexity and patchy experimental data. In concert with web-based databases of signaling pathways (URLs in Section 2.7), the methods described in this book form an emerging framework for a quantitative approach to studying cellular signaling.

2.6 GLOSSARY OF ABBREVIATIONS

PKC	Protein Kinase C
PKA	Protein Kinase A
CaMKII	Calcium Calmodulin Activated Protein Kinase Type II
MAPK	Mitogen-Activated Protein Kinase
AA	Arachidonic Acid
DAG	Diacylglycerol
cAMP	cyclic Adenosine MonoPhosphate
CaM	Calmodulin
GTP	Guanosine TriPhosphate
ACII, ACIII	Adenylyl Cyclases Type II, III
PLA_2	Phospholipase A2
EGF	Epidermal Growth Factor
EGFR	Epidermal Growth Factor Receptor

2.7 FURTHER READING

Modeling Signaling Pathways

Lauffenberger, D. A. and Linderman, J. J., *Receptors: Models for Binding, Trafficking and Signaling,* Oxford University Press, New York, 1993.

Thomas, R. and Thieffry, D., Developing a logical tool to analyse biological regulatory networks, in *Computing with Biological Metaphors*, Paton, R., Ed., Chapman and Hall, London, 1994.

Bhalla, U. S., The network within: Signaling Pathways, in *The Book of GENESIS*, 2nd ed., Bower, J. M. and Beeman, D., Eds., Springer-Verlag, New York, 1998.

Computation by Signaling Pathways

Bray, D., Protein molecules as computational elements in living cells, *Nature,* 376, 307, 1995.

Enzyme Basics

Stryer, L., *Biochemistry 4th Edition*, W. H. Freeman, New York, 1995, chap. 8.

Cell Biology Basics

Alberts, B., Bray, D., Lewis, J., Raff, M., Roberts, K., and Watson, J. D., *Molecular Biology of the Cell,* Garland Publishing, New York, 1983.

2.8 URLs

Kinetic Simulation Software

GENESIS/kinetikit: *http://www.bbb.caltech.edu/GENESIS*
XPP: *http://www.pitt.edu/~phase/*
Vcell: *http://www.nrcam.uchc.edu*

Models and Databases of Pathways

Signaling pathways, concentrations and rates from U.S. Bhalla and R.I. Iyengar, 1999: *http://www.ncbs.res.in/~bhalla/ltploop/* or *http://piris.pharm.mssm.edu/urilab/*
Genomics database of pathways: *http://www.genome.ad.jp/kegg/*
Cross-reference records of signalling molecules: *http://bioinfo.mshri.on.ca.*

REFERENCES

1. Michaelis, L. and Menten, M. L., Die Kinetik der invertinwirkung, *Biochem. Z.*, 49, 333, 1913.
2. Stryer, L., *Biochemistry*, 4th ed., W. H. Freeman, New York, 1995.
3. Hines, M., Efficient computation of branched nerve equations, *Int. J. Bio-Med. Comput.,* 15, 69, 1984.
4. Press, W. H., Flannery, B. P., Teukolsky, S. A., and Vetterling, W. T., *Numerical Recipes in C: The Art of Scientific Computing,* Cambridge University Press, Cambridge, 1988.
5. Mascagni, M. V. and Sherman, A. S., Numerical methods for neuronal modeling, in *Methods in Neuronal Modeling: From Ions to Networks,* Koch, C. and Segev, I., Eds., 2nd ed., MIT Press, Cambridge, 1998, 569.
6. MacGregor, R. J., *Neural and Brain Modeling*, Academic Press, San Diego, 1987.
7. Nishizuka, Y., Intracellular signaling by hydrolysis of phospholipids and activation of protein kinase C, *Science,* 258, 607, 1982.
8. Døskeland, S. O. and Øgreid, D., Characterization of the interchain and intrachain interactions between the binding sites of the free regulatory moeity of protein kinase I, *J. Biol. Chem.,* 259, 2291, 1984.
9. Hanson, P. I. and Schulman, H., Neuronal Ca^{2+}/Calmodulin-dependent protein kinases, *Annu. Rev. Biochem.,* 61, 559, 1992.
10. Huang, C.-Y. F. and Ferrel, J. E., Jr., Ultrasensitivity in the mitogen-activated protein kinase cascade, *Proc. Natl. Acad. Sci. U.S.A.*, 93, 10078, 1996.
11. Bhalla, U. S. and Iyengar, R. I., Emergent properties of networks of biological signaling pathways, *Science,* 283, 381, 1999.
12. Cobb, M. H., Boulton, T. G., and Robbins, D. J., Extracellular signal-regulated kinases: ERKs in progress, *Cell Regul.,* 2, 965, 1991.
13. Ahn, N. G., Seger, R., and Krebs, E.G., The mitogen-activated protein kinase activator, *Curr. Opin. Cell. Biol.,* 4, 992, 1992.
14. Sanghera, J. S., Paddon, H. B., Bader, S. A., and Pelech, S. L., Purification and characterization of a maturation-activated myelin basic protein kinase from sea star oocytes, *J. Biol. Chem.,* 265, 52, 1990.

15. Ortiz, J., Harris, H. W., Guitart, X., Terwillinger, R. Z., Haycock, J. W., and Nestler, E. J., Extracellular signal-regulated protein kinases (ERKs) and ERK kinase (MEK) in brain: regional distribution and regulation by chronic morphine, *J. Neurosci.,* 15, 1285, 1995.
16. Posner, I., Engel, M., and Levitzki, A., Kinetic model of the epidermal growth factor (EGF) receptor tyrosine kinase and a possible mechanism of its activation by EGF, *J. Biol. Chem.*, 267, 20638, 1992.
17. Alberts, B., Bray, D., Lewis, J., Raff, M., Roberts, K., and Watson, J. D., *Molecular Biology of the Cell*, Garland Publishing, New York, 1983.
18. Vanier, M. C. and Bower, J. M., A comparative survey of automated parameter-search methods for compartmental neural models, *J. Comput. Neurosci.,* 7, 149, 1999.
19. Biddlecome, G. H., Berstein, G., and Ross, E. M., Regulation of phospholipase C-β1 by Gq and m1 muscarinic cholinergic receptor. Steady-state balance of receptor-mediated activation and GTPase-activating protein-promoted deactivation, *J. Biol. Chem.,* 271, 7999, 1996.
20. Lin, L.-L., Wartmann, M., Lin, A. Y., Knopf, J. L., Seth, A., and Davis, R. J., $cPLA_2$ is phosphorylated and activated by MAP kinase, *Cell,* 72, 269, 1993.
21. Wahl, M. I., Jones, G. A., Nishibe, S., Rhee, S. G., and Carpenter, G., Growth factor stimulation of phospholipase C-γ1 activity. Comparative properties of control and activated enzymes, *J. Biol. Chem.,* 267, 10447, 1992.
22. De Koninck, P. and Schulman, H., Sensitivity of CaM kinase II to the frequency of Ca^{2+} oscillations, *Science,* 279, 227, 1998.
23. Bray, D., Protein molecules as computational elements in living cells, *Nature,* 376, 307, 1995.
24. De Bernardi, M. A. and Brooker, G., Single cell Ca/cAMP cross-talk monitored by simultaneous Ca/cAMP fluorescence ratio imaging, *Proc. Natl. Acad. Sci. U.S.A.,* 93, 4577, 1996.
25. Garrington, T. P. and Johnson, G. L., Organization and regulation of mitogen-activated protein kinase signaling pathways, *Curr. Opin. Cell Biol.,* 11, 211, 1999.
26. Schaechter, J. D. and Benowitz, L. I., Activation of protein kinase C by arachidonic acid selectively enhances the phosphorylation of GAP-43 in nerve terminal membranes, *J. Neurosci.,* 13, 4361, 1993.

3 Modeling Local and Global Calcium Signals Using Reaction-Diffusion Equations

Gregory D. Smith

CONTENTS

0-8493-2068-2/01/$0.00+$.50

3.1 CALCIUM SIGNALING IN NEURONS AND OTHER CELL TYPES

Chemical reaction and diffusion are central to quantitative neurobiology and biophysics. An obvious example is action potential-dependent chemical neurotransmission, a process that begins with influx of Ca^{2+} through presynaptic Ca^{2+} channels. As Ca^{2+} ions diffuse away from the mouth of voltage-gated plasma membrane Ca^{2+} channels and into the cytosolic milieu, "domains" of elevated intracellular Ca^{2+} concentration activate proteins associated with neurotransmitter release.[1] These Ca^{2+} domains are formed in the presence of ubiquitous Ca^{2+}-binding proteins of the presynaptic terminal. By binding and releasing free Ca^{2+}, endogenous Ca^{2+}-binding proteins and other "Ca^{2+} buffers" determine the range of action of Ca^{2+} ions, influence the time course of their effect, and facilitate clearance of Ca^{2+}.[2] Here and throughout this chapter, "Ca^{2+} buffer" refers to any Ca^{2+}-binding species, whether an endogenous Ca^{2+}-binding protein, an exogenous Ca^{2+} chelator (e.g., BAPTA or EGTA), a Ca^{2+} indicator dye, or molecules like ATP and phospholipids with non-specific but important Ca^{2+} buffering capacities (see Box 3.1).

While Ca^{2+} influx via voltage-gated Ca^{2+} channels of the plasma membrane is a major source of cytosolic Ca^{2+} in neurons, another source is the endoplasmic reticulum (ER), a continuous membrane-delimited intracellular compartment that plays an important role in Ca^{2+} signaling. In many neurons, the ER has integrative and regenerative properties analogous to those of the plasma membrane. For example, when metabotropic receptors of the plasma membrane are activated, they stimulate the production of the intracellular messenger, inositol 1,4,5-trisphosphate (IP_3),[3] which promotes Ca^{2+} release from intracellular stores by binding and activating IP_3

receptor Ca^{2+} channels (IP_3Rs) located on the ER membrane.[4] In addition to this IP_3-mediated pathway, a second parallel mechanism for Ca^{2+} release is subserved by ryanodine receptors (RyRs), intracellular Ca^{2+} channels (also on the ER) that are activated by cyclic ADP ribose. Importantly, both IP_3Rs and RyRs can be activated and/or inactivated by intracellular Ca^{2+}, leading to what has been called a "Ca^{2+} excitability" of the ER.[5,6] This Ca^{2+} excitability is the physiological basis for propagating waves of Ca^{2+}-induced Ca^{2+} release (CICR) that can be observed in neurons and other cell types via confocal microfluorimetry. For a review, see References 7, 8, 9, and 10.

Box 3.1

Representative Cytosolic Ca^{2+} Buffers

Name	k^+ ($\mu M^{-1}s^{-1}$)	k^- (s^{-1})	K (μM)[b]	D ($\mu m^2/s$)[c]	Ref.[a]
		Endogenous			
Troponin-C	90, 100	300, 7	3, 0.05–0.07	[d]	1
Sarcolemmal phospholipids	—	—	1100, 13	—	2
Calmodulin	500, 100	470, 37	0.9–2.0, 0.2–0.4	32	1
Calbindin-D_{28K}	20	8.6	0.4–1.0	27	3
Parvalbumin	6[e]	1	0.00037	36	1
		Exogenous			
EGTA	1.5	0.3	~0.2[f]	113	4
BAPTA	600	100	0.1–0.7	95	5
Fura-2	600	80–100	0.13–0.60	30–95[g]	5,10
Ca^{2+} Green-1	700	170	0.19–0.25	84	6
Ca^{2+} Green-1 dextran	~700	~170	0.24–0.35	20	7

[a] For a review, see References 7, 8, and 9.

[b] A range indicates that measurements were made under different experimental conditions; values for distinct binding sites are separated by a comma.

[c] In some cases estimated using the Stokes–Einstein relation.

[d] The troponin complex is immobile due to its association with actin.

[e] Even slower physiologically because magnesium must dissociate (with a time constant of approximately 1 sec) before Ca^{2+} can bind.[1]

[f] EGTA is strongly dependent on pH variation near 7.0.[7]

[g] As much as 65–70% of fura-2 may be immobilized.[10]

Source: Modified from Reference 21. With permission.

References

1. Falke, J., Drake, S., Hazard, A., and Peersen, O., Molecular tuning of ion binding to Ca^{2+} signaling proteins, *Q. Rev. Biophys.*, 27, 219, 1994.
2. Post, J. and Langer, G., Sarcolemmal Ca^{2+} binding sites in heart: I. Molecular origin in "gas-dissected" sarcolemma, *J. Membr. Bio.*, 129, 49, 1992.

(continued)

Box 3.1 (continued)

3. Koster, H., Hartog, A., Van Os, C., and Bindels, R., Calbindin-D28K facilitates cytosolic Ca^{2+} diffusion without interfering with Ca^{2+} signalling, *Cell Calcium*, 18, 187, 1995.
4. Tsien, R., New Ca^{2+} indicators and buffers with high selectivity against magnesium and protons: design, synthesis, and properties of prototype structures, *Biochem.*, 19, 2396, 1980.
5. Pethig, R., Kuhn, M., Payne, R., Adler, E., Chen, T., and Jaffe, L., On the dissociation constants of BAPTA-type Ca^{2+} buffers, *Cell Calcium*, 10, 491, 1989.
6. Eberhard, M. and Erne, P., Calcium binding to fluorescent calcium indicators: calcium green, calcium orange and calcium crimson, *Biochem. Biophys. Res. Commun.*, 180, 209, 1991.
7. Kao, J., Practical aspects of measuring $[Ca^{2+}]$ with fluorescent indicators, *Methods Cell Biol.*, 40, 155, 1994.
8. Baimbridge, K., Celio, M., and Rogers, J., Ca^{2+}-binding proteins in the nervous system, *Trends Neurosci.*, 15, 303, 1992.
9. Heizmann, C. and Hunziker, W., Intracellular Ca^{2+}-binding proteins: more sites than insights, *Trends Biochem. Sci.*, 16, 98, 1991.
10. Blatter, L. and Wier, W., Intracellular diffusion, binding, and compartmentalization of fluorescent Ca^{2+} indicators indo-1 and fura-2, *Biophys. J.*, 58, 1491, 1990.

Recent experimental evidence indicates that intracellular Ca^{2+} release events in neurons mediated by IP_3Rs and RyRs can be spatially localized, forming "localized Ca^{2+} elevations" analogous to Ca^{2+} domains formed during Ca^{2+} influx. To give just one example, repetitive activation of the synapse between parallel fibers and Purkinje cells evokes synaptic activity that produces IP_3 in discrete locations. In these neurons, subsequent IP_3-mediated Ca^{2+} release from intracellular stores is often localized to individual postsynaptic spines, or to multiple spines and adjacent dendritic shafts.[11] Spatially localized Ca^{2+} release also occurs in non-neuronal cell types. In fact, localized Ca^{2+} release events called Ca^{2+} "puffs" were first observed in the immature *Xenopus* oocycte, where Ca^{2+} puffs mediated by IP_3Rs can be evoked in response to both flash photolysis of caged IP_3 and microinjection of non-metabolizable IP_3 analogue.[12,13] Similarly, localized Ca^{2+} release events known as Ca^{2+} "sparks" are observed in cardiac myocytes.[14,15] Ca^{2+} sparks are mediated by RyRs located on the intracellular Ca^{2+} store of muscle cells, the sarcoplasmic reticulum (SR). During cardiac excitation-contraction coupling, Ca^{2+} sparks activated by Ca^{2+} influx through sarcolemmal Ca^{2+} channels are the "building blocks" of global Ca^{2+} responses that cause muscle contraction.[16]

The remainder of this chapter is organized as follows. First, we describe the governing equations for the reaction of Ca^{2+} with buffers and the buffered diffusion of Ca^{2+}. Numerical simulation of Ca^{2+} domains (and related analytical approximations) are covered in Section 3.3, where these methods are subsequently used to quantify the effect of indicator dye properties on the appearance of localized Ca^{2+} elevations such as Ca^{2+} sparks. In Section 3.4, reaction-diffusion equations are used to simulate propagating Ca^{2+} waves, the macroscopic properties of which are shown to depend on the discrete nature of Ca^{2+} release. Because of the many commonalities between Ca^{2+} regulatory mechanisms in neurons and myocytes, the mathematical and computational methods presented are readily applied to cellular neuronal models.

3.2 THEORETICAL FOUNDATION

Reaction-diffusion equations are often used to simulate the buffered diffusion of intracellular Ca^{2+}, an important process to include in biophysically realistic neuronal models. In this section, we first consider temporal aspects of Ca^{2+} signaling by writing and analyzing a system of ordinary differential equations (ODEs) for the reaction kinetics of Ca^{2+} interacting with Ca^{2+} buffer. Secondly, theoretical issues related to spatial aspects of Ca^{2+} signaling are explored by extending these equations to include diffusion of Ca^{2+} and buffer.

3.2.1 EQUATIONS FOR THE KINETICS OF CALCIUM BUFFERING

3.2.1.1 Single Well-Mixed Pool

We begin by assuming a single well-mixed pool (e.g., the cytoplasm of a neuron) where a bimolecular association reaction between Ca^{2+} and buffer takes place,

$$Ca^{2+} + B \underset{k^-}{\overset{k^+}{\longleftrightarrow}} CaB \tag{3.1}$$

In Equation 3.1, B represents free buffer, CaB represents Ca^{2+} bound buffer, and k^+ and k^- are association and dissociation rate constants, respectively. If we further assume that the reaction of Ca^{2+} with buffer follows mass action kinetics, we can write the following system of ODEs for the change in concentration of each species (see Chapter 2),

$$\frac{d[Ca^{2+}]}{dt} = R + J \tag{3.2}$$

$$\frac{d[B]}{dt} = R \tag{3.3}$$

$$\frac{d[CaB]}{dt} = -R \tag{3.4}$$

where the common reaction terms, R, are given by

$$R = -k^+[Ca^{2+}][B] + k^-[CaB] \tag{3.5}$$

and J represents Ca^{2+} influx. Both R and J have units of concentration per unit time; for example, if t is measured in seconds and $[Ca^{2+}]$ is measured in μM, then J has units of μM s^{-1}. If the current, i_{Ca}, through a collection of plasma membrane calcium channels generates a $[Ca^{2+}]$ increase in cellular volume, V, then the rate of Ca^{2+} influx is given by

$$J = \frac{\sigma}{V} = \frac{i_{Ca}}{2FV} = \left(5.1824\times10^{6}\right)\frac{i^{*}_{Ca}}{V^{*}}\,\mu M\,s^{-1} \tag{3.6}$$

where $F = 9.648 \times 10^4$ coul/mol is Faraday's constant; and V^* and i^*_{Ca} are volume and current measured in microliters and pA, respectively. Note that the reaction terms, R, appear in Equations 3.2 and 3.3 with the same sign, because association and dissociation of Ca^{2+} and buffer results in a commensurate loss or gain of both species. Conversely, R occurs with opposite sign in Equations 3.3 and 3.4. By defining the total buffer concentration, $[B]_T$, as $[B]_T \equiv [B] + [CaB]$, and summing Equations 3.3 and 3.4, we see that $d[B]_T/dt = 0$, i.e., total buffer concentration is constant. This implies that the Ca^{2+}-bound buffer concentration is always given by the conservation condition, $[CaB] = [B]_T - [B]$, and Equation 3.4 is superfluous.

Though we have simplified our problem by eliminating an equation, those that remain are nonlinear and a general analytical solution to Equations 3.2 and 3.3 is not known. Because of this, a computational approach is often taken in which these equations are numerically integrated. There will be several concrete examples of such simulations in this chapter; however, let us first consider two simplifications of Equations 3.2 and 3.3 that come about when buffer parameters are in select regimes: the so-called "excess buffer" and "rapid buffer" approximations.

In the excess buffer approximation (EBA),[17–19] Equations 3.2–3.4 are simplified by assuming that the concentration of free Ca^{2+} buffer, $[B]$, is high enough that its loss (via conversion into the Ca^{2+} bound form) is negligible. The EBA gets its name because this assumption of the unsaturability of Ca^{2+} buffer is likely to be valid when Ca^{2+} buffer is in excess, as might be the case when modeling the effect of high concentrations of exogenous Ca^{2+} chelator (e.g., EGTA) on an intracellular Ca^{2+} transient. Following a description of the EBA (see below), we present a complementary reduced formulation, the rapid buffer approximation (RBA).[20–22] In the RBA, Equations 3.2–3.4 are simplified by assuming that the reaction kinetics (association and dissociation rates) are rapid processes compared to the diffusion of Ca^{2+}. Although the derivations of the EBA and RBA presented here are heuristic, these approximations can be obtained in a more rigorous manner using the techniques of regular and singular perturbation of ODEs.[23,23B]

3.2.1.2 The Excess Buffer Approximation

We begin our derivation of the EBA by recalling that the association and dissociation rate constants for the bimolecular association reaction between Ca^{2+} and buffer can be combined to obtain a dissociation constant, K,

$$K = k^{-}/k^{+} \tag{3.7}$$

This dissociation constant of the buffer has units of μM and is the concentration of Ca^{2+} necessary to cause 50% of the buffer to be in Ca^{2+} bound form. To show this, consider the steady states of Equations 3.2–3.4 in the absence of any influx ($J = 0$). Setting the left hand sides of Equations 3.3 and 3.4 to zero, we solve for the equilibrium relations,

$$[B]_\infty = \frac{K[B]_T}{K + \left[Ca^{2+}\right]_\infty} \tag{3.8}$$

and

$$\left[CaB\right]_\infty = \frac{\left[Ca^{2+}\right]_\infty [B]_T}{K + \left[Ca^{2+}\right]_\infty} \tag{3.9}$$

where $[Ca^{2+}]_\infty$ is the "background" or ambient free Ca^{2+} concentration, and $[B]_\infty$ and $[CaB]_\infty$ are the equilibrium concentrations of free and bound buffer, respectively. In these expressions, K is the dissociation constant of the buffer, as defined by Equation 3.7. Note that higher values for K imply that the buffer has a lower affinity for Ca^{2+} and is less easily saturated.

If we assume that the buffer is unsaturable, that is, we assume that changes in $[Ca^{2+}]$ due to influx (J) give rise to negligible changes in free buffer concentration, then we can write $[B] \approx [B]_\infty$ and $[CaB] \approx [CaB]_\infty$. Substituting these values into Equation 3.2 gives,

$$\frac{d\left[Ca^{2+}\right]}{dt} \approx -k^+\left[Ca^{2+}\right][B]_\infty + k^-[CaB]_\infty + J \tag{3.10}$$

which can be rearranged to give,

$$\frac{d\left[Ca^{2+}\right]}{dt} \approx -k^+[B]_\infty\left(\left[Ca^{2+}\right] - \left[Ca^{2+}\right]_\infty\right) + J \tag{3.11}$$

a linear equation for $[Ca^{2+}]$ that is approximately valid in the excess buffer limit. Equation 3.11 expresses the fact that in a compartment where Ca^{2+} buffers are in excess (i.e., one in which Ca^{2+} influx does not significantly perturb the free and bound buffer concentrations), elevated $[Ca^{2+}]$ will exponentially decay to its steady state value, $[Ca^{2+}]_\infty$, with a time constant given by $\tau = 1/k^+ [B]_\infty$. Conversely, if the Ca^{2+} concentration is initially at the background level ($[Ca^{2+}]_{initial} = [Ca^{2+}]_\infty$), Ca^{2+} influx (J) into a compartment with unsaturable buffer results in an exponential relaxation of $[Ca^{2+}]$ to a new value,

$$\left[Ca^{2+}\right]_{final} = \left[Ca^{2+}\right]_\infty + J/k^+[B]_\infty. \tag{3.12}$$

Because we have assumed that the Ca^{2+} buffer is unsaturable, a single well-mixed compartment modeled using Equation 3.11 will never "fill up." In spite of this limitation, an exponentially decaying well-mixed pool is commonly used in neuronal models.[24]

3.2.1.3 The Rapid Buffer Approximation

The RBA is an alternative to the EBA that is based on entirely different considerations. If the equilibration time of a buffer is fast compared to the time-scale of changes in $[Ca^{2+}]$ expected due to Ca^{2+} influx,[20] we can make a quasi-steady-state approximation[25] and assume that changes in buffer concentration occur in such a manner that Ca^{2+} and buffer are essentially always in equilibrium (i.e., $R \approx 0$). The following equilibrium relations result,

$$[CaB] \approx \frac{[Ca^{2+}][B]_T}{K+[Ca^{2+}]} \tag{3.13}$$

and

$$[B] \approx \frac{K[B]_T}{K+[Ca^{2+}]}. \tag{3.14}$$

Using Equations 3.13 and 3.14, we can express the total (free plus bound) Ca^{2+} concentration, $[Ca^{2+}]_T$, as a simple function of $[Ca^{2+}]$,

$$[Ca^{2+}]_T \equiv [Ca^{2+}]+[CaB] = [Ca^{2+}] + \frac{[Ca^{2+}][B]_T}{K+[Ca^{2+}]} \tag{3.15}$$

Using Equations 3.13 and 3.15, two quantities can be defined that are of particular interest. The first is the "buffer capacity," i.e., the differential of bound Ca^{2+} with respect to free Ca^{2+},

$$\kappa = \frac{d[CaB]}{d[Ca^{2+}]} = \frac{K[B]_T}{(K+[Ca^{2+}])^2} \tag{3.16}$$

where the second equality is found by differentiating Equation 3.13. The second, which we will refer to as the "buffering factor," is the differential of free Ca^{2+} with respect to total Ca^{2+}, that is,

$$\beta = \frac{d[Ca^{2+}]}{d[Ca^{2+}]_T} = \frac{1}{1+\kappa} \tag{3.17}$$

which can be derived from Equation 3.15 by calculating $d[Ca^{2+}]_T/d[Ca^{2+}]$ and inverting the result. The significance of κ and β becomes apparent when Equations 3.2 and 3.4 are summed to give

$$d\left[Ca^{2+}\right]_T/dt = J \tag{3.18}$$

which using Equations 3.17 and 3.18 implies

$$\frac{d\left[Ca^{2+}\right]}{dt} = \frac{d\left[Ca^{2+}\right]}{d\left[Ca^{2+}\right]_T}\frac{d\left[Ca^{2+}\right]_T}{dt} = \beta J \tag{3.19}$$

where β is always some number between zero and one ($\beta \approx 1/100$ is not unreasonable, but the exact value depends on $[Ca^{2+}]$ and buffer parameters). Equation 3.19, the RBA for a single well-mixed compartment, quantifies the degree to which rapid buffers attenuate the effect of Ca^{2+} influx (J) on $[Ca^{2+}]$.

Although both β and κ are in general functions of $[Ca^{2+}]$, in certain circumstances this dependence can be weak (see Box 3.2). For example, when the $[Ca^{2+}]$ is high compared to the dissociation constant of the buffer ($[Ca^{2+}] >> K$), κ approaches zero and β approaches one, reflecting the fact that nearly saturated buffers will have little effect on $[Ca^{2+}]$. When $[Ca^{2+}]$ is low compared to the dissociation constant of the buffer ($[Ca^{2+}] << K$), κ and β are approximately constant.

Box 3.2

The Relationship Between κ, β, and D_{eff} in Low and High Ca^{2+} Limits[20,26]

Quantity and Formula	Low Ca^{2+} ($[Ca^{2+}] << K$)	High Ca^{2+} ($[Ca^{2+}] \to \infty$)
$\kappa = \dfrac{K[B]_T}{\left(K+\left[Ca^{2+}\right]\right)^2}$	$\approx \dfrac{[B]_T}{K}$	≈ 0
$\beta = \dfrac{1}{1+\kappa}$	$\approx \dfrac{K}{K+[B]_T}$	≈ 1
$D_{eff} = \beta\left(D_c + D_b\kappa\right)$	$\approx \dfrac{K}{K+[B]_T}\left(D_c + \dfrac{[B]_T}{K}D_b\right)$	$\approx D_c$

3.2.2 Reaction-Diffusion Equations for the Buffered Diffusion of Calcium

3.2.2.1 The Full Equations

To explore spatial aspects of Ca^{2+} signaling we extend Equations 3.2–3.5 to include multiple buffers and the diffusive movement of free Ca^{2+}, Ca^{2+}-bound buffer, and Ca^{2+}-free buffer. Assuming Fickian diffusion in a homogenous, isotropic medium, we write the following system of reaction-diffusion equations,[20]

$$\frac{\partial[Ca^{2+}]}{\partial t} = D_{Ca}\nabla^2[Ca^{2+}] + \sum_i R_i + J \tag{3.20}$$

$$\frac{\partial[B_i]}{\partial t} = D_{Bi}\nabla^2[B_i] + R_i \tag{3.21}$$

$$\frac{\partial[CaB_i]}{\partial t} = D_{CaBi}\nabla^2[CaB_i] - R_i \tag{3.22}$$

where the reaction terms, R_i, are given by

$$R_i = -k_i^+[Ca^{2+}][B_i] + k_i^-[CaB_i], \tag{3.23}$$

i is an index over Ca^{2+} buffers, and Equations 3.21 and 3.22 represent pairs of equations (two for each type of Ca^{2+} buffer that is being modeled).

Because Ca^{2+} has a molecular weight that is small in comparison to most Ca^{2+}-binding species, it is reasonable to assume that the $D_{Bi} \approx D_{CaBi} \equiv D_i$. Given this, Equations 3.21 and 3.22 can be summed to give,

$$\frac{\partial[B_i]_T}{\partial t} = \frac{\partial[CaB_i]}{\partial t} + \frac{\partial[B_i]}{\partial t} = D_i\nabla^2[CaB_i] + D_i\nabla^2[B_i] = D_i\nabla^2[B_i]_T, \tag{3.24}$$

where $[B_i]_T \equiv [B_i] + [CaB_i]$. Thus, providing that the $[B_i]_T$ profiles are initially uniform and there are no sources or sinks for Ca^{2+} buffer, the $[B_i]_T$ will remain uniform for all time. Thus, we write the following equations for the buffered diffusion of Ca^{2+},

$$\frac{\partial[Ca^{2+}]}{\partial t} = D_c\nabla^2[Ca^{2+}] + \sum_i R_i + J \tag{3.25}$$

$$\frac{\partial[B_i]}{\partial t} = D_i \nabla^2 [B_i] + R_i \tag{3.26}$$

where

$$R_i = -k_i^+ [Ca^{2+}][B_i] + k_i^- ([B_i]_T - [B_i]) \tag{3.27}$$

and R_i is expressed in terms of the free species, B_i, using $[CaB_i] = [B_i]_T - [B_i]$ (as in Section 3.2.1.1). Of course, if any of the buffer species being considered happens to be stationary, $D_i = 0$ and the Laplacian term does not enter into the right hand side of Equation 3.26.

3.2.2.2 The Reduced Equations and Effective Diffusion Coefficient

To give some intuition for the behavior of Equations 3.25–3.27, we follow Sections 3.2.1.2 and 3.2.1.3 and consider two reduced formulations. In the case of excess buffer, the equations for the buffered diffusion of Ca^{2+} become[17]

$$\frac{\partial[Ca^{2+}]}{\partial t} \approx D_c \nabla^2 [Ca^{2+}] - k^+ [B]_\infty ([Ca^{2+}] - [Ca^{2+}]_\infty) + J \tag{3.28}$$

while in the limit of rapid buffer, we have[20]

$$\frac{\partial[Ca^{2+}]}{\partial t} \approx \beta \left[(D_c + D_b \kappa) \nabla^2 [Ca^{2+}] - \frac{2 D_b \kappa}{K + [Ca^{2+}]} (\nabla [Ca^{2+}])^2 + J \right] \tag{3.29}$$

Although the derivation of Equation 3.29 is beyond the scope of this chapter,[20] it allows us to quantify the effect of rapid buffers on Ca^{2+} diffusion. For example, the pre-factor of the Laplacian term in Equation 3.29 can be identified as an effective diffusion coefficient, D_{eff}, given by,

$$D_{eff} \equiv \beta (D_c + D_b \kappa) \tag{3.30}$$

with κ and β defined as above. Note that D_{eff} is a monotonic increasing function of $[Ca^{2+}]$. When $[Ca^{2+}]$ is much higher than the dissociation constant of a buffer, the buffer is saturated and D_{eff} is approximately equal to the free Ca^{2+} diffusion coefficient, D_c (see Box 3.2). When $[Ca^{2+}]$ is low compared to the dissociation constant of a buffer, D_{eff} is approximately

$$D_{eff} \approx \frac{K}{K+[B]_T}\left(D_c + \frac{[B]_T}{K} D_b\right) \tag{3.31}$$

Equation 3.29 can be generalized to the case of multiple buffers, and buffer parameters can be constrained by requiring that D_{eff} be consistent with experiment.[20,27] When stationary buffers alone are under consideration ($D_b = 0$) and $[Ca^{2+}]$ is very low or high, Equation 3.19 is formally equivalent to the cable equation (see Chapter 8).[26] Furthermore, provided $D_b < D_c$ (a physiologically reasonable assumption), D_{eff} will necessarily be less than or equal to D_c, that is,

$$D_{eff} = \beta(D_c + D_b\kappa) \leq D_c \tag{3.32}$$

Thus, Ca^{2+} buffers (whether stationary or mobile) always reduce the effective diffusion coefficient for Ca^{2+}. However, Equation 3.29 shows that the effect of rapid Ca^{2+} buffers on Ca^{2+} transport is complicated by a non-diffusive contribution that is negative for all values of the Ca^{2+} gradient.[20,28,29]

3.3 MODELING CALCIUM DOMAINS AND LOCALIZED CALCIUM ELEVATIONS

3.3.1 Numerical Solution of the Full Equations

3.3.1.1 Geometry of Simulation and Initial and Boundary Conditions

To complete a reaction-diffusion formulation for the buffered diffusion of Ca^{2+}, a particular geometry of simulation must be specified and Equations 3.25–3.27 must be supplemented with boundary conditions and initial concentration profiles. In this section we will describe a numerical model of a Ca^{2+} domain (or local Ca^{2+} elevation) due to a point source for Ca^{2+} that represents either a single Ca^{2+} channel (or a tight cluster of channels).

The case of a single point source can be greatly simplified by assuming spherical symmetry. Position in a spherical polar coordinate system is specified by (r, θ, ϕ), where r is the radius (distance from the point source), θ is the declination, and ϕ is the azimuthal angle. In these coordinates, the Laplacian is given by

$$\nabla^2 = \frac{1}{r^2}\frac{\partial}{\partial r}\left(r^2 \frac{\partial}{\partial r}\right) + \frac{1}{r^2 \sin\theta}\frac{\partial}{\partial \theta}\left(\sin\theta \frac{\partial}{\partial \theta}\right) + \frac{1}{r^2 \sin^2\theta}\left(\frac{\partial^2}{\partial \phi^2}\right) \tag{3.33}$$

However, spherically symmetric concentration profiles will not have any functional dependence on the declination (ϕ) or azimuthal angle (θ), so Equation 3.33 simplifies to

$$\nabla^2 = \frac{1}{r^2}\frac{\partial}{\partial r}\left(r^2 \frac{\partial}{\partial r}\right) = \frac{\partial^2}{\partial r^2} + \frac{2}{r}\frac{\partial}{\partial r} \tag{3.34}$$

This assumption of spherical symmetry is justified when considering Ca^{2+} release from deep within a cell (where the elevated Ca^{2+} profile does not interact with boundaries) or Ca^{2+} influx through a plasma membrane Ca^{2+} channel (because hemispherical symmetry is equivalent to spherical symmetry with an adjusted source amplitude).[30]

A reasonable initial condition for this simulation is a uniform "background" Ca^{2+} profile of $[Ca^{2+}]_\infty = 0.1$ μM. We further assume that all buffers are initially in equilibrium with Ca^{2+} and we require buffer far from the source to remain in equilibrium with Ca^{2+} at all times,

$$\lim_{r \to \infty} [Ca^{2+}] = [Ca^{2+}]_\infty \tag{3.35}$$

and

$$\lim_{r \to \infty} [B_i] = [B_i]_\infty . \tag{3.36}$$

Near the source, we enforce the boundary conditions,

$$\lim_{r \to 0} \left(4\pi D_c r^2 \frac{\partial [Ca^{2+}]}{\partial r} \right) = \sigma \tag{3.37}$$

and

$$\lim_{r \to 0} \left(4\pi D_i r^2 \frac{\partial [B_i]}{\partial r} \right) = 0 \tag{3.38}$$

implying an influx of free Ca^{2+} at rate σ (expressed in units of μmoles/s and related to current by Faraday's law, $\sigma = i_{Ca} / z F$).

3.3.1.2 A Finite Difference Method for Solving the Full Equations

With the governing equations, geometry, and initial and boundary conditions specified, our model of a localized Ca^{2+} elevation is completely formulated. It remains to present a numerical scheme that will implement this model. For simplicity this section describes the forward Euler finite difference scheme for solving

Equations 3.25–3.27 with initial and boundary conditions as in Section 3.3.1.1. For review of more elaborate numerical methods (e.g., Crank–Nicolson-like implicit schemes) applied to neuroscience problems see References 24 and 31. Important issues that arise during the implementation of multidimensional numerical schemes are discussed in Reference 32.

We begin the description of the numerical scheme by choosing $J + 1$ mesh points

$$0 = r_0 < r_1 < r_2 < \ldots < r_{J-2} < r_{J-1} < r_J = R_{\max} \tag{3.39}$$

and writing a finite difference approximation to the Laplacian,

$$L\left(U_j^n\right) = \frac{1}{r_j^2\left(\Delta r_{j+1} + \Delta r_j\right)}\left[r_{j+1/2}^2\left(\frac{U_{j+1}^n - U_j^n}{\Delta r_{j+1}}\right) - r_{j-1/2}^2\left(\frac{U_j^n - U_{j-1}^n}{\Delta r_j}\right)\right](1 \le j \le J) \tag{3.40}$$

where U_j^n is an approximation to the function $u(r_j, t_n)$, u represents the concentration of Ca^{2+} or buffer, t_n is discrete time, and $t_n = n\Delta t$. In this expression, Δr_j is defined by $\Delta r_j = r_j - r_{j-1}$ for $1 \le j \le J$. Similarly, $r_{j+1/2}$ and $r_{j-1/2}$ are defined as $r_{j+1/2} = r_j + \Delta r_{j+1} / 2$ and $r_{j-1/2} = r_j - \Delta r_j / 2$.

Because r_j occurs in the denominator of Equation 3.40, this expression is unusable when $r_j = 0$. At the origin ($r_j = 0$, $j = 0$) a finite difference approximation to the Laplacian is given by,[33]

$$L\left(U_0^n\right) = \frac{3}{\Delta r^2}\left(2U_1^n - 2U_0^n\right) \tag{3.41}$$

where in agreement with Equation 3.38 we have used a reflective (or no flux) boundary condition

$$\left.\frac{\partial u}{\partial r}\right|_{r=0} = 0$$

in its discrete form, $U_{-1}^n = U_1^n$. This is intuitive when u represents buffer, because buffer cannot be transported via diffusion either into or out of the domain of the simulation. When u represents Ca^{2+}, Equation 3.37 is satisfied by incrementing the Ca^{2+} concentration at the origin by $J_{rel}\,\Delta t$ every time step in accordance with the time-dependence of σ and the volume that this mesh point represents (a sphere of radius $\Delta r/2$), and Faraday's law (see above). Thus,

$$J_{rel} = \frac{\sigma}{V} = \frac{i_{Ca}}{2FV} = \frac{6i_{Ca}^{*}pA}{2\left(9.648\times10^4 coul/mol\right)\pi\left(\Delta r^{*}\mu m\right)^3} = \frac{i_{Ca}^{*}(9897)\mu M\, s^{-1}}{\left(\Delta r^{*}\right)^3} \tag{3.42}$$

where i^*_{Ca} is measured in pA and Δr^* is measured in micrometers. The units of Equation 3.42 are reconciled by noting that 1 pA = 5.182×10^{-18} mol/s, 1 $\mu m^3 = 10^{-15}$ liters and A = coul/s.

Finally, if we impose an absorbing boundary condition for each species far from the source ($r = R_{max}$), then we can write the following forward Euler numerical scheme,

$$U_j^{n+1} = U_j^n + \Delta t\left[L\left(U_j^n\right) + R_j\right] \quad (0 \le j \le J) \tag{3.43}$$

where U_{J+1}^n is given by the equilibrium value, $[U]_\infty$, when needed for the Laplacian.

3.3.1.3 A Representative Simulation of a Localized Ca^{2+} Elevation

Color Figure 3.1* shows an example of a simulation of a local Ca^{2+} elevation performed using the numerical method just described. In this calculation, 250 μM stationary buffer was included in addition to 50 μM mobile buffer (both with K of 10 μM). A source amplitude of 5 pA corresponds to a cluster of IP_3Rs or RyRs (see legend for other parameters). Essentially, Color Figure 3.1 is a snapshot of the concentration profiles for each species in the problem after an elapsed time of 1 ms. Note the elevated $[Ca^{2+}]$ near the release site (*red line*). Because some of the free Ca^{2+} that enters the simulation at the origin reacts with buffer, the concentration of bound buffer (*solid green and blue lines*) is elevated near the source. Conversely, the concentration of free buffer (*dashed lines*) decreases near the source.

Color Figure 3.1 demonstrates an important difference between stationary and mobile Ca^{2+} buffers. By following the free and bound buffer profiles toward to source until these profiles cross, one finds the distance at which the buffer is 50% saturated. This distance is almost 0.3 μm for stationary buffer while it is much less, about 0.2 μm, for mobile buffer. In this simulation, stationary buffer is more easily saturated than mobile buffer, in spite of its fivefold higher concentration.

3.3.2 Estimates of Domain Calcium

3.3.2.1 Estimating Domain Ca^{2+} Using Steady-States to the Full Equations

If buffer and source parameters are appropriately specified, simulations such as Color Figure 3.1 can be used to estimate domain Ca^{2+}. Computer simulations suggest that Ca^{2+} domains approach a steady-state value very quickly after Ca^{2+} channels open, usually in less than a microsecond (see Box 3.3). This fact makes steady-state solutions of Equations 3.25–3.27 particularly relevant. Because numerical solution of steady-state profiles is computationally less intensive than solving time-dependent reaction-diffusion equations, this approach is common.[34,35,35B]

---

* Color Figure 3.1 follows page 140.

Assuming a single buffer and spherical symmetry, steady-states of Equations 3.25–3.27 will satisfy the following two coupled ODEs,

$$0 = D_c \frac{1}{r^2}\frac{d}{dr}\left(r^2 \frac{d[Ca^{2+}]}{dr}\right) - k^+[Ca^{2+}][B] + k^-([B]_T - [B]) \tag{3.44}$$

$$0 = D_b \frac{1}{r^2}\frac{d}{dr}\left(r^2 \frac{d[B]}{dr}\right) - k^+[Ca^{2+}][B] + k^-([B]_T - [B]) \tag{3.45}$$

with boundary conditions given by Equations 3.35–3.38.

If buffer is stationary ($D_b = 0$), the first term of Equation 3.45 is zero, implying the equilibrium relations, Equations 3.13 and 3.14. Substituting these into Equation 3.44 gives,

$$0 = \frac{d}{dr}\left(r^2 \frac{d[Ca^{2+}]}{dr}\right) \tag{3.46}$$

When this equation is integrated twice (using Equations 3.35 and 3.37, the boundary conditions for Ca^{2+}, to identify integration constants), we find that the steady-state Ca^{2+} profile is

$$[Ca^{2+}] = \sigma/4\pi D_c r + [Ca^{2+}]_\infty . \tag{3.47}$$

Because Equations 3.46 and 3.47 also arise in the absence of buffer ($[B_m]_T = 0$), we see that stationary buffers do not affect the steady-state Ca^{2+} profile in a Ca^{2+} domain.

In the following two sections, we return to Equations 3.44 and 3.45 without making the assumption that Ca^{2+} buffer is stationary. We consider these equations in the two asymptotic limits introduced in Sections 3.2.1.2 and 3.2.1.3. In both the excess and rapid buffer limits, analytical steady-state solutions for the Ca^{2+} and buffer profiles near an open Ca^{2+} channel have been derived that give insight into the effect of buffers on Ca^{2+} domains.

3.3.2.2 The Steady-State Excess Buffer Approximation

As in Section 3.2.1.2, we begin by assuming that the mobile buffer is unsaturable ($[B] \approx [B]_\infty$ for all r). Equation 3.44 then simplifies to,[17]

$$0 = D_c \frac{1}{r^2}\frac{d}{dr}\left(r^2 \frac{d[Ca^{2+}]}{dr}\right) - k^+[B]_\infty([Ca^{2+}] - [Ca^{2+}]_\infty) \tag{3.48}$$

The solution of this linear equation is the steady-state EBA,

$$[Ca^{2+}] = \frac{\sigma}{4\pi D_c r}\exp(-r/\lambda) + [Ca^{2+}]_\infty \tag{3.49}$$

where the length constant, λ, is given by

$$\lambda = \sqrt{\frac{D_c}{k^+[B]_\infty}} \tag{3.50}$$

Comparing Equations 3.47 and 3.49 we see that high concentrations of free mobile buffer ($[B]_\infty$) and large association rate constants (k^+) lead to restricted Ca^{2+} domains (see Box 3.4).[17,18,19,30] This helps us to understand the efficacy of rapid exogenous Ca^{2+} buffers (such as BAPTA) over buffers with slower kinetics (such as EGTA) during the experimental manipulation of Ca^{2+} domains.

Box 3.3 The Time Scale of Calcium Domain Formation

Example 1. Ignoring for the moment the effect of Ca^{2+} buffers, a quick calculation allows one to estimate the time scale for formation of a Ca^{2+} domain. Assuming spherical symmetry and a point source for Ca^{2+} turned on at time zero, the Ca^{2+} profile (in the absence of buffer) is given by[36,37]

$$[Ca^{2+}](r,t) = \frac{\sigma}{4\pi D_c r}\operatorname{erfc}\left(\frac{r}{2\sqrt{D_c t}}\right) + [Ca^{2+}]_\infty \tag{1}$$

Using erfc(0.005) $\approx$ 0.99, neglecting $[Ca^{2+}]_\infty$, and assuming D_c = 250 μm/s, calculate how much time must elapse for Ca^{2+} to achieve 99% of its steady-state value at a distance of 10 nm, 100 nm, or 1 μm from the source.

Example 2. Using the numerical solver for the full equations (Equations 3.25–3.27) for the buffered diffusion of Ca^{2+}, set $[B_i]_T = 0$ and confirm the values you obtained.

Example 3. Choose realistic buffer parameters (see Box 3.1) and repeat the simulation. Does Ca^{2+} buffer strongly affect the time it takes for $[Ca^{2+}]$ to achieve 99% steady-state?

3.3.2.3 The Steady-State Rapid Buffer Approximation

The steady-state RBA near an open Ca^{2+} channel can be derived by subtracting Equations 3.44 and 3.45 to give

$$0 = \frac{1}{r^2}\frac{d}{dr}\left[r^2\frac{d}{dr}\left(D_c[Ca^{2+}] - D_b[B]\right)\right] \tag{3.51}$$

which, similar to Equation 3.46, can be solved by integrating twice to obtain,

$$D_c[Ca^{2+}] - D_b[B] = A_1/r + A_2 \tag{3.52}$$

Using the boundary conditions, Equations 3.35–3.38, the integration constants, A_1 and A_2, are found,

$$D_c[Ca^{2+}] - D_b[B] = \frac{\sigma}{4\pi r} + D_c[Ca^{2+}]_\infty - D_b[B]_\infty \tag{3.53}$$

where $[B]_\infty$ is given by Equation 3.8. As in Section 3.2.1.3, we now assume that buffer kinetics are rapid and the equilibrium relation is approximately valid.[20] Using Equation 3.14 we substitute for $[B]$ in Equation 3.53 to give an implicit expression for $[Ca^{2+}]$,

$$D_c[Ca^{2+}] - D_b\frac{K[B]_T}{K + [Ca^{2+}]} = \frac{\sigma}{4\pi r} + D_c[Ca^{2+}]_\infty - D_b[B]_\infty \tag{3.54}$$

which using the quadratic formula gives the steady-state RBA,

$$[Ca^{2+}] = \frac{1}{2D_c}\left(-D_cK + \frac{\sigma}{4\pi r} + D_c[Ca^{2+}]_\infty - D_b[B]_\infty + \sqrt{\left(D_cK\frac{\sigma}{4\pi r} + D_c[Ca^{2+}]_\infty - D_b[B]_\infty\right) + 4D_cD_bK[B]_T}\right) \tag{3.55}$$

Box 3.4 Predicted Restriction of Calcium Domains

Example 1. Compare the steady-state RBA solution (Section 3.3.2.3, Equation 3.55) and the unbuffered Ca^{2+} domain (Equation 3.47). The difference,

$$\Delta[Ca^{2+}] = [Ca^{2+}]_{RBA} - [Ca^{2+}]_{No\ Buffer}, \tag{1}$$

is a negative valued function of r expressing the attenuation of domain Ca^{2+} by rapid mobile Ca^{2+} buffer. By calculating $\lim_{r\to 0} \Delta[Ca^{2+}]$, it can be shown that the decrease in domain Ca^{2+} due to the presence of a rapid mobile buffer near the source is,[22]

$$\Delta[Ca^{2+}] = -\frac{D_b}{D_c}[B]_\infty \tag{2}$$

This result suggests that the ability of a mobile buffer to decrease domain Ca^{2+} is dependent on the background Ca^{2+} concentration, because changes in $[Ca^{2+}]_\infty$ will affect $[B]_\infty$. The mathematically inclined reader may be interested in confirming this result.

(continued)

Box 3.4 (continued)

Example 2. Estimate the decrease in domain Ca^{2+} due to the presence of excess buffer. If

$$\Delta[Ca^{2+}] = [Ca^{2+}]_{\text{EBA}} - [Ca^{2+}]_{\text{No Buffer}},$$

it can be shown that

$$\lim_{r\to 0} \Delta[Ca^{2+}] = -\frac{\sigma}{4\pi D_c \lambda}$$

where λ is given by Equation 3.50 in Section 3.3.2.2. Confirm this result.

3.3.2.4 The Validity of the Steady-State Excess and Rapid Buffer Approximations

In the derivations of both the EBA and RBA presented above, assumptions have been made that may be more or less valid depending on the buffer parameters and source amplitude of interest. By comparing numerically calculated steady-state solutions of the full equations (Equations 3.44 and 3.45) to the analytical approximations given by the EBA and RBA (Equations 3.49 and 3.55), the conditions for the validity of both approximations can be investigated (see Box 3.4). As suggested by their names, the EBA tends to be valid when $[B]_T$ is large compared to the dissociation constant of the buffer, K. Conversely, the RBA tends to be valid when the association and dissociation rate constants of the buffer are large.

The derivations of the EBA and RBA presented in Sections 3.3.2.2 and 3.3.2.3 are heuristic, as were the derivations that first occurred in the literature.[17,22] Importantly, the EBA and RBA can be rigorously derived by nondimensionalizing Equations 3.44 and 3.45, identifying small dimensionless parameters, and performing perturbation analysis.[23,23B] Using this approach, the EBA and RBA have been shown to be the first order terms of two different asymptotic expansions of the steady-state to the full equations, Equations 3.25–3.27. Higher order terms have been derived, expanding the parameter regime over which each approximation is valid.[23] Such mathematical analysis and related numerical studies[21] have shown that the EBA and RBA are seldom in agreement. However, for many choices of buffer and source parameters, one or the other of these approximations is often valid (see Box 3.5).

The complementarity between the EBA and RBA can be understood in the following way. The EBA assumes unsaturable buffer, and as a consequence the buffer profile predicted by the EBA is unperturbed near the source, that is,

$$\lim_{r\to 0}[B]_{EBA} = [B]_\infty \tag{3.56}$$

Conversely, the RBA assumes local equilibrium between Ca^{2+} and buffer, Equation 3.14. Because $\lim_{r \to 0} [Ca^{2+}] = \infty$ (this follows from Equation 3.37), the buffer profile predicted by the RBA saturates near the source, that is,

$$\lim_{r \to 0} [B]_{RBA} = 0. \tag{3.57}$$

In general $[B]_\infty$ is not zero and either Equation 3.56 or 3.57 (but not both) may be in agreement with solutions of the full equations, Equations 3.44 and 3.45 (the correct result). Thus, for any given set of buffer and source parameters, we do not expect both the EBA and RBA to be simultaneously valid. Furthermore, Equations 3.56 and 3.57 suggest that the EBA is likely valid when the source amplitude is small, because the buffer profile will be relatively unperturbed near a weak source. Conversely, the RBA is likely valid when the source amplitude is large, because a strong source promotes buffer saturation. (See Box 3.5 and Reference 23.)

Box 3.5 The Validity of the EBA and RBA

The validity of the EBA and RBA can be determined by comparing these analytical approximations (Equations 3.49 and 3.55) to numerically calculated steady-state solutions of the full equations, Equation 3.44 and 3.45.

For example, Figure B3.1A shows a semilog-y plot of a simulated Ca^{2+} domain (see legend for parameters). The *solid line* is calculated using the full equations and the *dotted line* is the RBA, which is accurate enough that the two curves are indistinguishable. The free buffer profile is plotted in Figure B3.1B. Near the source, the RBA (*dotted line*) slightly overestimates the degree of buffer saturation (cf. *solid line*).

Figure B3.1C and D show similar calculations. Here buffer and source parameters are chosen so that the EBA is valid. The EBA (*dotted line*) slightly underestimates the Ca^{2+} domain and neglects some saturation of buffer predicted by the full equations (*solid line*). Note that the EBA Ca^{2+} domain is much smaller than the RBA domain. This reflects the restriction of the EBA Ca^{2+} domain by excess buffer (see Section 3.3.2.2) but also a reduced source amplitude in Figures B3.1C and D, compared to A and B.

Example 1. Using the numerical solver for the steady-state to the full equations, reproduce the free buffer profile shown in Figure B3.1B. Confirm that the steady-state RBA predicts this profile accurately.

Example 2. Now gradually decrease the source amplitude until the free mobile buffer near the source is approximately 50% saturated. With this reduced source amplitude, does the steady-state RBA accurately predict the free Ca^{2+} or buffer profiles? Does a strong source increase or decrease the validity of the steady-state RBA?

(continued)

Box 3.5 (continued)

Example 3. Using parameters as in Figures B3.1C and D, vary the source amplitude between 0.01 pA and 100 pA. For each value you use, compare the steady-state EBA to the steady-state of the full equations. Does a strong source increase or decrease the validity of the EBA?

Examples 2 and 3 suggest that the RBA and EBA can be understood as "strong" and "weak source" approximations, respectively, where strong and weak are defined by the ability of a Ca^{2+} source to locally saturate mobile buffer.[23]

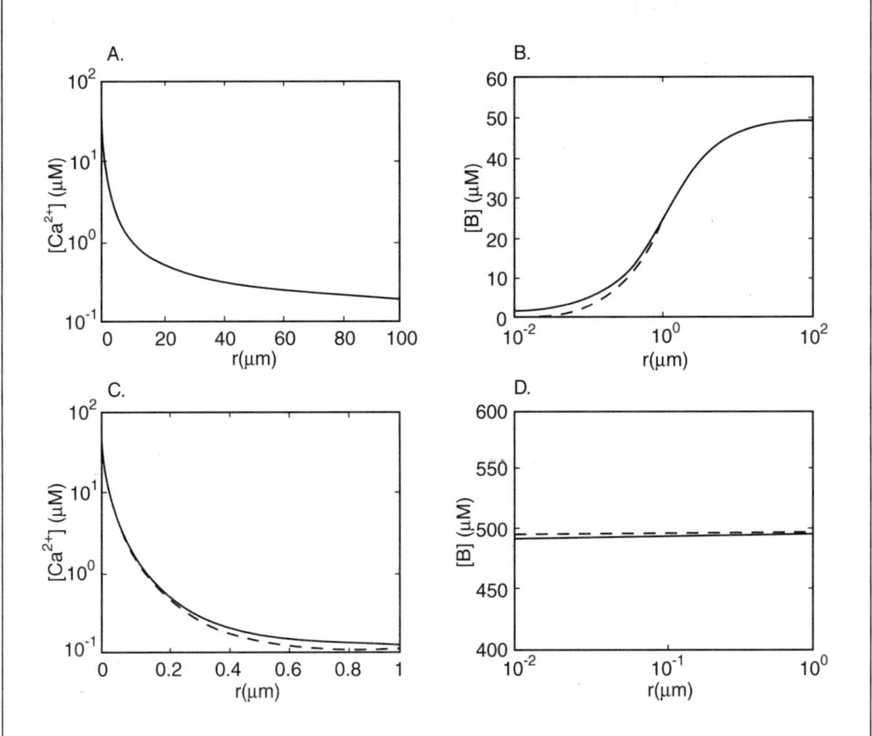

FIGURE B3.1 Comparison of RBA and EBA estimates (*dotted lines*) of steady-state Ca^{2+} and free buffer profiles to the numerically calculated exact result (according to Equations 3.44 and 3.45, *solid lines*). (A) Using parameters where the RBA is marginally valid, the Ca^{2+} profile (*solid line*) overlays the estimate given by the RBA (*dotted line, covered by solid line*). (B) The free buffer profile as estimated by the RBA (*dotted line*) completely saturates near the source (small r), although in the steady-state full equations predict free buffer, B, is only 96% saturated. (C,D) As A,B using a different set of parameters for which the EBA is marginally valid. In D the free buffer profile is relatively unperturbed from the background concentration of 500 μM. Parameters for (A,B): $iCa = 10$ pA, $[B]_T = 50$ μM, $k^+ = 100$ μM^{-1} s^{-1}, $k^- = 10^3$ s^{-1}, $D_c = 250$ μm^2/s, $D_b = 75$ μm^2/s, $[Ca^{2+}]_\infty = 0.1$ μM. Parameters changed for (C,D): $iCa = 0.1$ pA, $[B]_T = 500$ μM, $k^+ = 10$ μM^{-1} s^{-1}, $k^- = 10^2$ s^{-1}. Note that A and C are semilog-y plots while B and D are semilog-x plots.

3.3.3 Modeling the Interaction of Localized Calcium Elevations and Indicator Dyes

3.3.3.1 Including Indicator Dyes in Simulations of Localized Ca^{2+} Elevations

Although the analytical steady-state approximations discussed above give insight into the effect of Ca^{2+} buffers on Ca^{2+} domains, simulations of the dynamics of localized Ca^{2+} elevations such as Ca^{2+} sparks usually begin with time-dependent full or reduced equations for the buffered diffusion of Ca^{2+} (see Section 3.2.2). Simulations are most realistic when the binding of Ca^{2+} with indicator dye is explicitly accounted for in a reaction-diffusion formulation such as Equations 3.25–3.27. For example, in a numerical model of Ca^{2+} spark formation and detection in cardiac myocytes,[38] the Ca^{2+} indicator, fluo-3, was included among four endogenous Ca^{2+}-binding species, including calmodulin and troponin C as well as two additional low-affinity, high-capacity non-specific Ca^{2+}-binding sites provided by the SR and sacrolemmal membranes. Equations 3.25–3.27 were numerically solved giving concentration profiles for all the species in the problem, including Ca^{2+}-free and Ca^{2+}-bound fluo-3. Although more complex models of the relationship between indicator and fluorescence are available (e.g., see model of fura-2 fluorescence in Reference 39), Smith et al. (1998) assumed the fluorescence of fluo-3 is directly proportional to the concentration of bound indicator. In order to account for the optical blurring inherent in confocal (line scan) imaging, they calculated a spatially-averaged Ca^{2+}-bound indicator profile given by,

$$[CaB]_{avg}\left(x;Y_{offset},Z_{offset}\right)=$$

$$\iiint_V [CaB](x',y',z')G\left(x-x',Y_{offset}-y',Z_{offset}-z'\right)dx'dy'dz' \tag{3.58}$$

where G is a multidimensional Gaussian representing the point spread function of the confocal microscope and the integral is taken over the simulation volume. In this expression, and for the remainder of this section, we emphasize the idea that indicator dyes are mobile Ca^{2+} buffers by using $[B]$ and $[CaB]$ to represent the concentration of Ca^{2+}-free and Ca^{2+}-bound fluo-3, respectively. In Equation 3.58, Y_{offset} and Z_{offset} represent the degree to which the origin of the spark is out of register with the center of the point spread function, and x is distance along the line scan. In Smith et al. (1998), "smeared" Ca^{2+}-bound indicator dye profiles given by Equation 3.58 were normalized by the background fluorescence (proportional to $[CaB]_\infty$ given by Equation 3.9) and reported as $[CaB]_{avg}/[CaB]_\infty$.

Figure 3.1A uses these methods to simulate the time course of fluo-3 fluorescence in the presence of a 2 pA source for free Ca^{2+} of 15 ms duration. For other parameters see Reference 38. When 50 μM indicator is simulated (*solid line*), the spatially averaged fluorescence peaks around 2 times background and has a decay time constant of approximately 25 ms, similar to experiment. Figure 3.1D shows the Ca^{2+} and Ca^{2+}-bound indicator profiles (before spatial averaging) at a time of

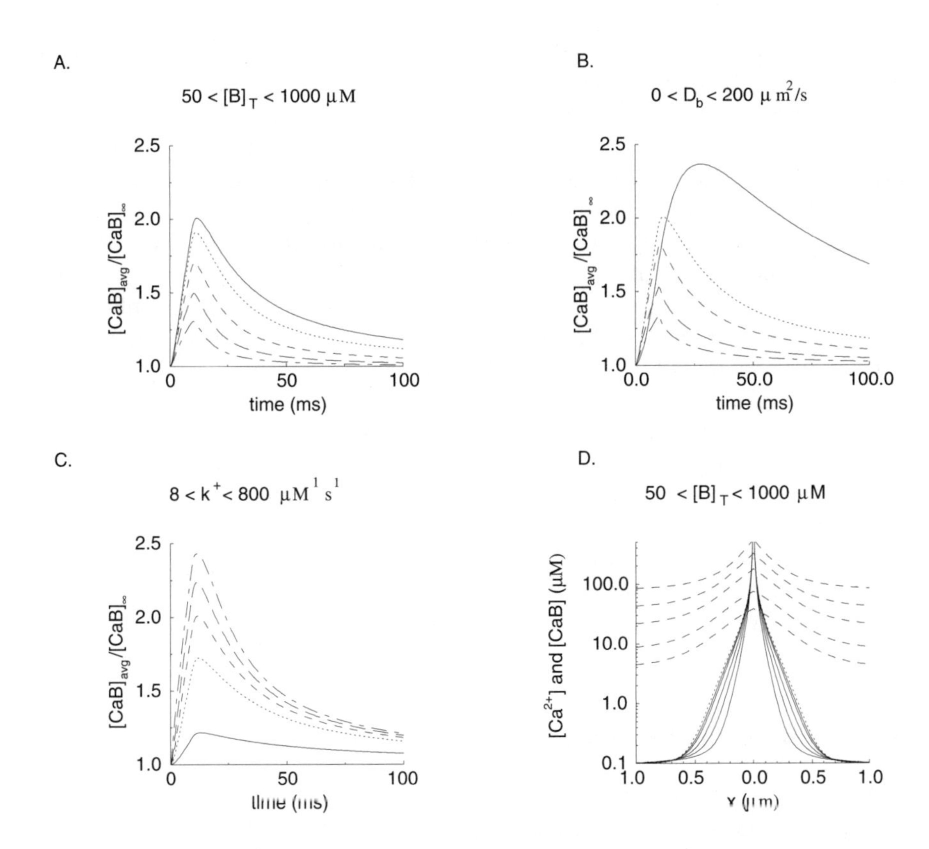

FIGURE 3.1 Effects of indicator dye parameters on Ca^{2+} spark properties. Source amplitude is 2 pA for 15 ms and simulated fluo-3 has K of 1.13 μM. (A,B,C) Time course of normalized, blurred fluorescence signal ($[CaB]_{avg}/[CaB]_{\infty}$), with Y_{offset} = 0.5 mm and Z_{offset} = 0 mm. (A) From top to bottom: indicator dye concentration ($[B]_T$) is 50, 100, 250, 500 and 1000 μM. Note that $[CaB]_{avg}/[CaB]_{\infty}$ is greater when there is less indicator. (B) From top to bottom, mobility of the dye, D_b, takes values of 0 (immobile), 20, 40, 100, and 200 μm²/s. (C) From bottom to top, the association rate constant k^+ of indicator is increased from 8 to 40, 80, 160 and 800 μM⁻¹s⁻¹ (K held constant by varying k^-). (D) Ca^{2+} (*solid and dotted lines*) and CaB (*dashed lines*) profiles at t =10 ms. Parameters as in A, except that $Y_{offset} = Z_{offset} = 0$. Uppermost CaB profile and lowermost Ca^{2+} profile use $[B]_T$ =1000 μM. *Dotted line* shows the Ca^{2+} profile with no fluo-3. Other parameters as in Reference 38. (Modified from Reference 38. With permission.)

10 ms, just before the source turns off. As expected, the indicator fluorescence is elevated where $[Ca^{2+}]$ is greatest.

3.3.3.2 Indicator Dye Properties and the Appearance of Localized Ca^{2+} Elevations

In addition to the simulation of a Ca^{2+} spark using standard parameters (solid line), Figure 3.1A presents a study of the effect of total concentration of indicator ($[B]_T$)

on the fluorescence time course during a Ca^{2+} spark. Adding higher concentrations of indicator ($[B]_T = 100 - 1000$ μM) has the counter-intuitive effect of decreasing the peak normalized fluorescence signal. This occurs because higher concentrations of indicator effect the free Ca^{2+} profile, causing it to be more restricted (see Figure 3.1D and Box 3.4). In addition, Figure 3.1A shows that high concentrations of indicator accelerate the decay of the simulated Ca^{2+} spark.

Changing indicator dye properties (such as mobility, D_b, and rate constants, k^+ and k^-) also have a profound effect on Ca^{2+} spark appearance (see Figure 3.1B and C). For example, the mobility of an indicator can decrease the peak amplitude of the Ca^{2+} spark, but improve the degree to which the peak of the spark is in register with the termination of Ca^{2+} release.[38] Thus, the use of immobile dyes (e.g., dextran-based Ca^{2+} indicators)[40] may have both advantages (greater fluorescence) and disadvantages (distorted kinetics). Similarly, spark brightness and width can be increased by using indicators with faster reaction kinetics. Because Ca^{2+} indicator dyes are mobile Ca^{2+} buffers, they are able to significantly perturb the underlying free Ca^{2+} profile, even when used at moderate concentrations (see Figure 3.1D). Figure 3.1 demonstrates that indicator dye properties are a primary determinant of the appearance of localized Ca^{2+} elevations.[38] Perhaps this is not surprising considering the ability of mobile buffers to perturb the time course of global Ca^{2+} transients.[20,39,41,42]

3.4 MODELING PROPAGATING CALCIUM WAVES

3.4.1 Localized Calcium Elevations Are the Building Blocks of Global Signals

Spatially localized Ca^{2+} elevations are important cellular signals that allow highly specific regulation of cellular function. Many cellular processes (including synaptic transmission, activity-dependent synaptic plasticity, and regulation of neuronal excitability) can be initiated by $[Ca^{2+}]$ changes in the absence of a global Ca^{2+} response. However, localized Ca^{2+} elevations are also of interest as the building blocks of global Ca^{2+} signals (such as Ca^{2+} oscillations and propagating waves of Ca^{2+}-induced Ca^{2+} release) that are observed in neurons and other cell types. For a review, see References 10 and 7, respectively.

The sections that follow describe a mathematical model for propagating Ca^{2+} waves in cardiac myocytes. The methods discussed are readily applied to simulations of neuronal Ca^{2+} waves, but note that the model presented reflects the fact that RyR-mediated Ca^{2+} release is responsible for Ca^{2+} waves in cardiac myocytes. Readers interested in modeling IP_3-mediated Ca^{2+} release may consult several early models of IP_3-mediated Ca^{2+} responses[43,44,45] as well as two models of the IP_3R[46,47] highly constrained by experimental data from single channel planar lipid bilayer experiments[4] and Ca^{2+} responses of the immature *Xenopus* oocyte during flash photolysis of caged IP_3.[48] Particularly useful starting points are the discussion of ligand-gated channels in Chapter 5 and a minimal model of IP_3R kinetics[49] that has been used in models of Ca^{2+} oscillations in pituitary gonadotrophs.[50,51,52] For a review see References 5 and 6.

3.4.2 Description of a Two-Compartment Model with Buffering

3.4.2.1 A Continuum, Two-Compartment Model

This section develops a model of propagating Ca^{2+} waves in cardiac myocytes by modifying the reaction-diffusion equations presented above in several ways. While Section 3.3 assumed spherical symmetry in calculations of Ca^{2+} domains and localized Ca^{2+} elevations, here we simplify a model with (potentially) three spatial dimensions by assuming that variations in Ca^{2+} and buffer concentration occur only in the longitudinal direction (see Figure 3.2). With this geometry in mind, we write a system of reaction-diffusion equations in which the $[Ca^{2+}]$ in the myoplasm and SR of a cardiac myocyte are distinguished,

$$\frac{\partial[Ca^{2+}]_{myo}}{\partial t} = D_{c,myo}\nabla^2[Ca^{2+}]_{myo} + \sum_i R_{i,myo} + J_{release} - J_{uptake} \tag{3.59}$$

$$\frac{\partial[B_i]_{myo}}{\partial t} = D_i\nabla^2[B_i]_{myo} + R_{i,myo} \tag{3.60}$$

$$R_{i,myo} = -k^+_{i,myo}[Ca^{2+}]_{myo}[B_i]_{myo} + k^-_{i,myo}\left([B_i]_{T,myo} - [B_i]_{myo}\right) \tag{3.61}$$

$$\frac{\partial[Ca^{2+}]_{SR}}{\partial t} = D_{c,SR}\nabla^2[Ca^{2+}]_{SR} + \sum_i R_{i,SR} - \frac{1}{V_{rel}}\left(J_{release} - J_{uptake}\right) \tag{3.62}$$

$$\frac{\partial[B_i]_{SR}}{\partial t} = D_{i,SR}\nabla^2[B_i]_{SR} + R_{i,SR} \tag{3.63}$$

$$R_{i,SR} = -k^+_{i,SR}[Ca^{2+}]_{SR}[B_i]_{SR} + k^-_{i,SR}\left([B_i]_{T,SR} - [B_i]_{SR}\right) \tag{3.64}$$

Equations 3.59–3.64 are the skeleton of a "two compartment" model for propagating Ca^{2+} waves that will be fleshed out below (see Figure 3.2 for diagram, buffer parameters are chosen following Reference 52B). These equations represent a "continuum model," because at every intracellular "point" a certain volume fraction is assigned to each compartment. Note the factor $V_{rel} = V_{SR}/V_{myo}$ in Equation 3.62, the relative volume occupied by the SR and myoplasm.

As mentioned in Section 3.2.2.2, the RBA[20] can be used to reduce Equations 3.59–3.64 to transport equations for $[Ca^{2+}]_{myo}$ and $[Ca^{2+}]_{SR}$ similar to Equation 3.29. The κ's and β's that occur in these reduced equations are similar to Equations 3.16 and 3.17, but generalized to account for multiple buffers.[28,52C] Regardless of whether full or reduced transport equations are used, the

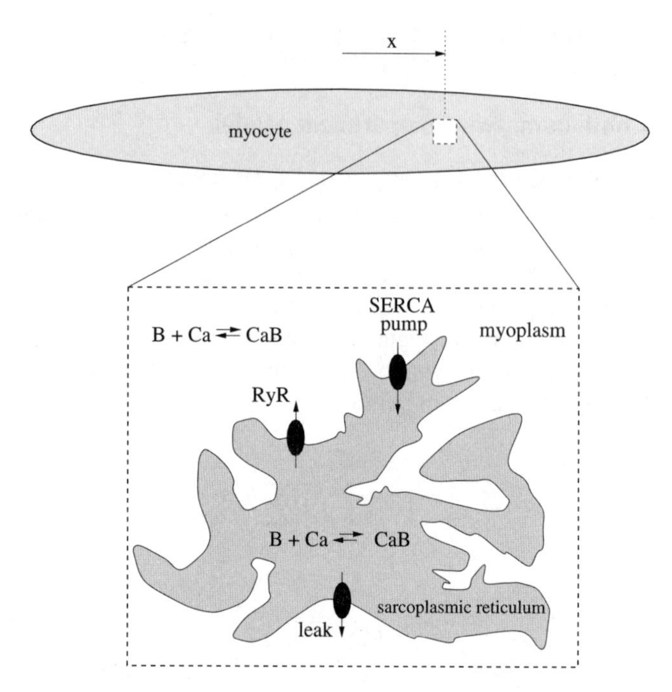

FIGURE 3.2 Diagram of compartments and fluxes for propagating Ca^{2+} wave simulation shown in Figure 3.4. Assuming no transverse variation in $[Ca^{2+}]$, two variables of interest are $[Ca^{2+}]_{myo}$ and $[Ca^{2+}]_{SR}$, both functions of time and longitudinal distance, x. Ca^{2+} is translated from the SR to the myoplasm (and back) via RyR-mediated Ca^{2+} release, a passive Ca^{2+} leak, and resequestration by Ca^{2+}-ATPases. The model also includes the important association of free Ca^{2+} with Ca^{2+}-binding species in each compartment, and longitudinal diffusion of Ca^{2+} and buffer in both myoplasm and SR.

dynamics of $[Ca^{2+}]_{myo}$ and $[Ca^{2+}]_{SR}$ are driven by the fluxes $J_{release}$ and J_{uptake} that are, in turn, functions of $[Ca^{2+}]_{myo}$ and $[Ca^{2+}]_{SR}$. The following two subsections elaborate on the mathematical form of these fluxes.

3.4.2.2 Ca²⁺ Release

The release flux includes two terms, $J_{release} = J_{RyR} + J_{leak}$, a passive leak with the form,

$$J_{leak} = v_{leak}\left(\left[Ca^{2+}\right]_{SR} - \left[Ca^{2+}\right]_{myo}\right) \tag{3.65}$$

as well as the flux due to CICR via the RyR

$$J_{RyR} = v_{RyR} f_0\left(\left[Ca^{2+}\right]_{SR} - \left[Ca^{2+}\right]_{myo}\right) \tag{3.66}$$

where f_O is the fraction of open Ca^{2+} release sites at any given time (see Chapter 5 for a general treatment of modeling the dynamics of voltage- and ligand-gated channels). The dynamics of f_O are implemented using a two-state model, $N \leftrightarrow R$, based on the six-state model release site shown in Figure 3.3. This two-state model is derived using the separation of time scales of activation and inactivation of Ca^{2+} release sites.[53,54] The non-refractory state, N, consists of two open and two closed states (C_0, C_1, O_0, O_1) that rapidly equilibrate and the refractory state, R, consists of two equilibrated closed states (C_2, C_3). The structure and rate constants of the model were chosen in agreement with experiment.[54,55,56]

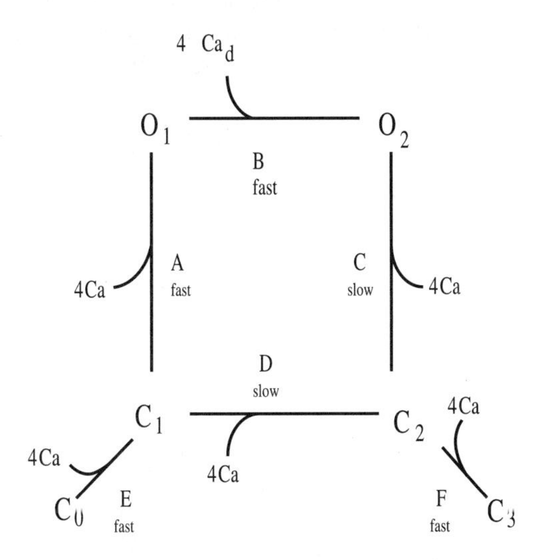

FIGURE 3.3 State diagram of the model Ca^{2+} release site used in Figure 3.4. Note the domain Ca^{2+}-mediated transition from state O_1 to O_2 (process B). The simulations of Figures 3.4 and 3.5 use $[Ca^{2+}]_d$ = 5 μM. Other release site parameters as in Table 1 of Reference 54. With permission.

3.4.2.3 Ca^{2+} Uptake

Uptake of Ca^{2+} due to Ca^{2+}-ATPases of the SR membrane can be modeled as an instantaneous function of $[Ca^{2+}]_{myo}$,[57]

$$J_{uptake} = \frac{v_{uptake}^{\max}\left(\left[Ca^{2+}\right]_{myo}\right)^m}{K_{uptake}^m + \left(\left[Ca^{2+}\right]_{myo}\right)^m} \tag{3.67}$$

where we assume Michaelis–Menten kinetics (see Chapter 2), K_{uptake} is 184 nM, m is 3.98, and $v_{uptake}^{\max} = 208\ \mu M^{-1} s^{-1}$ Athough plasma membrane Ca^{2+}-ATPases can be modeled in a similar manner, we neglect Ca^{2+} efflux here. Na^+/Ca^{2+} exchange (see References 58 and 59 for models) is not included, because of its minor role in regulating cellular Ca^{2+} transients in rat ventricular cardiac myocytes.[60]

3.4.2.4 Initial Conditions

The final step in our description of a two compartment, continuum model of propagating Ca^{2+} waves in cardiac myocytes is the specification of initial conditions. When the dynamics of the Ca^{2+} release flux, a passive leak, and resequestration by Ca^{2+} ATPases are combined in a single well-mixed pool model, the ODEs have a stable steady state with $[Ca^{2+}]_{myo} \approx 0.08$ μM, and $[Ca^{2+}]_{SR} \approx 28.34$ μM ($f_O \approx 0, f_N \approx 1$). Let's refer to these values as $[Ca^{2+}]^{\infty}_{myo}$ and $[Ca^{2+}]^{\infty}_{SR}$, respectively, and use them as spatially uniform initial conditions throughout the simulation. To trigger a propagating Ca^{2+} wave, $[Ca^{2+}]_{myo}$ will be elevated to $[Ca^{2+}]^{trigger}_{myo} \approx 0.2$ μM in the leftmost region of the model myocyte. If we wish to locally conserve total intracellular Ca^{2+}, we must decrease $[Ca^{2+}]_{SR}$ in this region according to,

$$\Delta\left[Ca^{2+}\right]_{T,myo} + V_{rel}\Delta\left[Ca^{2+}\right]_{T,SR} = 0 \tag{3.68}$$

where

$$\Delta\left[Ca^{2+}\right]_{T,myo} = \left[Ca^{2+}\right]^{trigger}_{T,myo} - \left[Ca^{2+}\right]^{\infty}_{T,myo} \tag{3.69}$$

$$\left[Ca^{2+}\right]_{T,myo} = \left[Ca^{2+}\right]_{myo} + \sum_i \frac{\left[Ca^{2+}\right]_{myo}\left[B_i\right]_{T,myo}}{K_{i,myo} + \left[Ca^{2+}\right]_{myo}} \tag{3.70}$$

and expressions similar to Equations 3.69 and 3.70 define $\Delta[Ca^{2+}]_{T,SR}$ and $[Ca^{2+}]_{T,SR}$. Note that Equation 3.70 is a generalization of Equation 3.15 for multiple buffers and that substituting Equation 3.70 into Equation 3.68 gives an implicit expression for $[Ca^{2+}]^{trigger}_{SR}$ in terms of $[Ca^{2+}]^{trigger}_{myo}$. To choose initial conditions in a consistent manner, Equation 3.68 must be numerically solved (e.g., using Newton's method).[61]

3.4.3 Simulations of a Propagating Calcium Wave

3.4.3.1 A Traveling Wave of Ca^{2+}-Induced Ca^{2+} Release

The solid lines of Figure 3.4A show a snapshot of ($t = 0.5$ s) of a simulated Ca^{2+} wave (see Section 3.4.2). The Ca^{2+} wave is propagating from left to right with a velocity of approximately 50 μm/s. Figure 3.4B shows the fraction of non-refractory release sites (f_N) as a function of position in the model myocyte. Most of the release sites are nonrefractory ($f_N \approx 1$) in the region of the myocyte being invaded by the propagating Ca^{2+} wave, and the decrease in f_N as the wave passes represents Ca^{2+}-inactivation. The beginnings of a slow recovery to the non-refractory state can be seen at the leftmost portions of Figure 3.4B. Note that the model predicts a rightward traveling depletion wave in the *SR* (see Figure 3.4C).

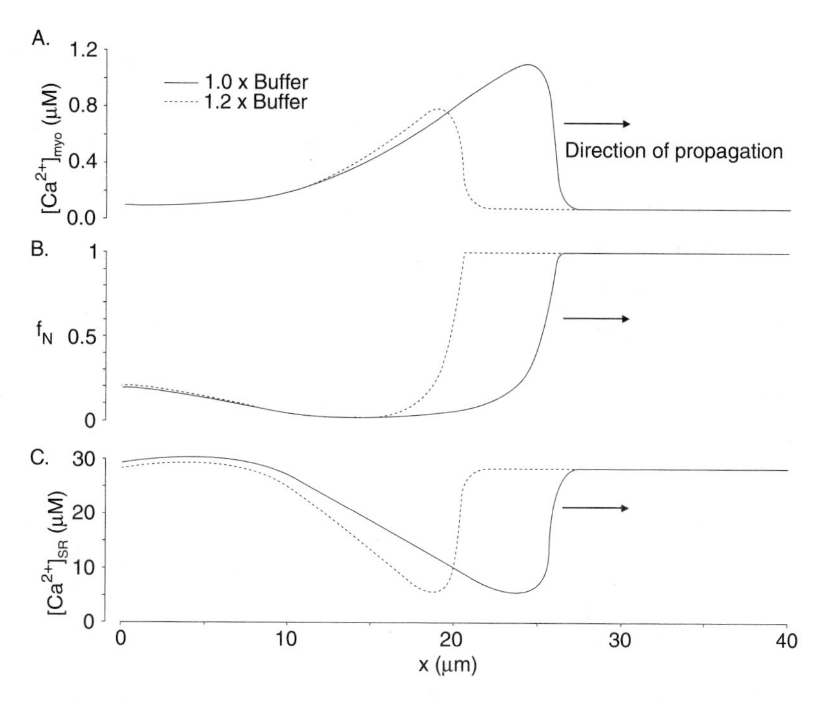

FIGURE 3.4 Simulations of propagating Ca^{2+} waves in a cardiac myocyte performed using a generalization of Equation 3.29 appropriate for a two compartment, continuum model with multiple buffers.[28] Solid lines show $[Ca^{2+}]_{myo}$, $[Ca^{2+}]_{SR}$, and fraction of non inactivated release sites (f_N) as a function of spatial position at $t = 0.3$ sec. *Dotted lines* show a similar calculation with the total concentration of each buffer increased by 20%. Ca^{2+} excitability of the model myocyte generates a trigger wave. Other wave phenomena are obtained when the homogenous system is oscillatory or bistable.[52C] Release site and cellular parameters as in Figure 3.3 and Reference 52B, respectively.

The dotted lines in Figure 3.4 are the result of a simulation identical to the one described above, except that $[B_i]_T$ for each buffer was increased by 20%. Comparison of the solid and dotted lines in Figure 3.4A shows that this increase in $[B_i]_T$ has the effect of decreasing the amplitude of the free Ca^{2+} signal at the peak of the wave (see Box 3.6). In the presence of additional buffer, the Ca^{2+} wave also travels at a slower velocity than the control wave (~40 μm/s).

The effect of increased $[B_i]_T$ on Ca^{2+} wave amplitude and velocity can be understood by considering the reduced equations for $[Ca^{2+}]_{myo}$ and $[Ca^{2+}]_{SR}$ (similar to Equation 3.29) that are obtained when the RBA is applied to Equations 3.59–3.64. Considering such equations, rapid Ca^{2+} buffers can be shown to have three distinct effects.[28] First, in a two compartment model, the release and uptake fluxes are scaled by a factor of β (different for the myoplasm and *SR*). Thus, when $[B_i]_T$ is increased in the myoplasm, β_{myo} decreases, and the effect of *SR* Ca^{2+} release on $[Ca^{2+}]_{myo}$ is minimized. Secondly, when $[B_i]_T$ is increased, the effective diffusion coefficient for Ca^{2+}, D_{eff}, is reduced as discussed in Section 3.2.2.2. This decrease in D_{eff} may contribute to a slower propagation velocity of the wave. Thirdly, Equation 3.29 indicates that the effect of rapid Ca^{2+} buffers on wave speed is complicated by a non-diffusive sink effect that allows mobile Ca^{2+} buffers to quickly clear Ca^{2+} gradients.[20,28,29,62]

Box 3.6 The Effect of Rapid Calcium Buffers on Calcium Wave Propagation

The CICR wave simulation can be used to investigate the effect of rapid Ca^{2+} buffers on Ca^{2+} wave propagation.

Example 1. Using parameter values from Color Figure 3.1, include in the myoplasmic compartment two Ca^{2+} buffers, one stationary (B_s) and one mobile (B_m). Notice that both the stationary and mobile buffer have the same dissociation constant ($K_s = K_m = 10$ µM), and that the diffusion coefficient for the mobile buffer is much less than that of free Ca^{2+} ($D_b = 75$ µm²/s << $D_c = 250$ µm²). For simplicity, use the same parameters in the *SR* compartment. Set $[B_s]_T$ = 250 µM and $[B_m]_T$ = 50 µM (in both the myoplasm and *SR*) and confirm that you are able to initiate a propagating Ca^{2+} wave.

Example 2. Now vary the fraction of stationary versus mobile buffer in the myoplasmic compartment from 100% stationary to 100% mobile, making sure that $[B_s]_T + [B_m]_T = 300$ µM remains constant. What is the effect of buffer mobility on the amplitude of the Ca^{2+} wave? What is the effect on the propagation velocity of the wave? Attempt to rationalize your observations by considering the functional dependence of D_{eff} on $[B_s]_T$ and $[B_m]_T$ (see Section 3.2.2.2, Equation 3.30).

Example 3. Now keeping the fraction of stationary versus mobile buffer in the myoplasmic compartment fixed (5/6 stationary and 1/6 mobile), increase $[B_i]_T$. In computer simulations of Ca^{2+} waves in the immature *Xenopus* oocyte, the width of trigger waves remains relatively constant as $[B_i]_T$ is increased, while the wave amplitude and wave speed both decrease.[28] Do you observe similar effects?

Example 4. In Example 3, did you remember to choose the initial condition for $[Ca^{2+}]_{SR}$ (in the region where the Ca^{2+} wave is initiated) in accordance with Equation 3.68? When the total myoplasmic buffer is increased, $[Ca^{2+}]_{SR}^{trigger}$ will decrease if $[Ca^{2+}]_{myo}^{trigger}$ is fixed and initial conditions are chosen to locally conserve total Ca^{2+}.

Example 5. What are the consequences on wave amplitude and velocity when $[B_i]_T$ in the *SR* is increased?

3.4.3.2 Spark-Mediated Propagating Ca^{2+} Waves

Experimental observations of propagating Ca^{2+} waves in cardiac myocytes[14,15] indicate that the simulations in the previous section may be oversimplified, because the Ca^{2+} release flux described by Equation 3.66 does not account for the discrete nature of Ca^{2+} release. In this section we account for the inhomogeneous nature of the myoplasm in cardiac myocytes by letting the maximum conductance for the *RyR* be a periodic function of spatial position. That is, we redefine v_{ryr} in Equation 3.66 so that it takes values of either one or zero, where values of one represent the localization of *RyRs* to the junctional *SR* (near T-tubules and Z-lines). All other parameters in the problem remain spatially uniform.

Because in this section we wish to highlight the consequences of discrete Ca^{2+} release for the macroscopic properties of Ca^{2+} waves, we account for buffering in the simplest way possible (using constant valued effective diffusion coefficients for Ca^{2+} in the myoplasm and *SR*) and constant buffering factors for both compartments. Admittedly, these assumptions are made as a matter of convenience; nevertheless, we write

$$\frac{\partial[Ca^{2+}]_{myo}}{\partial t} = \hat{D}_{eff,myo}\nabla^2[Ca^{2+}]_{myo} + \hat{\beta}_{myo}\left(J_{release} - J_{uptake}\right) \tag{3.71}$$

$$\frac{\partial[Ca^{2+}]_{SR}}{\partial t} = \hat{D}_{eff,SR}\nabla^2[Ca^{2+}]_{SR} - \frac{\hat{\beta}_{SR}}{V_{rel}}\left(J_{release} - J_{uptake}\right) \tag{3.72}$$

where the hats indicate constant quantities. If we further assume that $\hat{D}_{eff,myo} = \hat{D}_{eff,SR}$, total cell Ca^{2+} (weighted by volume fraction) is locally conserved, and we can use this fact to eliminate the equation for $[Ca^{2+}]_{SR}$. That is, if we define

$$[Ca^{2+}]_T = [Ca^{2+}]_{myo} + V_{rel}[Ca^{2+}]_{SR} \tag{3.73}$$

Equations 3.71 and 3.72 imply that $[Ca^{2+}]_T$ is a constant and thus $[Ca^{2+}]_{SR}$ can be expressed as,

$$[Ca^{2+}]_{SR} = \frac{[Ca^{2+}]_T - [Ca^{2+}]_{myo}}{V_{rel}} \tag{3.74}$$

Using Equations 3.65 and 3.66, we now rewrite the Ca^{2+} release flux as

$$J_{release} = \left(v_{leak} + v_{RyR} f_o\right)\left(\frac{[Ca^{2+}]_T - [Ca^{2+}]_{myo}}{V_{rel}} - [Ca^{2+}]_{myo}\right) \tag{3.75}$$

Figure 3.5 shows a snapshot of a simulated spark-mediated Ca^{2+} wave based on Equations 3.67, 3.71, and 3.75. Elevated $[Ca^{2+}]_{myo}$ in the center initiates two propagating waves, one moving left and the other right. A depletion wave in the SR is also observed (a consequence of Equation 3.74).

The simulation shown in Figure 3.5 differs from that in Figure 3.4 in several ways. As described above, we have for simplicity assumed constant (and equivalent) effective diffusion coefficients in the myoplasm and *SR*. This assumption, made primarily for convenience, allowed us to eliminate the transport equation for $[Ca^{2+}]_{SR}$. In this respect, Figure 3.5 is less realistic than Figure 3.4, which accurately accounts for the effects of Ca^{2+} buffers. On the other hand, Figure 3.4 assumes a homogenous Ca^{2+} release flux, while Figure 3.5 explicitly models the elementary Ca^{2+} release events, Ca^{2+} sparks, that are the building blocks of propagating Ca^{2+} waves. An array of 50 spatially discrete Ca^{2+} release sites are included in Figure 3.5 (note that the crest of the wave has a distinct sculpted form). Color Figure 3.2* shows a "waterfall" plot of the same simulation. The reciprocal of the slope of the wave front is the wave speed, approximately 70 µm/s, in agreement with experiment.

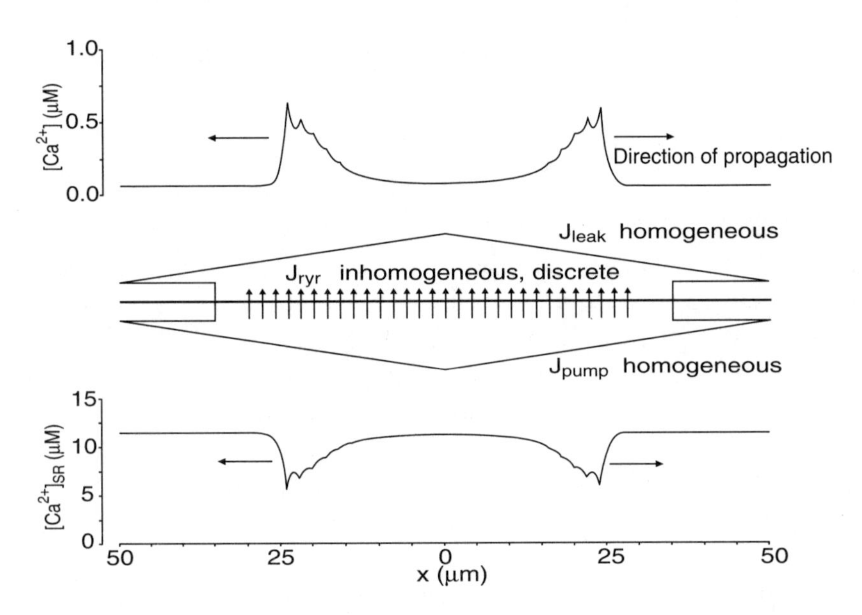

FIGURE 3.5 Snapshot of two spark-mediated Ca^{2+} waves 200 ms after initiation with elevated $[Ca^{2+}]$ at center. (A,C) Profiles of $[Ca^{2+}]_{myo}$ and $[Ca^{2+}]_{SR}$ as a function of spatial position. (B) Diagram shows Ca^{2+} fluxes in the model. A passive Ca^{2+} leak, J_{leak} (*upward large arrow*), and active Ca^{2+} uptake, J_{pump} (*downward large arrow*), are modeled as homogenous fluxes. The Ca^{2+} release flux, J_{ryr} (*small upward arrows*), is inhomogeneous and takes nonzero values only at discrete locations separated by approximately 2 µm, the intra-sarcomeric distance. Release site and cellular parameters as in Figure 3.3 and Table 2 of Reference 54, respectively. (From Reference 54. With permission.)

ACKNOWLEDGMENTS

This work is supported in part by a National Institute of Health Intramural Research Training Assistantship and NEI National Research Service Award EY06903-01.

* Color Figure 3.2 follows page 140.

Box 3.7 The Velocity of Spark-Mediated Calcium Wave

The velocity of conventional reaction-diffusion trigger waves is proportional to the square root of the diffusion coefficient for the activating species (here Ca^{2+}). Luther's equation expresses this relationship,[63,64]

$$v \propto \sqrt{\frac{D}{\tau_{chem}}} \quad (1)$$

where τ_{chem} is a time constant associated with the autocatalytic process leading to the reaction- diffusion wave. However, the velocity of simulated spark-mediated waves has been reported[54,65] to be proportional to the diffusion coefficient, that is,

$$v \propto \frac{D}{d} \quad (2)$$

where d is the inter-site separation. Thus, it appears that Ca^{2+} release site separation influences the macroscopic properties of spark-mediated Ca^{2+} waves. (For detailed analysis of the transition from spark-mediated waves to continuous reaction-diffusion waves, see Reference 66.)

Example 1. Using the spark-mediate Ca^{2+} wave simulation, vary the diffusion coefficient and confirm the proportionality given by Equation 2. Also confirm Equation 2 using the minimal "Fire-Diffuse-Fire" model[65,66] of Ca^{2+} wave propagation (see Reference 66 for description of model).

Example 2. Consider the velocity of wave propagation as a function of the inter-site separation. First, fix the effective diffusion coefficient, D, and vary the inter-site separation distance, d. (Note that when d is changed, v_{ryr} is adjusted so that the maximum possible Ca^{2+} release rate per unit length remains constant.) Using $D = 30\ \mu m^2/s$, begin with $d \geq 2\ \mu m$ and confirm that the velocity decreases linearly as the site separation distance gets larger. Next, decrease the site separation, d, until the velocity of propagation saturates. When $d = 0$ there is no inter-site separation and the velocity you observe should be the velocity of the continuum reaction-diffusion wave (v_{cont}).

Example 3. Consider the point at which the saturation of propagation velocity became evident. When $D/d \approx v_{cont}$, the velocity of the Ca^{2+} wave should be nearly v_{cont}.[65] Conversely, Ca^{2+} wave propagation is likely saltatory when $D/d << v_{cont}$. Confirm these predictions.

REFERENCES

1. Neher, E., Vesicle pools and Ca^{2+} microdomains: new tools for understanding their roles in neurotransmitter release, *Neuron*, 20, 389, 1998.
2. Stanley, E. F., Calcium entry and the functional organization of the presynaptic transmitter release site, in *Excitatory Amino Acids and Synaptic Transmission*, 2nd ed., Wheal, H. V. and Thomson, A. M., Eds., Academic Press, New York, 1995.
3. Berridge, M. J., Inositol trisphosphate and calcium signaling, *Nature*, 361, 315, 1993.

4. Bezprozvanny, I., Watras, J., and Ehrlich, B. E., Bell-shaped calcium-response curves of Ins(1,4,5)P3- and calcium-gated channels from endoplasmic reticulum of cerebellum, *Nature*, 351, 751, 1991.
5. Keizer, J., Li, Y. X., Stojilkovic, S., and Rinzel, J., InsP3-induced Ca^{2+} excitability of the endoplasmic reticulum, *Mol. Biol. Cell*, 6, 945, 1995.
6. Li, Y. X., Keizer, J., Stojilkovic, S. S., and Rinzel, J., Ca^{2+} excitability of the ER membrane: an explanation for IP_3-induced Ca^{2+} oscillations, *Am. J. Physiol.*, 269, C1079, 1995.
7. Clapham, D. E., Calcium signaling, *Cell*, 80, 259, 1995.
8. Berridge, M. J., Elementary and global aspects of calcium signalling, *J. Physiol.* (London), 499, 291, 1997.
9. Berridge, M. J., Bootman, M. D., and Lipp, P., Calcium—a life and death signal, *Nature*, 395, 645, 1998.
10. Berridge, M. J., Neuronal calcium signaling, *Neuron*, 21, 13, 1998.
11. Finch, E. and Augustine, G., Local calcium signalling by inositol-1,4,5-trisphosphate in Purkinje cell dendrites, *Nature*, 396, 753, 1998.
12. Yao, Y., Choi, J., and Parker, I., Quantal puffs of intracellular Ca^{2+} evoked by inositol trisphosphate in Xenopus oocytes, *J. Physiol.* (London), 482, 533, 1995.
13. Parker, I., Choi, J., and Yao, Y., Elementary events of InsP3-induced Ca^{2+} liberation in Xenopus oocytes: hot spots, puffs and blips, *Cell Calcium*, 20, 105, 1996.
14. Cheng, H., Lederer, W. J., and Cannell, M. B., Calcium sparks: elementary events underlying excitation-contraction coupling in heart muscle, *Science*, 262, 740, 1993.
15. Cheng, H., Lederer, M. R., Xiao, R. P., Gomez, A. M., Zhou, Y. Y., Ziman, B., Spurgeon, H., Lakatta, E. G., and Lederer, W. J., Excitation-contraction coupling in heart: new insights from Ca^{2+} sparks, *Cell Calcium*, 20, 129, 1996.
16. Fabiato, A., Simulated calcium current can both cause calcium loading in and trigger calcium release from the sarcoplasmic reticulum of a skinned canine cardiac Purkinje cell, *J. Gen. Physiol.*, 85, 291, 1985.
17. Neher, E., Concentration profiles of intracellular Ca^{2+} in the presence of diffusible chelator, *Exp. Brain Res.*, 14, 80, 1986.
18. Neher, E., Usefulness and limitations of linear approximations to the understanding of Ca^{2+} signals, *Cell Calcium*, 24, 345, 1998.
19. Naraghi, M. and Neher, E., Linearized buffered Ca^{2+} diffusion in microdomains and its implications for calculation of $[Ca^{2+}]$ at the mouth of a calcium channel, *J. Neuroscience*, 17, 6961, 1997.
20. Wagner, J. and Keizer, J., Effects of rapid buffers on Ca^{2+} diffusion and Ca^{2+} oscillations, *Biophys. J.*, 67, 447, 1994.
21. Smith, G. D., Wagner, J., and Keizer, J., Validity of the rapid buffering approximation near a point source for Ca^{2+} ions, *Biophys. J.*, 70, 2527, 1996.
22. Smith, G. D., Analytical steady-state solution to the rapid buffering approximation near an open Ca^{2+} channel, *Biophys. J.*, 71, 3064, 1996.
23. Smith, G. D., Dai, L., Miura, R., and Sherman, A., Asymptotic analysis buffered Ca^{2+} diffusion near a point source, *SIAM J. Appl. Math.*, in press.

23B. Lin, C. C. and Segel, L. A., *Mathematics Applied to Deterministic Problems in the Natural Sciences*, Society for Industrial and Applied Mathematics, Philadelphia, 1988.

24. De Schutter, E. and Smolen, P., Calcium dynamics in large neuronal models, in *Methods in Neuronal Modeling: From Ions to Networks*, 2nd ed., Koch, C. and Segev, I., Eds., MIT Press, Cambridge, 1998.
25. Edelstein-Keshet, L., *Mathematical Models in Biology*, Random House, New York, 1988, 275.

26. Zador, A. and Koch, C., Linearized models of calcium dynamics: formal equivalence to the cable equation, *J. Neurosci.*, 14, 4705, 1994.
27. Allbritton, N. L., Meyer, T., and Stryer, L., Range of messenger action of calcium ion and inositol 1,4,5-trisphosphate, *Science*, 258, 1812, 1992.
28. Jafri, M. S. and Keizer, J., On the roles of Ca^{2+} diffusion, Ca^{2+} buffers, and the endoplasmic reticulum in IP_3-induced Ca^{2+} waves, *Biophys. J.*, 69, 2139, 1995.
29. Sneyd, J., Dale, P. D., and Duffy, A., Traveling waves in buffered systems: applications to calcium waves, *SIAM J. Applied Math.*, 58, 1178, 1998.
30. Stern, M. D., Buffering of Ca^{2+} in the vicinity of a channel pore, *Cell Calcium*, 13, 183, 1992.
31. Mascagni, M. and Sherman, A. S., Numerical methods for neuronal modeling, in *Methods in Neuronal Modeling: From Ions to Networks*, 2nd ed., Koch, C. and Segev, I., Eds., MIT Press, Cambridge, 1998.
32. Morton, K. W. and Mayers, D. F., *Numerical Solution of Partial Differential Equations: An Introduction*, Cambridge University Press, Cambridge, 1994.
33. Smith, G. D., *Numerical Solution of Partial Differential Equations: Finite Difference Methods*, 3rd ed., Clarendon Press, Oxford, 1985.
34. Roberts, W. M., Spatial calcium buffering in saccular hair cells, *Nature*, 363, 74, 1993.
35. Roberts, W. M., Localization of calcium signals by a mobile calcium buffer in frog saccular hair cells, *J. Neurosci.,* 14, 3246, 1994.
35B. Bertram, R., Smith, G. D., and Sherman, A., A modeling study of the effects of overlapping Ca^{2+} microdomains on neurotransmitter release, *Biophys. J.*, 76, 735, 1999.
36. Crank, J., *The Mathematics of Diffusion*, 2nd ed., Clarendon Press, Oxford, 1975.
37. Carslaw, H. S. and Jaeger, J. C., *Conduction of Heat in Solids*, 2nd ed., Clarendon Press, Oxford, 1959.
38. Smith, G. D., Keizer, J., Stern, M., Lederer, W. J., and Cheng, H., A simple numerical model of Ca^{2+} spark formation and detection in cardiac myocytes, *Biophys. J.*, 75, 15, 1998.
39. Blumenfeld, H., Zablow, L., and Sabatini, B., Evaluation of cellular mechanisms for modulation of calcium transients using a mathematical model of fura-2 Ca^{2+} imaging in Aplysia sensory neurons, *Biophys. J.*, 63, 1146, 1992.
40. Horne, J. H. and Meyer, T., Elementary calcium-release units induced by inositol trisphosphate, *Science*, 276, 1690, 1997.
41. Sala, F. and Hernandez-Cruz, A., Calcium diffusion modeling in a spherical neuron. Relevance of buffering properties., *Biophys. J.*, 57, 313, 1990.
42. Gabso, M., Neher, E., and Spira, M. E., Low mobility of the Ca^{2+} buffers in axons of cultured Aplysia neurons, *Neuron*, 18, 473, 1997.
43. Meyer T. and Stryer, L., Molecular model for receptor-stimulated calcium spiking, *Proc. Natl. Acad. Sci. U.S.A.,* 85, 5051, 1988.
44. Goldbeter, A., Dupont, G., and Berridge, M. J., Minimal model for signal-induced Ca^{2+} oscillations and for their frequency encoding through protein phosphorylation, *Proc. Natl. Acad. Sci. USA*, 87, 1461, 1990.
45. Atri, A., Amundson, J., Clapham, D., and Sneyd, J., A single-pool model for intracellular calcium oscillations and waves in the Xenopus laevis oocyte, *Biophys. J.*, 65, 1727, 1993.
46. Keizer, J. and De Young, G. W., Two roles of Ca^{2+} in agonist stimulated Ca^{2+} oscillations, *Biophys. J.*, 61, 649, 1992.

47. De Young, G. W. and Keizer, J., A single-pool inositol 1,4,5-trisphosphate-receptor-based model for agonist-stimulated oscillations in Ca^{2+} concentration, *Proc. Natl. Acad. Sci. U.S.A.*, 89, 9895, 1992.
48. Parker, I. and Ivorra, I., Inhibition by Ca^{2+} of inositol trisphosphate-mediated Ca^{2+} liberation: a possible mechanism for oscillatory release of Ca^{2+}, *Proc. Natl. Acad. Sci. U.S.A.*, 87, 260, 1990.
49. Li, Y. X. and Rinzel, J., Equations for InsP3 receptor-mediated $[Ca^{2+}]i$ oscillations derived from a detailed kinetic model: a Hodgkin–Huxley like formalism, *J. Theor. Biol.*, 166, 461, 1994.
50. Li., Y. X., Rinzel, J., Keizer, J., and Stojilkovic, S. S., Calcium oscillations in pituitary gonadotrophs: comparison of experiment and theory, *Proc. Natl. Acad. Sci. U.S.A.*, 91, 58, 1994.
51. Li, Y. X., Rinzel, J., Vergara, L., and Stojilkovic, S. S., Spontaneous electrical and calcium oscillations in unstimulated pituitary gonadotrophs, *Biophys. J.*, 69, 785, 1995.
52. Li, Y. X., Stojilkovic, S. S., Keizer, J., and Rinzel, J., Sensing and refilling calcium stores in an excitable cell, *Biophys. J.*, 72, 1080, 1997.
52B. Jafri, M. S., Rice, J. J., and Winslow, R. L., Cardiac Ca^{2+} dynamics: the roles of ryanodine receptor adaptation and sarcoplasmic reticulum load, *Biophys. J.*, 74, 1149, 1998.
52C. Jafri, M. S. and Keizer, J., Diffusion of inositol 1,4,5-trisphosphate but not Ca^{2+} is necessary for a class of inositol 1,4,5-trisphosphate-induced Ca^{2+} waves, *Proc. Natl. Acad. Sci. U.S.A.*, 91, 9485, 1994.
53. Keizer, J. and Levine, L., Ryanodine receptor adaptation and Ca^{2+}-induced Ca^{2+} release-dependent Ca^{2+} oscillations, *Biophys. J.*, 71, 3477, 1996.
54. Keizer, J. and Smith, G. D., Spark-to-wave transition: saltatory transmission of Ca^{2+} waves in cardiac myocytes, *Biophys. Chem.*, 72, 87, 1998.
55. Gyorke, S. and Fill, M., Ryanodine receptor adaptation: control mechanism of Ca(2+)-induced Ca^{2+} release in heart, *Science*, 260, 807, 1993.
56. Gyorke, S., Velez, P., Suarez-Isla, B., and Fill, M., Activation of single cardiac and skeletal ryanodine receptor channels by flash photolysis of caged Ca^{2+}, *Biophys. J.*, 66, 1879, 1994.
57. Bassani, J. W., Bassani, R. A., and Bers, D. M., Relaxation in rabbit and rat cardiac cells: species-dependent differences in cellular mechanisms, *J. Physiol.* (London), 476, 279, 1994.
58. Gabbiani, F., Midtgaard, J., and Knopfel, T., Synaptic integration in a model of cerebellar granule cells, *J. Neurophysiol.*, 72, 999, 1994.
59. DiFrancesco, D. and Noble, D., A model of cardiac electrical activity incorporating ionic pumps and concentration changes, *Philos. Trans. R. Soc. London Ser. B Biol. Sci.*, 307, 353, 1985.
60. Balke, C. W., Egan, T. M., and Wier, W. G., Processes that remove calcium from the cytoplasm during excitation-contraction coupling in intact rat heart cells, *J. Physiol.* (London), 474, 447, 1994.
61. Press, W., Teukolsky, S., Vetterling, W., and Flannery, B., *Numerical Recipes in C: The Art of Scientific Programming*, 2nd ed., Cambridge University Press, Cambridge, 1992.
62. Koch, C., *Biophysics of Computation: Information Processing in Single Neurons*, Oxford University Press, Oxford, 1999, 248.
63. Jaffe, L., Classes and mechanisms of Ca^{2+} waves, *Cell Calcium*, 14, 736, 1993.
64. Murray, J. D., *Mathematical Biology*, Springer-Verlag, New York, 1990, 277.

65. Keizer, J., Smith, G. D., Ponce-Dawson, S., and Pearson, J., Saltatory propagation of Ca^{2+} waves by Ca^{2+} sparks, *Biophys. J.*, 75, 595, 1998.
66. Ponce-Dawson, S., Keizer, J., and Pearson, J. E., Fire-diffuse-fire model of dynamics of intracellular calcium waves, *Proc. Natl. Acad. Sci. U.S.A.*, 96, 6060, 1999.

4 Monte Carlo Methods for Simulating Realistic Synaptic Microphysiology Using MCell

Joel R. Stiles and Thomas M. Bartol

CONTENTS

0-8493-2068-2/01/$0.00+$.50

4.1 INTRODUCTION

The function of the nervous system can be investigated at widely different scales, and computational approaches can range from space-filled atomic resolution (e.g., molecular dynamics simulations of protein structure) to space-independent methods (e.g., information theory applied to spike trains). At the level of cellular and subcellular signaling, models can be tailored to different questions that require different levels of structural realism. For example, neuronal excitation can be addressed using isopotential compartmental models (Chapters 5–9), biochemical networks of signaling pathways might be simulated using well-mixed assumptions (Chapter 2), and certain types of reaction-diffusion problems can be handled using spatial simplifications and boundary conditions that allow effectively one-dimensional (1-D) analytic and/or finite difference approaches (Chapter 3).

At some level of reaction-diffusion problems, however, realistic 3-D cell structures become critical components of quantitative modeling. For example, if identification of pre- and postsynaptic factors that contribute to synaptic function, variability, and crosstalk are of interest,[3,8,11,15,22,23,33] then incorporation of actual ultrastructure into the model cannot be avoided. Similarly, if quantitative simulations of realistic Ca^{+2} dynamics in and around dendritic spines are the goal, then the shapes, sizes, and other biophysical properties of real intra- and extracellular spaces will have to be included in models. Thus, a realistic, fully 3-D approach is required for quantitative modeling of *microphysiology*, i.e., cellular physiology that includes *3-D ultrastructural organization* (intra- and extracellular diffusion spaces, site densities and distributions of macromolecules), *movements of mobile molecules* (signaling ligands, ions, gases, free water), and *biochemical reactions* (transition paths, reaction rates).

In this chapter we illustrate *Monte Carlo methods* for realistic 3-D modeling of synaptic microphysiology. We begin with a discussion of the steps required to create a microphysiological model, and then briefly compare the two alternative computational paradigms that can be used to simulate the model, *finite element* versus *Monte Carlo* (Section 4.2). Although the general power and increased realism of Monte Carlo methods have been recognized for many years,[9,13,21] inadequate computer resources precluded many applications until the recent explosion in speed, memory, and storage capacity. We will describe in this chapter MCell, Monte Carlo simulator of cellular microphysiology. MCell is freely available and can be obtained for a wide variety of UNIX, Windows, and MacIntosh environments, see *http://www.mcell.cnl.salk.edu* or *http://www.mcell.psc.edu*. With MCell highly realistic synaptic simulations can be run on present-day workstations given suitably optimized algorithms and run-time conditions. In addition, very large-scale projects can be run on increasingly accessible parallel processing resources if necessary.

The definition and control of model input and output parameters, simulation operations, and critical run-time optimizations are covered in Section 4.3. The theoretical foundation of simulation operations and numerical accuracy is discussed in Section 4.4. Finally, Sections 4.5 and 4.6 illustrate the use of MCell's diffusion, reaction, and optimization features to simulate quantal current generation[1,5] together with neurotransmitter exocytosis[30–32] at a realistic neuromuscular junction[33] and neuronal cell body.

4.2 GENERAL BACKGROUND ON DESIGN AND SIMULATION OF MICROPHYSIOLOGICAL MODELS

4.2.1 Model Design

Quantitative microphysiological modeling is in its infancy, largely because realistic 3-D models necessarily entail a "scaled-up" computational approach compared to simpler models, and can rapidly escalate into supercomputing territory. In essence, such modeling encompasses four steps, each of which can require considerable computing resources and expertise: reconstruction, model visualization and design, simulation, and visualization and analysis of results. The third step is the primary focus of this chapter; the third and fourth together can involve large-scale parameter fitting and sensitivity analyses (i.e., the sensitivity of a model output quantity to the value of one or more input parameters, Chapter 1, and see also Reference 32).

To simulate a realistic microphysiological system, a representation of the relevant cellular and subcellular structures must first be generated at very high resolution. It is not necessary (nor presently possible) to include the 3-D structure of each molecule, but their positions in space with respect to diffusion boundaries (i.e., cell and organelle membranes) must either be known and included in the model, or predicted by comparison of simulation results to experimental data. The requisite model of cellular structures thus requires a highly accurate 3-D reconstruction, most likely at the EM rather than light level. While light level methods for single cells are fairly well established (Chapter 6), EM level reconstruction of intra- and extracellular spaces for use with simulations is new ground, and is heavily dependent on large-scale computer graphics algorithms. Some of the issues introduced by the need for high resolution and accuracy are outlined in Reference 33, and a general treatment of surface representation and visualization can be found in Reference 28, which also includes the *Visualization Toolkit* software. Another extremely useful software package for model visualization and multidimensional data analysis is IBM DataExplorer (OpenDX, which includes extensive documentation and tutorials; http://www.research.ibm.com/dx).

4.2.2 Simulations Based on Microphysiological Models

Regardless of the methods used to create a microphysiological model, it must be designed so that it subsequently can be imported into a simulation program. At this level of detail and realism, an interactive link between model design and simulation is not a trivial problem, nor is a general solution presently available. For the simulation methods covered in this chapter, the present state of the art is a combination of the above graphics software and a Model Description Language (MDL) covered in Section 4.3.

The simulation program itself can be based on one of two general numerical paradigms, finite element (FE) or Monte Carlo (MC). Both methods have been utilized heavily in computational chemistry and physics but have not yet become commonplace in computational biology, again largely due to the effort involved in first creating the model to be simulated. As outlined below, the FE approach is a 3-D extension of familiar methods based on differential equations (Section 1.1), while the MC approach is altogether different. With a set of equations one predicts the idealized average behavior of a system, while with MC methods one uses random numbers and probabilities to simulate individual cases of the system's behavior. This use of random numbers to "throw dice" and make decisions led Ulam and Von Neumann to coin the "Monte Carlo" appellation in the days of the Manhattan Project.[21] Specifics of MC methods vary according to the problems being investigated, but the results invariably include some form of quantitative stochastic noise that reflects the underlying probabilistic algorithms. In some situations this noise is an undesired drawback and must be reduced by averaging across multiple simulations, but in other cases (especially biology) the variability itself may be of great interest (e.g., synaptic current variability) and may contain useful information.

FE Methods — With FE simulations, a 3-D space is subdivided into contiguous volume elements, or *voxels*. Well-mixed conditions are assumed within each voxel, and differential equations that describe mass action kinetics are used to compute fluxes between and reaction rates within each voxel.[29,34] The methods for solution of the equations are essentially no different than has been described in preceding chapters for problems of reduced dimensionality. High numerical accuracy can be achieved by finely subdividing both space and time, i.e., by using fine spatial and temporal granularity.

As long as the voxels of a FE model are arranged in a regular 3-D grid, their implementation in a simulation can be relatively straightforward, but with realistic cellular structures the grid must be extremely fine and/or must be irregular in shape. With an extremely fine grid the voxels are small and numerous, and the computational expense can grow to be very large (e.g., hydrodynamics and turbulence simulations). With irregular voxels the design of the grid itself becomes an additional large-scale problem, and the dependence of numerical accuracy on grid properties becomes more difficult to assess.

As voxel size decreases, the product of voxel volume and reactant concentration is likely to yield only fractional amounts of molecules. Mass action equations still predict the average behavior of the system, but the stochastic nature and sometimes non-intuitive variability of interactions on the molecular scale are ignored. In principle, some degree of stochastic behavior can be incorporated into equation-based methods,[2,10,18] but at this stage the alternative MC modeling approach becomes highly appealing.

MC Methods — For 3-D reaction-diffusion problems MC methods replace voxels and sets of differential equations with stochastic molecular events simulated directly within the reconstructed volume of tissue. Individual ligand molecules diffuse by means of *random walk movements*, which reproduce net displacements that in reality would arise from Brownian motion. Movement trajectories can be reflected from arbitrary surfaces that represent cell and organelle membranes, and thus *a quantitative simulation of diffusion in complex spatial locales is obtained without the use of voxels*. In addition, reaction transitions such as ligand binding, unbinding, and protein conformational changes can be simulated probabilistically on a molecule-by-molecule basis, using random numbers to throw the dice and test each different possibility against *a corresponding MC probability*. Tests for binding are performed each time a diffusing ligand molecule "hits" an available binding site, and successful tests for unbinding are followed by newly generated random walk movements. *Hence, combined reaction and diffusion can be simulated in any arbitrary 3-D space*. The MC modeling approach thus is very general, and in a sense is easier to implement for complex structures than the FE approach. In addition, MC simulations reproduce the stochastic variability and non-intuitive behavior of discontinuous, realistic 3-D microenvironments that contain finite numbers of molecules.

4.3 GENERAL MONTE CARLO SIMULATIONS: SPECIFIC ISSUES OF DESIGN AND EXECUTION

4.3.1 A Matter of Scale

While sets of differential equations can be evaluated numerically in a variety of ways (Andrew Huxley first computed the Hodgkin–Huxley equations for action potential generation using a mechanical hand-cranked calculator), and models based simply on coupled differential equations can be designed and run within equation-solving software environments (e.g., Matlab or Mathematica), the scope and characteristics of general MC simulations dictate a very different approach. Until recently, MC simulation programs in neuroscience were tailored specifically to a single problem and a simplified structure.[1,5–7,11,15,35] For *fully arbitrary* and *realistic* reaction-diffusion simulations, on the other hand, one must first generate a high-resolution 3-D reconstruction and populate it with molecular constituents (as outlined briefly in Section 4.2.1), and then a representation of the complete model (e.g., 5–50 Mbytes for a synaptic reconstruction) must be read and simulated. If the simulation program itself is based on computationally "naïve" MC algorithms, the time required to run large-scale simulations will be overwhelming (say, months to years) on even the fastest existing computers. With highly optimized algorithms, however, the same simulation might literally run in minutes on a workstation (Section 4.5).

Here we illustrate the design and simulation of realistic MC models using MCell and its Model Description Language (MDL). Both are presently unique tools — MCell as a highly optimized simulation program, and the MDL as a high-level user interface and link between the steps of reconstruction, model design, simulation, and output of results. We outline the underlying concepts and operations in detail so that: (1) the non-programmer can begin to run complex simulations with a minimum of effort; (2) an understanding of MC methods precludes major mistakes in the choice of input parameter values which may lead to profoundly degraded numerical accuracy; and

(3) one could, in principle at least, write a general MC simulation program from scratch (although we expect the 20+ years of aggregate theoretical, programming, and development work will dissuade even the most adventuresome).

4.3.2 Input and Output Parameters

Input parameters for a MC simulation include those that define the microphysiological environment (*simulation objects*, e.g., surfaces, positions of molecules, etc.), together with those that specify how the simulation is actually run (e.g., the length of the MC time-step Δt, the number of time-step iterations, and run-time optimizations). For MCell in particular, simulations are designed and controlled using its MDL user interface, in essence a simple programming language designed for generality, readability, and easy use by programming novices. The MDL and MCell's program flow are summarized in Figure 4.1 and Boxes 4.1 and 4.2. Specific examples follow in Sections 4.5 and 4.6.

Random Number Seed — In a sense, the "first" input parameter for a MC simulation is a *seed value* for generation of random numbers. This seed value determines the values of the random numbers used in the simulation (Section 4.3.4), and hence determines the outcome of decisions made during the simulation. When an MCell simulation is started, the seed value is specified together with the name of an MDL *input file*.* As shown schematically in Figure 4.1, the input files are parsed (read and interpreted), the specified simulation conditions are initialized, and then time-step iterations begin.

Arbitrary Surfaces — To be used with MC random walk and binding algorithms, *arbitrary surfaces* are created from *polygon meshes*. Meshes that represent reconstructed pre- and postsynaptic membranes can easily contain of order 10^6 individual polygons (*mesh elements*, MEs; Figures 4.1, 4.3, and 4.4), and each ME may be subdivided to obtain discrete *effector tiles* (ETs) that can be used to model stationary molecules (see below).** It is important to note that meshes generated for use with MC simulations must be exact, i.e., must "*hold water*." If vertices shared by adjacent MEs do not agree exactly, then random walk movements will pass through the gaps no matter how small. Reconstructions created just for visualization purposes or for use with compartmental equation-based models do not share this requirement, and typically are inadequate for MC simulations.

Surface Properties and Ligand Molecules — In order to simulate an impermeable membrane, individual MEs of a surface must be *reflective* to diffusing ligand molecules. On the other hand, it is often necessary to place stationary molecules (i.e., ETs) on surfaces that do not constitute diffusion boundaries. For example, the acetylcholinesterase (AChE) molecules in the example of Section 4.5 are located on a *transparent* surface that represents the synaptic basal lamina within the synaptic

* The specified input file itself often makes reference to other additional MDL files that contain objects and parameter definitions to be included in the simulation (see INCLUDE_FILE keyword statement in Box 4.2).

** With MCell, spatial dimensions are given in μm, and polygon meshes are created from a POLYGON_LIST object (Box 4.2). Each ME can have any number of vertices but must be convex (all internal angles less than 180°) and exactly planar. Triangular MEs are used most often (guaranteed convex and planar), and a ME *must* be triangular if it contains ETs.

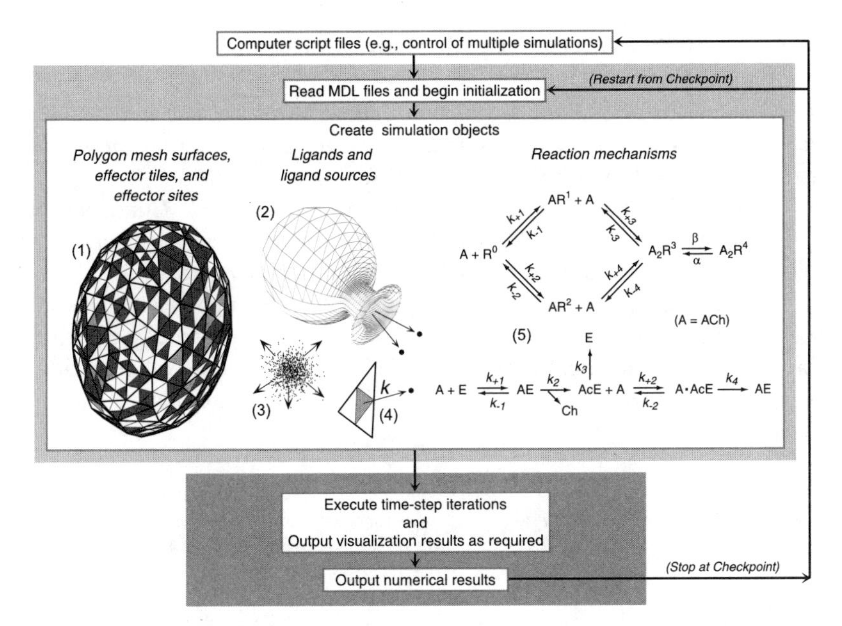

FIGURE 4.1 General Overview of MCell Simulations. Several examples of simulation objects include: (1) ovoid polygon mesh (similar to the simplified nerve cell body used in Section 4.6); (2) complex polygon mesh (synaptic vesicle and exocytotic fusion pore, see Section 4.5); (3) spherical source of ligand molecules; (4) surface that creates ligand molecules at rate k; and (5) reaction mechanisms for acetylcholine (A), acetylcholine receptors (R), and acetylcholinesterase (E, Section 4.5). Objects (1) and (4) illustrate barycentric subdivision of mesh elements (triangles with heavy black outlines) to create effector tiles (smaller triangles) that cover the surface completely and may contain effector sites (shaded triangles) in different chemical states (light vs. dark shading). Program flow and operations during initialization (light gray box) and time-step iterations (dark gray box) are detailed in Figures 4.2A and 4.2B, respectively. MDL, MCell Model Description Language.

cleft. Ligand molecules (acetylcholine, ACh) thus are able to diffuse through the basal lamina unless they encounter and bind to an AChE site. Section 4.5 also shows how transparent surfaces may be used to sample the concentration of diffusing molecules (ACh and choline, Ch, obtained after hydrolysis of ACh by AChE) in restricted regions of space. Each type of ligand molecule used in a simulation is minimally defined by a diffusion coefficient value (D_L),* which, together with the value of Δt, determines the distribution of step lengths used for random walk movements (Sections 4.4.1 and 4.4.2).

* Each type of ligand molecule is defined by two user-specified input parameters, an arbitrary name and a diffusion coefficient value ($cm^2 \cdot s^{-1}$; see DEFINE_LIGAND object in Box 4.2). Multiple predefined (see SPHERICAL_RELEASE_SITE object in Box 4.2) or arbitrary initial distributions of diffusing molecules can be used. An arbitrary initial distribution can be created within a closed polygon mesh (to shape the distribution), and then the bounding mesh can be removed (or changed from reflective to transparent) in order to initiate outward diffusion (see Section 4.3.3). In addition, different temporal patterns can be designed to trigger the release of different amounts and types of ligand molecules (see DEFINE_RELEASE_PATTERN object in Box 4.2, and Sections 4.5 and 4.6).

Box 4.1 Introduction to the MCell Model Description Language (MDL)

MDL files consist of user *comments*, user-defined *variables*, *keywords* (capitalized), *keyword statements*, and *keyword statement blocks* that have *subordinate* statements enclosed in braces (see Box 4.2 for general syntax).

User *comments* are begun with "/*" and ended with "*/". Comments can be nested (comments within comments), and are ignored when the file is parsed.

User *variables* are defined by equating an arbitrary name to a value. Names are typically given in lower case to distinguish them from upper case keywords. Types of variables include text, text expressions, numerical values, numerical expressions, and numerical arrays (elements may be values or expressions). Numerical expressions may include a variety of standard math functions. Examples:

```
user_id = "qwerty"                        /* definition of a text variable named user_id */
output_file = "run_001." & user_id        /* text expression named output_file with elements
                                             joined by the & operator; result would be
                                             "run_001.qwerty" */
k_plus = 1.5e8                            /* definition of a numerical variable named k_plus */
new_k_plus = k_plus*SQRT(3)               /* numerical expression named new_k_plus, uses the
                                             square root operator */
location = [0, 1, 0]                      /* definition of numerical array named location */
```

Keyword statements and *blocks* are used to:

- *Define Values for Simulation Parameters.* As illustrated in Box 4.2, two statements are required for every simulation, to define the time-step value and the number of time-step iterations. Many optional statements can be used to define other parameters and run-time optimizations.
- *Define Logical Objects.* Logical objects have no physical location, and specify sets of input parameters for different types of ligand molecules, ligand release patterns, and chemical reaction mechanisms. Logical object definitions begin with a keyword that contains the word DEFINE.
- *Design Templates for Physical Objects.* Physical objects have a location in space, and include various types of surfaces and ligand release sites. The user specifies an object name, and the first keyword describes the object. Physical objects are initially invoked as templates that can be modified in various ways, and only exist in a simulation if instantiated (see below).
- *Design Metaobject Templates.* Physical object templates can be grouped into metaobjects, which in turn can be grouped into unlimited levels of higher order metaobjects.

(continued)

Box 4.1 (continued)

- *Instantiate Physical Objects and Metaobjects*. Creates actual simulation objects from templates.
- *Output Data for Visualizations and Animations*. See page 97 and Box 4.2.
- *Output Reaction Data Statistics*. See Box 4.2.
- *Output Other Data*. Uses syntax and formatting similar to the C programming language. Allows arbitrary file creation and write operations, printing of messages to the command line window, and conversion of numerical values to text variables.

Box 4.2 General MDL File Organization

The left column shows an abbreviated version of an MDL file, as explained briefly in the right column. MDL keywords are capitalized, and italics indicate names, values, or expressions that would be supplied by the user. Subordinate statements within statement blocks have been omitted, and their positions are indicated by ellipsis marks. When the simulation is started, the file is read (parsed) from top to bottom. Some calculations for initialization are performed while parsing, so there is some order-dependence to the file layout.

MDL file	Explanation
/* comment to describe purpose of MDL file */	Comments can appear anywhere to document the file.
variable_name_1 = *text_expression* *variable_name_2* = *numerical_expression* *variable_name_3* = *numerical_array*	User-defined variables can appear anywhere between statement blocks.
INCLUDE_FILE = *text_expression*	Include files can appear anywhere between statement blocks and should be documented consistently for submission to an online repository (see MCell web sites).
/* Required keyword statements */	
TIME_STEP = *numerical_expression*	Value given in seconds.
ITERATIONS = *numerical_expression*	Total number of time-step iterations.
/* Optional keyword statements */	
EFFECTOR_GRID_DENSITY = *numerical_expression*	Global value for barycentric tiling (tiles · μm^{-2}).
PARTITION_X = [*numerical array*]	Positions along x-axis to insert spatial partitions.
PARTITION_Y = [*numerical array*]	Positions along y-axis to insert spatial partitions.
PARTITION_Z = [*numerical array*]	Positions along z-axis to insert spatial partitions.
CHECKPOINT_INFILE = *text_expression*	Name of checkpoint file to read during initialization.
CHECKPOINT_OUTFILE = *text_expression*	Name of checkpoint file to write before stopping.
CHECKPOINT_ITERATIONS = *numerical_expression*	Number of checkpoint iterations to run before stopping

(continued)

Box 4.2 (continued)

/* Optional logical object definitions */	
DEFINE_LIGAND {...}	Requires user-specified name and diffusion coefficient.
DEFINE_RELEASE_PATTERN {...}	Timing and amount of ligand release.
DEFINE_REACTION {...}	Chemical reaction mechanism associated with ESs.
/* Optional physical object templates */	
name_1 BOX {...}	For simple structures and/or ligand sampling. May include ESs.
name_2 POLYGON_LIST {...}	For complex polygon mesh structures. May include ESs.
name_3 SPHERICAL_RELEASE_SITE {...}	Spherical distribution of ligand molecules (arbitrary diameter). May be associated with a release pattern.
/* Optional metaobject templates */	
name_4 OBJECT {...}	Hierarchical groups of physical objects and/or other metaobjects.
/* Instantiation statement(s) */	
INSTANTIATE *name_5* OBJECT {...}	Create an instance of a physical object. May include geometric transformations.
/* Optional visualization and reaction statistics output */	
VIZ_DATA_OUTPUT {...}	See example models (Sections 4.5 and 4.6).
REACTION_DATA_OUTPUT {...}	See example models (Sections 4.5 and 4.6).

Effector Sites and Reaction Mechanisms — Individual ligand-binding or other stationary molecules (e.g., receptors, transporters, or enzymes) are simulated by using *effector sites* (ESs) on MEs of surfaces. *Each ES is an ET that has a chemical reaction mechanism (see below) associated with it.* Effector sites used in conjunction with arbitrary polygon mesh surfaces and reaction mechanisms constitute one of the most powerful and unique features of MCell simulations. When the user specifies that a ME contains a particular type of ES at some density, σ_{ES} (sites $\cdot$ μm^{-2}), MCell:

1. Covers the ME with a triangular grid of ETs using a method called barycentric subdivision (Figure 4.1). The number of ETs (N_{ET}) depends on the area of the ME (A_{ME}) and a global input parameter, the effector grid density σ_{EG} (tiles $\cdot$ μm^{-2}).* With barycentric tiling, the area of an ET (A_{ET}) is given exactly by A_{ME}/N_{ET}, and approximately by $A_{EG} = 1/\sigma_{EG}$. A_{ME}, on the other hand, depends only on how the surface was constructed.
2. Uses random numbers to decide which of the available ETs are ESs (the ratio $(\sigma_{ES} \cdot A_{ME})/(N_{ET})$ gives the probability that an ET is an ES). Note that *different types of ESs* (e.g., different subclasses of glutamate receptors) *can be intermixed on the same ME.*

Since ESs occupy positions on surfaces in a real 3-D space, reaction mechanisms** used in MC simulations can encompass the *polarity* (directionality) of

* See EFFECTOR_GRID_DENSITY keyword statement in Box 4.2.

** See DEFINE_REACTION object in Box 4.2, and Sections 4.5 and 4.6.

ligand binding and unbinding, as well as the typical *state diagram* and associated *rate constants* that define transitions between different chemical states (Figure 4.1). For example:

1. If an ES on a reflective polygon mesh is used to model a receptor protein on the plasmalemma, binding and unbinding could be defined to occur on the extracellular side of the surface.
2. An ES used as a transporter protein could bind molecules on one side of the surface and unbind on the other side.
3. An ES that represents an enzyme localized on an intra- or extracellular scaffold (transparent surface) could bind and unbind from either side.

With MCell's MDL, rate constants are input as conventional bulk solution values, in units of ($M^{-1} \cdot s^{-1}$) for *bimolecular associations* (i.e., ligand *binding*, k_{+n} values in Figure 4.1), or (s^{-1}) for *unimolecular transitions* (i.e., ligand *unbinding* [k_- values], *transformation* [k_2], *destruction* [k_3 and k_4], or *de novo production* [k], as well as ES conformational changes [α and β]). When the simulation is initialized (Section 4.3.4), bimolecular and unimolecular rate constants are converted into MC probabilities (p_b and p_k values, respectively; Sections 4.4.3–4.4.5).

Output of Visualization and Reaction Data — Since diffusion and all reaction transitions (events) occur on a molecule-by-molecule basis in a MC simulation, it is possible to track the statistics of each and every kind of event in a space- and/or time-dependent manner (e.g., the number of ESs in a particular state, the number of ligand molecules in a sampled volume, fluxes, transitions). In addition, it is often necessary to visualize snap-shots of a simulation, or create an animation from successive snap-shots. Thus, the amount of output information can be enormous. As illustrated in Sections 4.5 and 4.6, MCell's MDL includes unique facilities for selective filtering, formatting, and timing of output.*

4.3.3 Maximizing Flexibility and Efficiency Through Checkpointing

Given the nature of realistic MC simulations, computer time can be lengthy even with highly optimized algorithms. Premature termination of a simulation, with concomitant loss of results (e.g., because one's present allotment of time on a multi-user computer is used up), can be avoided through the use of *checkpointing*. As indicated by the feedback loop in Figure 4.1, checkpointing is a general technique that allows a running program to be stopped and restarted at specified checkpoints (hence the name).**

---

* See VIZ_DATA_OUTPUT and REACTION_DATA_OUTPUT in Box 4.2. Supported formats for visualization output include: IBM DataExplorer (open source, http://www.research.ibm.com/dx), Pixar RenderMan (proprietary, http://www.pixar.com), Blue Moon Rendering Tools (shareware, adheres to the RenderMan interface standard, http://www.bmrt.org), rayshade (open source, ftp://graphics.stanford.edu/pub/rayshade), povray (open source, http://www.povray.org), irit (open source, http://www.cs.technion.ac.il/~irit).

** At each checkpoint, all the information required to restart is saved in a *checkpoint file*. With MCell simulations, each checkpoint (given as a number of time-step iterations) and checkpoint file name is specified with a keyword statement in the MDL input file (Box 4.2).

One simple use of checkpointing is to subdivide a lengthy simulation into a sequence of shorter runs, and thereby avoid premature termination. A more powerful adaptation is to introduce one or more parameter changes when the simulation restarts. It is also possible to reuse intermediate results saved at different checkpoints. In this way computation time can be reduced dramatically, and simulation conditions can branch from a common point into parallel tracks in which one or more input parameters are varied. Examples of changes that can be incorporated into an MCell checkpoint sequence* include: existing surfaces can be removed and new surfaces can be added; existing surfaces can be moved (e.g., to simulate an expanding exocytotic fusion pore that joins a synaptic vesicle to a presynaptic membrane[30]); surfaces can be changed from reflective to transparent or absorptive, etc.; new ligand molecules and effector sites can be added; new reaction mechanisms can be added; existing reaction mechanisms can be modified; the simulation time-step can be changed; the type and amount of information to be output as results can be changed.

4.3.4 Initialization, Optimized Random Number Generation, and Time-Step Events

Initialization — Figure 4.2A summarizes how an MCell simulation is initialized before time-step execution begins. In essence, initialization includes: (1) instantiation (creation) of pre-existing objects from preceding checkpoint data (if any); (2) instantiation of new simulation objects; and (3) set-up of runtime optimizations. As outlined below and demonstrated in Section 4.5.2, the user must specify runtime optimizations efficiently to run large-scale realistic simulations, because this can decrease the required computer time by *orders of magnitude*.

Random Numbers — Since random numbers are used to make decisions even during initialization (e.g., to place ESs on MEs), a stream of available values is set up when initialization begins (Figure 4.2A). A computer's "random" numbers in reality are "pseudorandom" because they are computed as a deterministic stream using a mathematical algorithm and an initial seed value (for background information on various algorithms, see References 14 and 19). Each number in the stream is actually a sequence of binary bits, and the best algorithms return not just statistically uncorrelated numbers, but also uncorrelated bits within each number. Although most programming languages include built-in functions that compute pseudorandom numbers, the underlying algorithms generally are not adequate to produce uncorrelated bits. MCell therefore uses a self-contained, 64-bit cryptographic-quality algorithm that has been tested at the bit level, and also includes 3000 pre-defined seed values. The advantages are: speed, because the bits in each "single" random number can be subdivided to obtain multiple values for the computational price of one (a random number can be split and used to pick both a distance and direction for a random walk movement, Section 4.4.2); and reproducibility, because

* Modifications to input parameters in successive MDL files used for a checkpoint sequence can be set up by the user in advance, and/or the MDL can be used to introduce incremental parameter changes. In addition, users familiar with operating system script files (e.g., UNIX shell scripts or DOS batch files) can use them for an additional level of flexibility and automation (Figure 4.1).

a given seed produces the same stream of values, and therefore identical simulation results, regardless of the computer platform or operating system.

Time-Step Events — During each simulation time-step, many decisions must be made between different possible events that can occur to *reactant* molecules, where *reactant* includes all existing ligand molecules, bound ligand-ES complexes, and any unbound ESs currently in a chemical state that can undergo a unimolecular reaction transition. A list of all reactants is created when a simulation is initialized (Figure 4.2A), and subsequent time-step events and program flow are summarized in Figure 4.2B. Most computation time for a simulation is consumed by diffusing molecules, because their movement trajectories, or *rays* (a radial distance l_r and direction ξ), must be traced through space to determine whether the ray intersects a ME of a surface. If not, the ligand molecule is simply placed at the endpoint of the ray for the duration of the time-step. If so, the final result depends on the presence or absence of a reactive ES at the point of intersection (P_I), and/or the properties of the ME, as outlined in Figure 4.2B. If the ligand molecule is neither absorbed by the surface nor retained at P_I because of a bimolecular reaction, then its movement must be continued. After passing through or reflecting from the ME (transparent or reflective ME, respectively), the search for an intersection begins again and this process of *ray marching* continues until the molecule is either absorbed, reacts with an ES, or travels a total distance l_r and remains at the resulting endpoint of motion for this time-step.

4.3.5 Additional Run-Time Optimizations: Spatial Partitions and Fast Realistic Random Walk Methods

Spatial Partitions — As illustrated in Figures 4.2 and 4.3, MCell's ray marching algorithms also include a run-time optimization called *spatial partitions*. In the *absence* of partitions, every ME must be checked for a potential intersection every time a diffusing molecule moves. Thus, the computer time for a simulation is roughly proportional to the total number of MEs, i.e., a simulation with 100,000 MEs would require ~1000-fold more computer time than one with 100, and a simulation of one synaptic current at a reconstructed synapse might run for weeks or months. With spatial partitions, however, the number of MEs has very little impact on computer time, and simulations of currents in realistic synaptic structures can run in minutes (Section 4.5.2).

Spatial partitions are transparent planes (Figure 4.3) that the user places along the *x*, *y*, and/or *z* axes* to create spatial subvolumes (SSVs). When the simulation is initialized, the partitions are inserted and the MEs included within each resulting SSV are determined (Figures 4.2A and 4.3). The user's goal is to arrange a sufficient number of partitions so that each SSV includes no more than a few MEs. During time-step iterations, the SSV in which each ligand molecule currently resides is always known (Figure 4.2), so the search for intersections during ray marching can be limited to those MEs included in the ligand molecule's current SSV, as well as

* See PARTITION_X, PARTITION_Y, and PARTITION_Z keyword statements in Box 4.2.

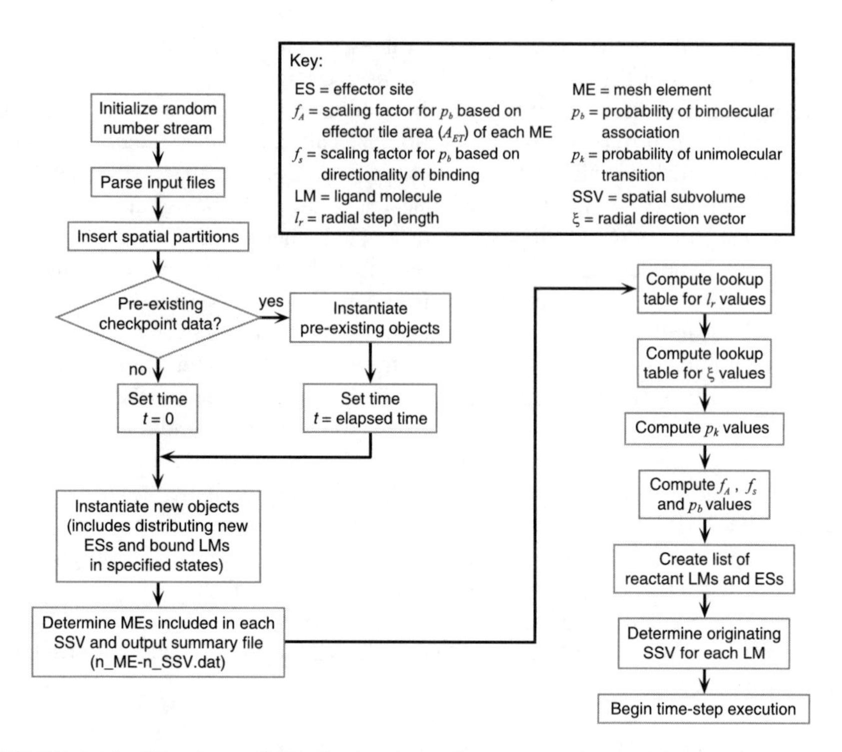

FIGURE 4.2A Flowchart of Initialization Operations. In general, initialization includes set-up of: (1) random numbers; (2) pre-existing and new objects; (3) run-time optimizations such as spatial partitions and random walk look-up tables; (4) calculation of probabilities and scaling factors for reaction transitions; and (5) a reactant list that is traversed during the first time-step (Figure 4.2B). To help the user optimize the placement and number of spatial partitions (Sections 4.3.5 and 4.5.2), MCell writes an output file (n_ME-n_SSV.dat) that can be used to create a frequency histogram for the number of mesh elements included in spatial subvolumes (see Figure 4.5B).

the boundaries of the SSV itself. If a ligand molecule passes through an SSV boundary, ray marching simply continues in the new SSV using an updated list of MEs and boundaries (Figure 4.2B).

As simulations grow in structural complexity, and therefore include an increasing number of MEs, the number of spatial partitions can be increased to keep the average number of MEs per SSV about constant. Under these conditions, *MCell's execution speed will also remain nearly constant, i.e., independent of the number of MEs.* Since partitions influence only the simulation's execution speed, and not the random number stream, net ligand displacements, nor reaction decisions, *simulation results are identical regardless of partitioning.*

Random Walk — While most of the computational expense incurred for diffusing molecules arises from ray marching, another significant component is generation of the random walk rays themselves. The simplest (and least realistic) random walk algorithm uses a constant step length and chooses a positive or negative x, y,

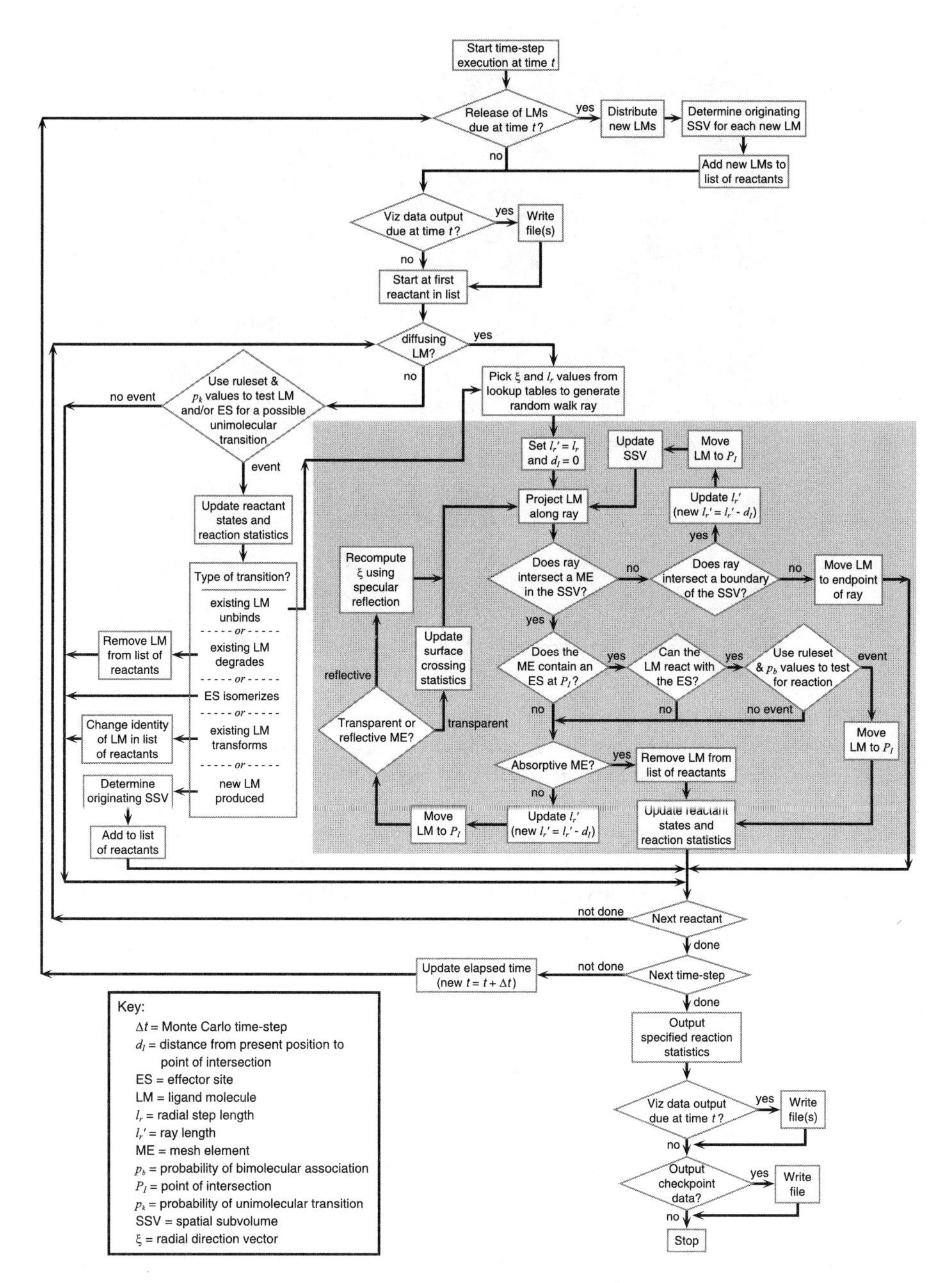

FIGURE 4.2B Flowchart of Time-Step Operations. During each time-step, a list of valid reactants is traversed. Depending on each reactant's identity, random numbers are used either to generate a random walk movement and test for possible bimolecular associations (gray box), or to test for possible unimolecular transitions. Additional operations may include release of new ligand molecules and output of visualization data. Specified reaction statistics and checkpoint data are written to output files after the last iteration.

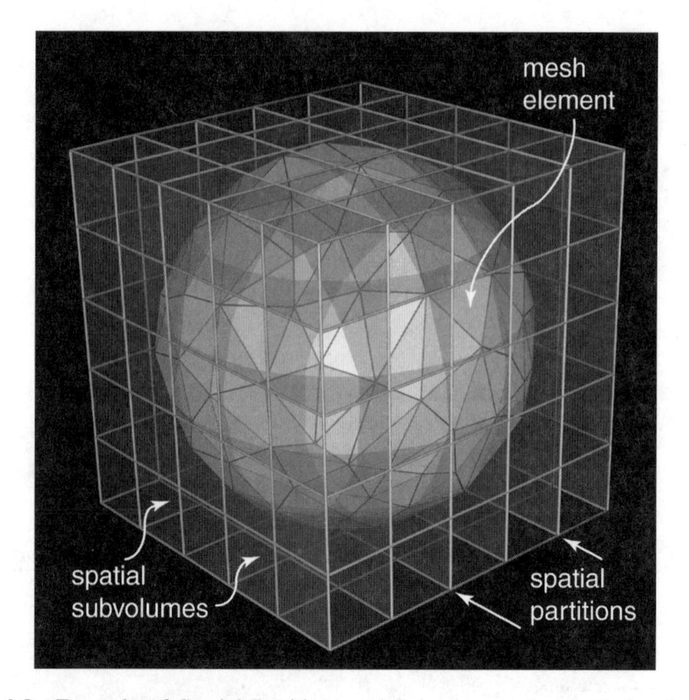

FIGURE 4.3 Example of Spatial Partitions and Subvolumes. A spherical polygon mesh surface is shown with spatial partitions (transparent planes) along the *x*, *y*, and *z* axes. Spatial subvolumes (cuboids) are created between the partitions, and under optimal conditions each subvolume includes (wholly or partly) no more than a small number of mesh elements (Section 4.3.5). The search for collisions between diffusing ligand molecules and mesh elements (ray tracing and marching, Figure 4.2B) then can be restricted to individual subvolumes, dramatically increasing execution speed (Section 4.5.2). As shown for clarity, each partition extends just beyond the dimensions of the sphere. In an actual simulation, however, partitions extend to the bounds of available space ($\sim 10^{151}$ μm $\cong \infty$), and may be placed anywhere along each axis.

and *z* displacement for each movement.[9] Movements are generated rapidly, but molecules occupy positions on a 3-D Cartesian lattice. For accurate simulation of diffusion, the granularity of the lattice must be very fine compared to the size of any restrictive structures through which the molecules move. Fine granularity requires a very small step length, and hence a very small simulation time-step. The smaller the time-step, the more time-step iterations are required, and thus the advantage of rapidly generated movements is negated.

The most realistic random walk algorithm computes a random radial direction and step length for each movement. Because the step lengths are obtained from a theoretical diffusion distribution (Section 4.4.1) and the molecules do not move along a lattice, accurate simulation of fluxes through restrictive structures can be obtained with a longer time-step than is required with simpler algorithms. However, the trigonometric and numerical integration operations required for each movement are very time-consuming, so the overall execution speed is very slow. MCell achieves the same maximal realism and accuracy with the speed of the simplest algorithm by using extensive look-up tables for equally probable radial step lengths (l_r) and direction vectors (ξ). The values in each table are computed only once when the simulation is initialized (Figure 4.2A), and thereafter are chosen as required (Figure 4.2B) using a subdivided random number (Section 4.3.4). MCell's random walk algorithms are outlined more fully in the following section, as is the calculation of p_b and p_k values used in testing for possible reaction transitions during time-step iterations (Figure 4.2B).

4.4 THEORETICAL FOUNDATION

4.4.1 ANALYTIC DIFFUSION THEORY FOR NET RADIAL DISPLACEMENTS

An actual diffusing ligand molecule initially at point P_1 at time ($t = 0$) has some thermal velocity and undergoes Brownian movements (i.e., collides with water molecules) on the sub-ps timescale at room temperature.[4] After many collisions accumulate over a longer time interval ($t = \Delta t$), the difference between the molecule's ending position P_2 and initial position P_1 yields a net radial displacement (distance r) in a random direction. To calculate the theoretical distribution of net radial displacements (that will be used to generate random walk step lengths), we start from Fick's second law, i.e., the diffusion equation for concentration C at any point in space as a function of time:

$$\left(\frac{\partial C}{\partial t}\right)_{x,y,z} = D_L\left[\left(\frac{\partial^2 C}{\partial x^2}\right)_t + \left(\frac{\partial^2 C}{\partial y^2}\right)_t + \left(\frac{\partial^2 C}{\partial z^2}\right)_t\right] \tag{4.1}$$

If a concentration gradient exists in only one dimension (ϑ), whether linear (x, y, or z) or radial (r), Equation 4.1 reduces to:

$$\left(\frac{\partial C}{\partial t}\right)_{\vartheta} = D_L\left(\frac{\partial^2 C}{\partial \vartheta^2}\right)_t, \quad \vartheta = x, y, z \text{ or } r \tag{4.2}$$

which can be solved analytically for certain sets of boundary conditions. For a point source of M molecules and a time interval Δt, the solution is[9]

$$C(r, \Delta t) = \frac{M}{\left(4\pi D_L \Delta t\right)^{3/2}} e^{-r^2/4D_L\Delta t} \tag{4.3}$$

Equation 4.3 gives the theoretical ligand concentration (molecules per unit volume) as a continuous function of time and radial distance (and appears in similar form in Chapter 3). Multiplying Equation 4.3 by the volume of a spherical shell

($4\pi r^2\, dr$) gives the amount of ligand (N_r, number of molecules) contained in the shell at time Δt:

$$N_r = \frac{M}{(4\pi D_L \Delta t)^{3/2}} e^{-r^2/4D_L\Delta t}(4\pi r^2\, dr) \tag{4.4}$$

Dividing both sides of Equation 4.4 by the total amount of ligand (M) gives the fractional amount (N_r/M) in the shell, or the *fractional probability* (p_r) of a net radial displacement between r and (r + dr) *for a single diffusing molecule*. Hence:

$$p_r = \frac{1}{(4\pi D_L \Delta t)^{3/2}} e^{-r^2/4D_L\Delta t}(4\pi r^2\, dr) \tag{4.5}$$

The mean radial displacement ($\bar{l}_r$) is obtained from the expectation value of r, i.e., by integrating across all probability-weighted values of r, and is:

$$\bar{l}_r = 2\sqrt{\frac{4D_L \Delta t}{\pi}} \tag{4.6}$$

If P_1 is defined as the origin of a Cartesian coordinate system and a set of radial displacements in random directions is evaluated, the average (±) x, y, and z components ($\Delta\bar{x}$, $\Delta\bar{y}$, and $\Delta\bar{z}$) for the set are all equal to ($\bar{l}_\perp$), where:

$$\bar{l}_\perp = \sqrt{\frac{4D_L \Delta t}{\pi}} \tag{4.7}$$

Because of radial symmetry, the same result is obtained for (±) $\Delta\bar{x}$, $\Delta\bar{y}$, and $\Delta\bar{z}$ no matter how the Cartesian axes are rotated around P_1. In general, then, $\bar{l}_\perp$ is the average value of (±) $\Delta\psi$, where ψ is any arbitrary axis that passes through P_1. For a plane at any arbitrary location in space, there exists an axis ψ that is perpendicular (normal) to the plane. Thus, the average increase or decrease in distance between a diffusing molecule and the plane (measured with respect to ψ) is given by $\bar{l}_\perp$. Comparison of Equations 4.6 and 4.7 shows that $\bar{l}_\perp$ is equal to $\bar{l}_r/2$.

4.4.2 Optimized Brownian Dynamics Random Walk

The probability distribution function of Equation 4.5 can be rewritten for a dimensionless parameter s, where s is defined as $r/\sqrt{4D_L\Delta t}$:

$$p_s = \frac{4}{\sqrt{\pi}} s^2 e^{-s^2}\, ds \tag{4.8}$$

When an MCell simulation begins, Equation 4.8 is integrated to obtain a large set of equally probable values of s (1024 under default conditions), and these values are stored in a look-up table. Thereafter, each time a radial step length l_r is needed for a random walk movement, a random number is used to pick a value of s from the table. To convert from s to l_r, the chosen value of s is multiplied by the scaling factor $\sqrt{4D_L\Delta t}$ (which has units of distance).

A second look-up table is used for an extensive set of equally probable radial directions (>130,000 by default). These directions are stored as the x, y, and z components of unit vectors (ξ) that radiate out from a point and are spaced evenly from each other. An initial subset is computed to fill one octant of a unit sphere, and then is reflected to fill the remaining seven octants. This symmetrical replication of the initial subset guarantees the complete absence of directional bias, which is critical because even a tiny net bias can accumulate over thousands of time-steps to produce substantial drift. Once random numbers have been used to pick l_r and ξ for a random walk ray, the x, y, and z components of ξ are multiplied by l_r. The results then are added to the (x, y, z) values of the molecule's initial position (P_1) to find a possible endpoint of motion (P_2). Finally, the ray is traced and marched as required (Figure 4.2B).

As introduced briefly in Section 4.3.5, simpler random walk algorithms can be designed in many ways, but the MCell method has the following advantages (which also distinguish it from our own earlier algorithm used with MC simulations):[5]

1. Realism — movements are radially symmetrical during each individual time-step, and under default conditions are essentially free of discernible spatial granularity.
2. Speed — no penalty is incurred for realism, because numerous optimizations make it as fast or faster than simpler, less realistic methods.
3. Numerical accuracy — extensive sets of radial distances and directions are available for each movement, so concentration gradients are reproduced accurately within distances as small as twice the mean radial step length $\bar{l}_r$ (Equation 4.6). Thus, to simulate diffusion through a constriction like an exocytotic fusion pore, the time-step Δt need only be chosen so that $\bar{l}_r$ is about twofold smaller than the radius of the pore.[30,33]
4. Adjustability — although the default numbers of radial step lengths (1024) and directions (~130,000) are sufficient for almost any conceivable circumstance, the user can increase or decrease either value, or even replace the look-up table for direction vectors with directions chosen at random for each movement (full machine precision, slower execution).

4.4.3 General Overview of Monte Carlo Probabilities

Monte Carlo binding and unbinding probabilities were first introduced in simulations of a simplified vertebrate neuromuscular junction that contained regular rectangular junctional folds, and therefore simple rectangular grids for binding sites.[5] The simulation program was soon merged with another developed independently at about the same time,[33] and after extensive generalization the earliest version of MCell was created;[30,31] (see References 26 and 33 for a more complete historical overview). Simulation of highly realistic rather than simplified models is now possible after much additional optimization and development. Here we present detailed general derivations of the MC probabilities for bimolecular (p_b) and unimolecular (p_k) reaction transitions (see Chapter 2, Box 2.1 for an overview of reaction mechanisms), as required for the new methods and capabilities. This is important but difficult ground for newcomers, and luckily the practical "take-home" message can be summarized simply: *For a given set of input conditions, simulation results will converge to the correct answer as a single input parameter, the time-step* Δt*, decreases.* Unlike FE simulations, there is no separate parameter (i.e., voxel size and shape) that determines the simulation's spatial granularity. Instead, *spatial granularity decreases hand-in-hand with decreasing temporal granularity, because random walk distances become smaller as* Δt *becomes smaller.*

Each time that a reactant molecule is tested for possible chemical transitions (Figure 4.2B), the value of a single random number (κ) is compared to the relevant MC probability value computed during initialization (Figure 4.2A). Each probability value is subdivided into fractional amounts if the reactant has more than one available transition path. For example, in the reaction mechanisms of Figure 4.1:

1. A random walk ray for a diffusing A molecule might intersect ("hit") an ES in the R^0 state, and A then would be tested for binding to either of two independent binding sites defined by parallel transition paths and the bulk solution rate constants k_{+1} and k_{+2}. Each of the two rate constants would be used to compute a fraction of p_b, the MC probability for bimolecular association (see below). In general, each ES can represent a molecule with an *arbitrary number of binding sites.*
2. A ray might hit an ES in the E state, which has a single available binding site, and hence a single rate constant would be used to compute p_b.
3. An ES in the A_2R^3 state might be tested simultaneously for unbinding from either binding site, or isomerization to the (A_2R^4) state. The three rate constants k_{-3}, k_{-4}, and β would each be used to compute a fraction of p_k, the MC probability for a unimolecular transition.

4.4.4 The MC Probability of Unimolecular Transitions

Derivation — As illustrated in Figures 4.1 and 4.2B, MCell simulations presently can include five different types of unimolecular transition. In each case (e.g., the $A_2R^3 \rightarrow A_2R^4$ isomerization in Figure 4.1), the transition is governed by a first-order rate constant k that has units of inverse time (s^{-1}). In general, if initial state S^0 can undergo one of n different possible transitions:

$$S^1 \xleftarrow{k_1} S^0 \begin{array}{l} \xrightarrow{k_2} S^2 \\ \quad\vdots \\ \xrightarrow{k_n} S^n \end{array} \tag{4.9}$$

and the reaction proceeds for some time Δt, the total probability (p_{kt}) that a *single molecule in the* S^0 *state undergoes any transition* is given by (Box 4.3):

$$p_{kt} = 1 - \exp\left[-\left(\sum_{1}^{n} k_i\right)\Delta t\right] \quad (4.10)$$

and the fractional probabilities of each individual transition are:

$$p_{k1} = p_{kt} \cdot \frac{k_1}{\sum_{1}^{n} k_i}, \ldots p_{kn} = p_{kt} \cdot \frac{k_n}{\sum_{1}^{n} k_i}; \quad \sum_{1}^{n} p_{ki} = p_{kt} \quad (4.11)$$

Implementation, Validation, and Accuracy — Since p_{kt} is the probability of any transition during Δt, the probability of no transition is $(1 - p_{kt})$. The decision between all possible events (including no transition) is made with maximal run-time efficiency by comparing the value of a single random number $(0 \leq \kappa \leq 1)$ to the cumulative set of probabilities $(p_{k1}, p_{k1} + p_{k2}, \ldots, p_{kt}, 1)$.*

MCell results for unimolecular transitions can be verified with a simple simulation of Equation 4.9, i.e., one that starts with a set of ESs in state S^0, and then tallies:** (1) the number of all transitions from S^0 per time-step Δt, to verify the expected exponential distribution of lifetimes for the S^0 state (with a mean value of τ, Box 4.3); and (2) the number of ESs in each state after each time-step, to verify the expected proportions (Equation 4.11). For high numerical accuracy, the value of Δt used for the simulation must be small compared to τ. Of course, the MCell output will also include stochastic noise, the magnitude of which will depend on the absolute number of ESs. Such noise can be reduced by averaging across multiple simulations run with different random number seeds, and will decrease by a factor $1/\sqrt{n_s}$, where n_s is the number of simulations.

4.4.5 The Momte Carlo Probability of Bimolecular Associations

Derivation — For a simulation of bimolecular association between ligand A and receptor R with n possible binding sites:

$$A + R \begin{array}{l} \xrightarrow{k_{+1}} AR^1 \\ \quad \vdots \\ \xrightarrow{k_{+n}} AR^n \end{array} \quad (4.12)$$

fractional values of p_b are used to test for binding to any one of the available sites each time a random walk ray hits an ES in state R (Figure 4.2B). The total number

* If $\kappa \leq p_{k1}$, then the first possible transition occurs. If $p_{k1} < \kappa \leq (p_{k1} + p_{k2})$, then the second possible transition occurs, and so on. If $\kappa > p_{kt}$, then no transition occurs.

** See REACTION_DATA_OUTPUT in Box 4.2 and COUNT statements in Sections 4.5 and 4.6.

of times that a *particular* ES is hit during a time-step Δt can range from zero to any (+) integer value, depends on the local *concentration-dependent* flux of A molecules into the ES, and shows stochastic variability across successive trials. The *average* number of hits per time-step (N_H) thus is a (+) *real number that approaches zero as either* Δt *or the concentration of A approaches zero.*

Box 4.3 Derivation of the MC Probability for Unimolecular Transitions

If the reaction of Equation 4.9 (Section 4.4.4) proceeds for any time t from an initial concentration $(S^0)_o$, the total probability (p_{kt}) that a *single molecule in the* S^0 *state undergoes a transition* is given by the *fraction of* $(S^0)_o$ *that undergoes any transition during time t*:

$$p_{kt} = \frac{\left(S^1\right)_t + \left(S^2\right)_t + \ldots\left(S^n\right)_t}{\left(S^0\right)_o} = 1 - \frac{\left(S^0\right)_t}{\left(S^0\right)_o} \tag{1}$$

The general rate equation depends only on time and (S^0):

$$\begin{aligned} -d\left(S^0\right) &= d\left(S^1\right) + d\left(S^2\right) + \ldots d\left(S^n\right) \\ &= \left(k_1 + k_2 + \ldots k_n\right)\left(S^0\right)dt = \left(\sum_1^n k_i\right)\left(S^0\right)dt \end{aligned} \tag{2}$$

and hence can be integrated directly to obtain p_{kt}. From Equation 4.2:

$$\int_{\left(S^0\right)_o}^{\left(S^0\right)_t} \frac{d\left(S^0\right)}{\left(S^0\right)} = -\left(\sum_1^n k_i\right)\int_0^t dt \tag{3}$$

and the solution is

$$\frac{\left(S^0\right)_t}{\left(S^0\right)_o} = \exp\left[-\left(\sum_1^n k_i\right)t\right] \tag{4}$$

From Equation 4 the lifetime of S^0 is exponentially distributed with a mean value

$$\tau = 1 / \sum_1^n k_i \tag{5}$$

From Equations 4 and 1:

$$p_{kt} = 1 - \exp\left[-\left(\sum_1^n k_i\right)t\right] \tag{6}$$

as given in Equation 4.10 (Section 4.4.4).

Since p_b is the probability that binding *occurs* at any one of the available sites after a *single* hit, the probability that binding *does not occur* after a *single* hit is $(1 - p_b)$. The probability that binding has not occurred *after a total of N_H hits* is $(1 - p_b)^{N_H}$, and thus the total probability (p_{bt}) that binding *has occurred* during Δt, i.e., *after any one of the N_H hits,** is given by:

$$p_{bt} = 1 - (1 - p_b)^{N_H} \tag{4.13}$$

In order for the binding kinetics of a MC simulation to be quantitatively correct, *the average instantaneous binding rate must equal the idealized binding rate predicted by mass action kinetics*. Therefore, the value of p_{bt} must equal a corresponding probability (p_t) calculated from mass action rate equations. For a short interval of time Δt:

$$p_t \cong \zeta = \left(\sum_1^n k_{+i}\right)(A)_o \,\Delta t \tag{4.14}$$

where $(A)_o$ is the local concentration of ligand molecules around a single R molecule at the beginning of Δt (Box 4.4). To calculate p_b for use in MC simulations, Equation 4.13 thus is set equal to Equation 4.14 (see Box 4.5):

$$1-(1-p_b)^{N_H} = p_t \cong \zeta = \left(\sum_1^n k_{+i}\right)(A)_o \,\Delta t \tag{4.15}$$

and then Equation 4.15 must be solved for p_b. Doing so directly yields a ligand concentration-dependent expression for p_b, which would make p_b a space- and time-dependent parameter specific to each ES during a running MC simulation. The resulting computational cost would be staggering. Fortunately, however, as Δt approaches 0, p_b, and N_H must also approach 0, and under such conditions the term $(1 - p_b)^{N_H}$ in Equation 4.15 approaches $(1 - N_H \cdot p_b)$.** After substitution and rearrangement:

$$p_b = \left(\sum_1^n k_{+i}\right)\frac{(A)_o \,\Delta t}{N_H} \quad ; \text{ for small } \Delta t \tag{4.16}$$

---

* A concrete analogy to this probability problem goes as follows: Roll a six-sided die 3 (i.e., N_H) times. What is the probability that a 1 *is not* obtained on *any* of the three trials? The probability of rolling a 1 (i.e., p_b) is 1/6, and so the probability of *not* rolling a 1 is (1 – 1/6 = 5/6) for the first trial *and* the second trial *and* the third trial. Thus, the probability that a 1 is not obtained within three rolls is $(5/6) \cdot (5/6) \cdot (5/6)$, or $(5/6)^3$, i.e., $(1 - p_b)^{N_H}$. The probability that a 1 *will be* rolled (i.e., binding will occur) within three trials therefore must be $1 - (5/6)^3$, or $1 - (1 - p_b)^{N_H}$ as given in Equation 4.13.

** At the limit of $\Delta t = 0$, ligand molecules make no movements at all, and hence no hits can occur. With $N_H = 0$, both $(1 - p_b)^{N_H}$ and $(1 - N_H \cdot p_b)$ evaluate to unity. For small non-zero values of N_H, the agreement between the two terms is especially close if p_b is also small (and p_b, like N_H, decreases as Δt decreases). For example, if both N_H and p_b are 0.1, the two terms agree to within 0.05%.

As mentioned above, the term N_H in Equation 4.16 is determined by the local concentration-dependent flux (J) of A molecules into one ES. For small Δt, J (and therefore N_H) is directly proportional to $(A)_o$, and after substitution of a final expression for N_H (Box 4.6) into Equation 4.16:

$$p_b = \left(\sum_1^n k_{+i}\right)\frac{1}{2(N_a)(A_{ET})}\left(\frac{\pi\Delta t}{D_L}\right)^{1/2} \tag{4.17}$$

where N_a is Avogadro's number, A_{ET} is the area of an ET, and D_L is the diffusion coefficient of ligand A. The expression for p_b in Equation 4.17 is independent of free ligand concentration, and therefore can be implemented very efficiently in simulations.

Box 4.4 Bimolecular Associations: Derivation of ζ

If the reaction of Equation 4.12 (Section 4.4.5) proceeds for some time t from initial concentrations $(A)_o$, $(R)_o$, and $(AR)_o$, the mass action probability (p_t) that a *single R molecule becomes bound* is given by the fraction of $(R)_o$ that becomes bound, i.e.,

$$p_t = \frac{\sum_1^n\left[(AR^i)_t - (AR^i)_o\right]}{(R)_o}, \tag{1}$$

or simply

$$\frac{\sum (AR^i)_t}{(R)_o} \tag{2}$$

if

$$\sum = \sum_1^n \text{ and } \sum (AR^i)_o = 0 \tag{3}$$

(as used below). The general rate equation for production of the n bound states is

$$\partial\left[\sum (AR^i)\right] = -\partial(A) = -\partial(R) = \left(\sum k_{+i}\right)(A)(R)\partial t \tag{4}$$

(continued)

Box 4.4 (continued)

but Equation 4 cannot be integrated directly to obtain (AR)*t* because (A), (R), and (AR) are functions of space as well as time. In Equation 4, the quantity ($D_A + D_R$), i.e., the sum of the diffusion coefficients for A and R, is implicitly included in the values of k_+, together with the sizes and shapes of the molecules, and the activation energy for each binding reaction. If at least one of the diffusion coefficients is large but the k_+ values are small (e.g., because the activation energy is large), then the rate of reaction is not "diffusion-limited," i.e., the solution is always "well-mixed" because the rate of binding is slow compared to the rate of diffusion. Under such conditions, appreciable spatial concentration gradients do not form as binding proceeds, so the partial differentials of Equation 4 can be replaced with ordinary differentials:

$$d\left[\sum\left(AR^i\right)\right] = -d(A) = -d(R) = \left(\sum k_{+i}\right)(A)(R)dt \qquad (5)$$

Since the concentration terms in Equation 5 are independent of space, by definition they are *equally valid* for the bulk solution and at the *local level in the vicinity of single molecules*. Equation 5 can be integrated to determine $(AR)_t$:

$$\int_{\sum\left(AR^i\right)_o}^{\sum\left(AR^i\right)_t} d\left[\sum\left(AR^i\right)\right] = \left(\sum k_{+i}\right)\int_o^t (A)(R)dt =$$

$$\left(\sum k_{+1}\right)\int_o^t \left(A_o - \sum AR^i\right)\left(R_o - \sum AR^i\right)dt \qquad (6)$$

After integration, final analytic expressions for p_t are

$$p_t = \frac{\sum\left(AR^i\right)_t}{(R)_o} = \frac{1}{(R)_o}\left(\frac{\left(b^2+q\right)\left(\exp\left[\left(\sum k_{+1}\right)\cdot t\sqrt{-q}\right]-1\right)}{2\left(b+\sqrt{-q}-\left(b-\sqrt{-q}\right)\exp\left[\left(\sum k_{+i}\right)t\sqrt{-q}\right]\right)}\right) \qquad (7)$$

$$\left(\text{if } (A)_o \neq (R_o)\ ;\ \text{where } b = -\ (A_o + R_o) \text{ and } q = 4(A_o)(R_o) - b^2\right)$$

or

$$p_t = \frac{\sum\left(AR^i\right)_t}{(R)_o} = \frac{1}{(R)_o}\left(\frac{\left(\sum k_{+i}\right)(A)_o^2 t}{1+\left(\sum k_{+i}\right)(A)_o t}\right) = \frac{\left(\sum k_{+i}\right)(A)_o t}{1+\left(\sum k_{+i}\right)(A)_o t} \qquad (8)$$

$$\left(if (A)_o = (R)_o\right)$$

(continued)

Box 4.4 (continued)

From Equations 7 and 8, the probability (p_t) that a single R molecule becomes bound during an arbitrarily long interval of time (t) depends on all the k_+ values, $(A)_o$, and $(R)_o$; $p_t = 0$ for $t = 0$, and for $t = \infty$, $p_t = 1$ if $(A)_o \geq (R)_o$, or $p_t = (A)_o/(R)_o$ if $(A)_o < (R)_o$.

If the interval of time is very short, so that $t = \Delta t$ and $(AR^i)_{\Delta t}$ is much less than both $(A)_o$ and $(R)_o$, then

$$\left(A_o - \sum\left(AR^i\right)_{\Delta t}\right) \cong (A)_o \text{ and } \left(R_o - \sum\left(AR^i\right)_{\Delta t}\right) \cong (R)_o. \tag{9}$$

Equation 4.3 then becomes:

$$\int_{\sum\left(AR^i\right)_o}^{\sum\left(AR^i\right)_{\Delta t}} d\left[\sum\left(AR^i\right)\right] \cong \left(\sum k_{+i}\right)(A)_o(R)_o \int_o^{\Delta t} dt \tag{10}$$

and after integration:

$$\sum\left(AR^i\right)_{\Delta t} \cong \left(\sum k_{+i}\right)(A)_o(R)_o \Delta t \tag{11}$$

Thus, for a short interval of time Δt:

$$p_t = \frac{\sum\left(AR^i\right)_{\Delta t}}{(R)_o} \cong \zeta = \left(\sum k_{+i}\right)(A)_o \Delta t \tag{12}$$

where $(A)_o$ is both the instantaneous local concentration of ligand molecules in the immediate vicinity of a single R molecule, and the average bulk solution ligand concentration.

If the MEs of a surface have different shapes and sizes (which is generally the case for reconstructions), the value of A_{ET}, and hence p_b, is different for each ME. The exact values of A_{ET} are determined at the time of barycentric subdivision, and are slightly smaller than the approximate value, A_{EG} (which itself is the inverse of σ_{EG}, the global effector grid density; Section 4.3.2 and Box 4.2). A factor

$$f_A = \frac{A_{EG}}{A_{ET}} = \frac{1}{\sigma_{EG} \cdot A_{ET}} \tag{4.18}$$

is calculated for each ME during initialization (Figure 4.2A), and a final expression for p_b is:

$$p_b = \sum_1^n p_{bi} = \sum_1^n \left(f_{si} \cdot k_{+i}\right) \cdot X = f_{s1} \cdot k_{+1} \cdot X + \ldots + f_{sn} \cdot k_{+n} \cdot X;$$
$$X = \left(\frac{f_A \cdot \sigma_{EG}}{2 \cdot N_a}\right)\left(\frac{\pi \Delta t}{D_L}\right)^{1/2} \tag{4.19}$$

where the term f_{si} is used to account for the polarity of binding to each of the n possible binding sites (as specified in the reaction mechanism for the ligand molecule and ES; see Sections 4.5 and 4.6). If binding can occur after a hit to either side of the ET, then both "poles" of the ES are valid surface area for binding and the value of f_{si} is unity. If binding can only occur after a hit to one particular side of the ET, then only the "positive" or "negative" pole of the ES is valid surface area, and f_{si} is 0 if the invalid pole is hit, or 2 if the valid pole is hit. The value of 2 in the latter case compensates for the apparent twofold reduction in N_H.

Box 4.5 Bimolecular Associations: Equating p_{bt} to p_t

Equation 12 in Box 4.4 was derived under conditions in which concentration gradients are absent at all times. If this is not true, the expression for p_t still holds in small local regions where the concentration change is negligible, but on the macroscopic scale Equation 4 in Box 4.4 must be evaluated numerically using either finite element or MC methods. With a finite element approach, space is subdivided into small volume elements (voxels), and the contents of each voxel are assumed to be well mixed. The rate of reaction within each voxel is computed using an ordinary differential equation like Equation 5 in Box 4.4, and the flux between voxels is computed from concentration differences and the values of D_A and D_R. With MCell's MC algorithms, ligand fluxes and binding depend only on random walk movements and intersections with individual ESs, and hence are *always local* on the spatial scale of the random walk and the temporal scale of Δt (see Box 4.6). The binding probability p_b is calculated by equating p_{bt} to $p_t \cong \zeta$ (Equation 4.15, Section 4.4.5), so the MC binding rate satisfies the requirements of mass action kinetics at all points in space. If, for a particular geometric arrangement of ESs, the effective rate of ligand diffusion happens to be fast compared to the rate of binding, the formation of concentration gradients during binding will be negligible and the MCell simulation will match the analytic prediction of Equations 7 or 8 in Box 4.4. If the relative rate of diffusion is not fast, however, the rate of binding in the simulation will depend on both space and time, as idealized in Equation 4 in Box 4.4 (rather than Equations 7 or 8) for an infinite number of molecules.

Box 4.6 Bimolecular Associations: Derivation and Use of N_H to Compute p_b

In principle, the instantaneous flux of ligand molecules into a surface ("hits" per unit time) depends on the net velocity (v, distance per unit time) of the molecules *toward* the surface, the surface's area (A_s), and the instantaneous ligand concentration (A) adjacent to the surface:

$$\frac{\text{hits}}{\partial t} = J = \frac{1}{2}(N_a)(v)(A_s)(A) \tag{1}$$

(continued)

Box 4.6 (continued)

where N_a is Avogadro's number, and the factor (1/2) accounts for the fraction of molecules that have a net velocity *away* from the surface. The number of hits during an interval of time t is obtained by integrating Equation 4.1:

$$(\text{number of hits})_t = \int_0^t J\,\partial t = \frac{1}{2}(N_a)(v)(A_s)\int_0^t (A)\partial t \quad (2)$$

In an MCell simulation, ESs occupy ETs on MEs, and each side (front and back) of an ET has some area, A_{ET}. Thus, A in Equation 2 is $2 \cdot A_{ET}$. The average radial distance traveled by each ligand molecule during a time-step Δt is $\bar{l}_r$ (Equation 4.6, Section 4.4.1), but the average net displacement ($\Delta\bar{\psi}$) toward or away from any ME (a portion of a plane) is $\bar{l}_\perp$ (Equation 4.7, Section 4.4.1). Therefore, half of all ligand molecules move toward the ME with an apparent velocity of $v = \bar{l}_\perp/\Delta t$. Substituting into Equation 2:

$$N_H = \frac{1}{2}(N_a)(\bar{l}_\perp/\Delta t)(2 \cdot A_{ET})\int_0^t (A)\partial t = (N_a)(\bar{l}_\perp/\Delta t)(A_{ET})\int_0^t (A)\partial t \quad (3)$$

As Δt decreases, the distance $\bar{l}_\perp$ decreases, and hence the population of ligand molecules that can reach the ET becomes increasingly restricted to the local region of space adjacent to the ET. This reduces the sampling of any static or changing concentration gradients in the region. Therefore, as Δt decreases, the flux of A into the ET is determined by a concentration (A) that approaches the instantaneous concentration $(A)_o$ adjacent to the surface, and Equation 4.3 becomes:

$$\begin{aligned} N_H &= (N_a)(\bar{l}_\perp/\Delta t)(A_{ET})(A)_o\int_0^{\Delta t} dt \\ &= (N_a)(\bar{l}_\perp/\Delta t)(A_{ET})(A)_o\Delta t; \quad \text{for small } \Delta t \end{aligned} \quad (4)$$

The $(A)_o$ term in Equation 4 has the same meaning as $(A)_o$ in Equation 4.16 (Section 4.4.5), and thus the two terms cancel when Equation 4 is substituted into Equation 4.16 to obtain p_b:

$$p_b = \left(\sum_1^n k_{+i}\right)\frac{1}{(N_a)(\bar{l}_\perp/\Delta t)(A_{ET})} \quad (5)$$

After Equation 4.7 (Section 4.4.1) is used to substitute for $\bar{l}_\perp$ in Equation 5:

$$p_b = \left(\sum_1^n k_{+i}\right)\frac{1}{2(N_a)(A_{ET})}\left(\frac{\pi\Delta t}{D_L}\right)^{1/2} \quad (6)$$

as given in Equation 4.17 (Section 4.4.5). Thus, p_b depends on $\sqrt{\Delta t}$ (see Box 4.7).

Box 4.7 Bimolecular Associations: Dependence of p_b on $\sqrt{\Delta t}$

The dependence of p_b on $\sqrt{\Delta t}$ arises because ζ (Equation 4.15, and the numerator of Equation 4.16, Section 4.4.5) is directly proportional to Δt, but N_H (denominator of Equation 4.16, Section 4.4.5) is proportional to $\sqrt{\Delta t}$. If both ζ and N_H were linearly dependent on Δt, p_b would be independent of the simulation time-step. The dependence of N_H on $\sqrt{\Delta t}$ arises because the apparent velocity of ligand molecules toward ESs is inversely proportional to $\sqrt{\Delta t}$. Over time Δt, where Δt is longer than the time between actual Brownian collisions (sub-ps scale at room temperature), a real diffusing molecule follows some tortuous path between a starting position P_1 and ending position P_2. The *total* path length is determined by the molecule's real thermal velocity and scales directly with Δt, but the *average radial distance* ($\bar{l}_r$) between P_1 and P_2 scales with $\sqrt{\Delta t}$, as does the *average axial displacement* ($\bar{l}_\perp$) measured with respect to any plane's normal axis ψ (see discussion of Equations 4.6 and 4.7, Section 4.4.1). Since a random walk approximation of Brownian motion replaces the molecule's actual tortuous path with straight-line radial movement between P_1 and P_2, the apparent velocity of motion toward any plane (distance traveled per unit time;

$$v = \bar{l}_\perp / \Delta t = \sqrt{4D_L / (\pi \Delta t)}\ )$$

is less than the real thermal velocity and changes as $1/\sqrt{\Delta t}$.

Implementation, Validation, and Accuracy — Since p_b gives the probability that binding occurs to any one of n binding sites, $(1 - p_b)$ gives the probability that no binding occurs. As discussed previously for unimolecular transitions, the decision between all possible events is made by comparing the value of a single random number ($0 \le \kappa \le 1$) to the cumulative set of probabilities (p_b, $p_{b1} + p_{b2}$, ..., p_b, 1).

Equation 4.19 shows how p_b depends on the value(s) of k_+ and three additional user-specified input parameters; p_b is directly proportional to k_+ and σ_{EG}, inversely proportional to $\sqrt{D_L}$, and directly proportional to $\sqrt{\Delta t}$, (Box 4.7). The parameter σ_{EG} is usually set to a fixed value such as the maximum packing density of receptor proteins in plasma membrane (e.g., 10–15 × 10^3 μm^{-2}), and k_+ and D_L ordinarily are constrained by experimental or theoretical estimates. The user thus can vary Δt to adjust the value of p_b, which is one determinant of a simulation's numerical accuracy.

In general, Δt should be chosen so that $p_b << 1$, or, as a rule of thumb, not greater than ~0.5 to obtain errors less than 1–2% (relative amounts of bound and unbound reactant states). The user must be aware that an inappropriately long time-step can cause p_b to exceed 1, in which case numerical errors can be quite large in favor of unbound reactants. If k_+ is increased and/or D_L is decreased appreciably (e.g., during a search of parameter space to fit experimental data), the resulting increase in p_b may require a concomitant decrease in Δt to maintain accuracy. Also, if the free ligand concentration, and therefore N_H, varies in time and space (e.g., during ligand release from a synaptic vesicle), then (in principle) the numerical accuracy varies in time and space for a given value of p_b. This follows from the approximation used between Equations 4.15 and 4.16, i.e., that $(1 - p_b)^{N_H}$ approaches $(1 - N_H \cdot P_b)$ as N_H and p_b decrease with decreasing Δt. For reasonable values of Δt, the overall effect of free ligand concentration on accuracy is likely to be very small. If necessary, however, checkpointing can be used to change Δt during the course of the simulation. Such an adaptive time-step would be shorter during those times when free ligand concentration is high, thus maintaining accuracy.

The validity of Equation 4.19 and MCell's binding algorithms can be tested by comparing a simple simulation of Equation 4.12 to the time course of binding predicted by mass action rate theory (Equations 7 or 8 in Box 4.4). For example, if the ligand molecules and ESs are distributed uniformly to simulate well-mixed conditions, the MCell results will converge to the analytic expectation as Δt is decreased. On the other hand, a simulation run with arbitrary non-uniform ligand molecule and/or ES distributions will converge to results that accurately reproduce the influence of diffusion on the time course of binding.[33] In general, this type of result cannot be obtained analytically, and is the typical aim of MCell simulations (see Box 4.5).

4.4.6 The Heuristics of Reversible Reactions and Multiple Binding Sites

If a reaction mechanism includes only irreversible transitions, and ESs can bind only a single ligand molecule, each reactant can undergo no more than one transition per Δt. Thus, the MCell reaction algorithms (Figure 4.2B) are strictly first-order under such conditions. With reversible reactions and/or multiple binding sites per ES, however, the algorithms are higher-order, i.e., individual molecules may undergo multiple "sub-Δt" transitions during each time-step. Thus, some degree of hidden reversibility is introduced. For example, an ES initially in the AR^1 state (Figure 4.1) might unbind to reach the R^0 state at some point during a time-step, and then later during the same time-step might bind 1 or 2 ligand molecules to begin the next iteration in the AR^1, AR^2, or A_2R^3 state.

In principle, first-order and higher-order algorithms converge to the same result as Δt decreases, but if the higher-order approach can be suitably balanced for complex cyclic reactions, its advantage is higher numerical accuracy for a given value of Δt. Different higher-order approaches can be tested by simulating simple and complex reactions at equilibrium, and then comparing the fractional amounts of each reactant to analytic predictions or to a finite difference simulation of the corresponding rate equations. As indicated in Figure 4.2B, MCell's algorithms use an optimized set of rules to make decisions regarding sub-Δt transitions, and additional details can be found in Reference 33.

4.5 EXAMPLE MODEL: ACETYLCHOLINE EXOCYTOSIS AND MINIATURE ENDPLATE CURRENT GENERATION AT A REALISTIC NEUROMUSCULAR JUNCTION

4.5.1 Logistics of Design, Input, and Output

Medium- to large-scale reconstructions and simulations of single synapses or neuropil can be used to investigate pre- and postsynaptic factors that contribute to synaptic variability. As one example, Figure 4.4A shows a realistic medium-scale model of rat diaphragm neuromuscular junction. The model consists of pre- and postsynaptic membrane surfaces, a transparent basal lamina layer that follows the postsynaptic contour inside the cleft space, and 30 ACh vesicles arrayed above the openings into six junctional folds that have highly variable topology. The size and time course of simulated miniature endplate currents (mEPCs) varies significantly from one release site to another.[33]

As illustrated for the postsynaptic membrane in Figure 4.4B, each of the surfaces in the model is a polygon mesh that has been optimized for MCell simulations and visualization, i.e., each ME is a nearly equilateral triangle.[33] It is important that the front and back faces of each ME are oriented consistently so that the polarity of ESs on the surface is also consistent. Figure 4.4C shows how the right-hand-rule is applied to MEs as they are imported from the MDL file, to distinguish the front from the back face. In this model, the total number of MEs for nerve, muscle, and basal lamina surfaces exceeds 54,000. For large-scale synaptic reconstructions presently in development, this number can easily increase by more than an order of magnitude.

Figure 4.4D shows the MEs that comprise the muscle membrane, together with glyphs that indicate ESs for acetylcholine receptors (AChRs) on the surface (~91,000). As discussed in Section 4.3.2, each ME has been covered with ETs (not shown) using barycentric subdivision and a global effector grid density (σ_{EG}) that in this case is 10,000 μm^{-2}. The position and area of each ET ($A_{ET} \cong 1/\ \sigma_{EG} = 100\ nm^2$) is indicated approximately by the position and size of the AChR glyphs (8.5 nm diameter). The membrane surface is actually composed of three different meshes that fit together *exactly* (Section 4.3.2), i.e., a top, middle and bottom portion with respect to the depth of the folds (muscle-membrane.mdl), and AChR ESs are present on the top two parts at different site densities. On the tops of the folds, the site density is 7250 μm^{-2} (~66,000 AChRs), and in the middle region is about 70% less (~25,000 AChRs). Thus, about 70% or 20% of all ETs are occupied by ESs in the top and middle regions, respectively, as can be seen in Figure 4.4D by the relative spacing of glyphs located either above or down inside the fold. Aside from AChR ESs, the model also includes ~59,000 AChE sites distributed throughout the basal lamina (1800 μm^{-2}), and ~5800 choline (Ch) reuptake sites (ChRs) in the nerve membrane (1000 μm^{-2}). The ChRs bind Ch molecules inside the cleft (where they

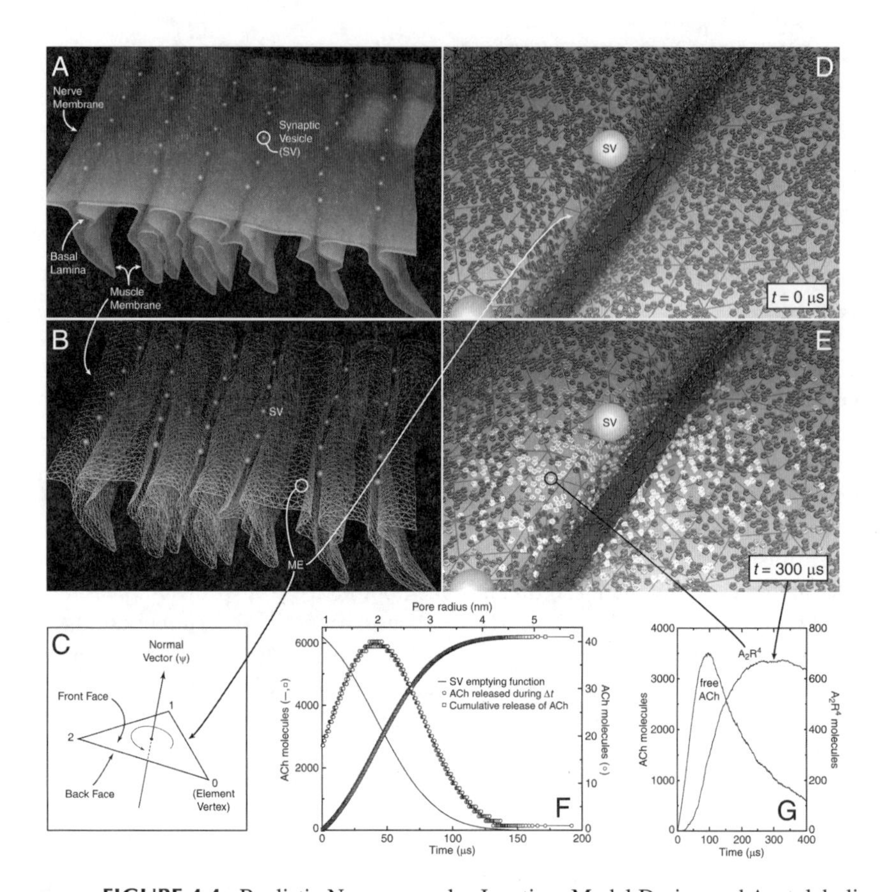

FIGURE 4.4 Realistic Neuromuscular Junction: Model Design and Acetylcholine (ACh) Exocytosis. (A) Model surfaces, and possible locations of ACh release indicated by synaptic vesicles. (B) Optimized polygon mesh for the muscle membrane surface. The entire mesh is composed of individual mesh elements (MEs) that are nearly equilateral triangles. (C) Each ME is defined by its vertex positions and an ordered list of connections between the vertices. This diagram shows how the right-hand-rule is applied to the list of connections to determine the front and back faces of the ME. The first connection extends from vertex 0 to vertex 1 and so on (vertex numbering is arbitrary), and the normal vector (ψ) defined by the right-hand-rule passes through the ME from back to front. (D) Closer view of muscle membrane with nerve and basal lamina surfaces removed. Individual MEs are subdivided into effector tiles, and each tile may contain an effector site as indicated by glyphs that represent acetylcholine receptors (AChRs, see text). (E) Snap-shot of muscle membrane during the simulation, showing bound AChRs (lighter gray) close to the point of ACh release (for more detailed views and explanation, see Color Figures 4.1–4.3). (F) The time course of SV emptying during ACh exocytosis is shown by the solid curve (see text). The corresponding time course of ACh appearance in the cleft can be simulated using incremental release during time-steps (circles), to obtain the desired cumulative release (squares). (G) Example of output from one simulation, showing the amount of free ACh in the entire synaptic cleft, and the number of AChRs in the double-bound open conformation (A_2R^4 state). Panels A, B, D, and E were rendered with IBM DataExplorer.

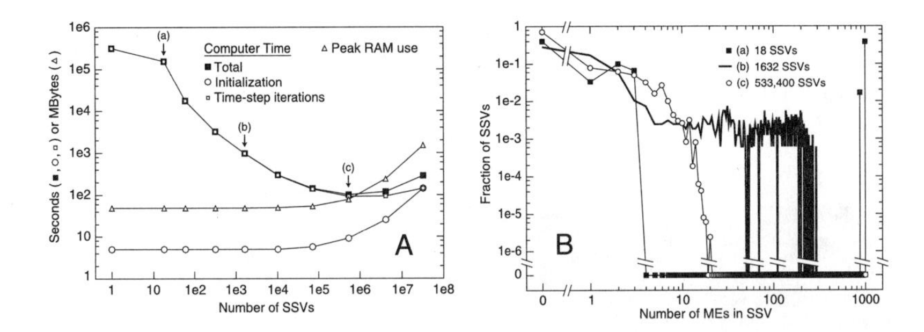

FIGURE 4.5 Relationship Between Computer Time, Memory (RAM) Use, and Spatial Partitioning. (A) Ten simulations were run using increasing numbers of spatial partitions to create increasing numbers of spatial subvolumes (SSVs; abcissa). For each simulation, the log-log plot shows the: (1) time required for initialization of SSVs and all other simulation objects (circles); (2) time required to execute the time-step iterations (open squares); (3) total run time (closed squares); and (4) peak memory use during the simulation. (B) Frequency distributions (log-log scales) for mesh elements (MEs) contained in SSVs, shown for three of the partitioning conditions (labeled a, b, and c) used in A. With few partitions and SSVs (a), many of the SSVs contain 900 or more MEs, making ray tracing and ray marching (Figure 4.2B) extremely inefficient and execution speed extremely slow. With an optimal number and size (see text) of SSVs (c), most of the SSVs contain fewer than 10 MEs, and no SSV contains more than 20. Thus, execution speed is faster by orders of magnitude.

are produced by ACh hydrolysis) and release them on the opposite (intracellular) side of the nerve membrane. The site densities and distributions used for AChR and AChE ESs are from EM autoradiographic measurements,[24,25,27] while the density for ChRs is an illustrative guess.

Exocytosis of ACh from one synaptic vesicle (SV) was simulated using a time course of vesicle emptying (Figure 4.4F) predicted by rapid fusion pore expansion.[30] An explicit model of the vesicle and expanding fusion pore can be simulated with MCell using checkpointing (Section 4.3.3), but requires a sub-ns time-step as the ACh diffuses out through the constrictive pore. As detailed elsewhere[31.33] and shown in Figure 4.4F, the same time course of release can be obtained with a time-step on the μs scale if the actual vesicle and pore are replaced by incremental releases of ACh directly within the cleft under the (missing) pore (release_sites.mdl). Aside from allowing a much longer time-step and thus far fewer iterations, this method also side-steps the need for checkpointing, and hence the simulation runs as fast as if all the ACh were released instantaneously.

The reaction mechanisms for AChR and AChE ESs are as shown in Figure 4.1. The illustrative reaction for Ch reuptake is simply reversible binding followed by rate-limiting translocation and release across the nerve membrane (reaction_mechanisms.mdl). To obtain the predicted size and time course of an mEPC, the number of AChR ESs in the A_2R^4 state is output for each time-step. This is shown for the rising phase and peak in Figure 4.4G, together with the amount of free ACh in the entire cleft. To visualize a snap-shot of the simulation (e.g., Figure 4.4E), the positions of each ME, ES, and ligand molecule must be output along with current state information for the ESs and ligand molecules. For an animation that shows the spatial evolution of the mEPC (i.e., "saturated disk" formation),[5,16,17,26,33] state information must be output repeatedly for multiple time-steps. Figures 4.4 and 4.6 and Color Figures 4.1–4.6* show examples of 3-D renderings done with IBM DataExplorer or Pixar RenderMan.

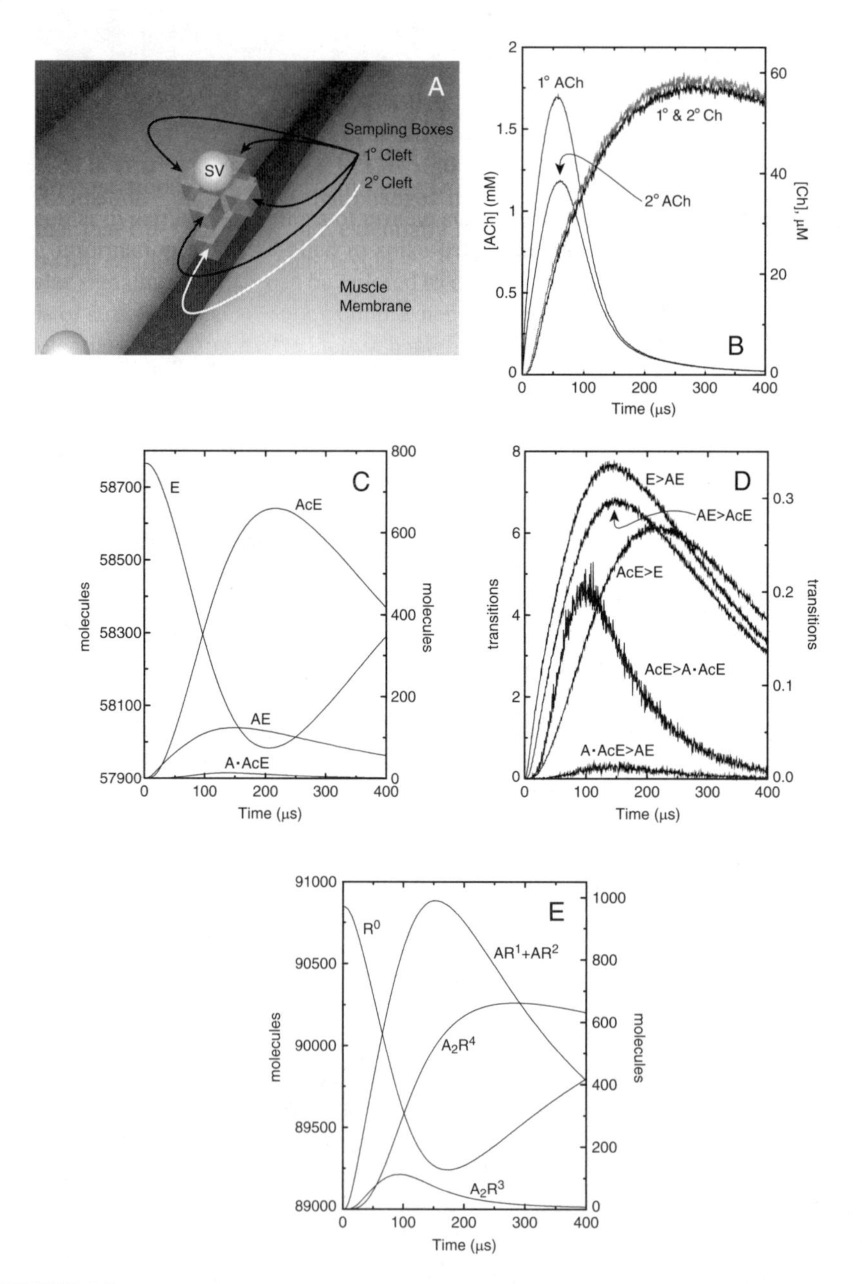

FIGURE 4.6

(continued)

* Color Figures 4.1–4.6 follow page 140.

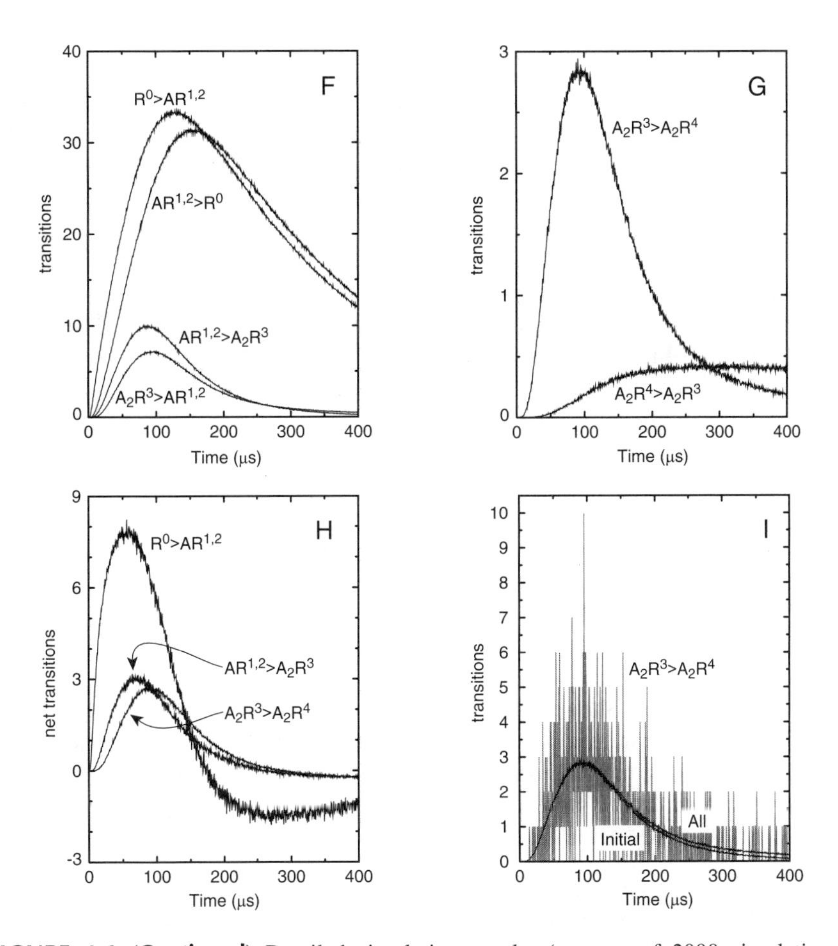

FIGURE 4.6 (Continued) Detailed simulation results (average of 2000 simulations). (A) Positions of transparent boxes used to sample acetylcholine (ACh) and choline (Ch) concentrations during the simulations. (B) Time course of ACh and Ch concentration sampled in primary and secondary cleft. (C and E) Amounts of unbound and bound AChE and AChR intermediates (states are labeled as in Figure 4.1). (D, F, and G) Selected examples of absolute transition rates (number of transitions from starting to ending state per time-step; $\Delta t = 0.5$ μs). Results are summed for the two single-bound AChR states, as indicated by the $AR^{1,2}$ label. (H) Examples of net transition rates (forward transitions minus reverse transitions per time-step) for AChR states. Note the change from positive to negative values during the mEPC rising phase or at time of peak amplitude (cf. A_2R^4 curve in E). (I) Detail of AChR opening transitions. Noisy results from one simulation are shown for comparison with the averaged curves labeled "All" and "Initial." "All" indicates *all transitions*, and is the same curve as shown previously in G on an expanded scale. "Initial" indicates a *subset of all transitions*, i.e., $A_2R^3 > A_2R^4$ transitions that *were not preceded by occupation of the A_2R^4 state during the previous time-step.* In terms of single channel kinetics, "Initial" transitions are those that initiate a burst of openings, versus those that occur within a burst after a "flicker" closure. Here the difference between All and Initial is small because the channel opening probability ($\beta/(\beta + k_{-1} + k_{-2})$, Figure 4.1) is small and flickering is rare, but for other channels and/or rate constant values the difference can be large. The time course of the mEPC is determined by the time course of Initial openings, and direct prediction of such transition subtypes can only be obtained from Monte Carlo simulations.

4.5.2 Run-Time Logistics

Using a present-day workstation to simulate one complete mEPC in the absence of spatial partitions (i.e., where the entire model space constitutes a single SSV), the computer memory and time requirements are approximately 50 MBytes and 1 week, respectively (Figure 4.5A).* As the number of spatial partitions (which require memory) is increased in all three dimensions, the resulting SSVs contain fewer MEs (Figure 4.5B), and therefore the efficiency of ray marching (Section 4.3.5) is increased substantially. As a result, computer time drops precipitously and memory use increases gradually, until optimal conditions are obtained with ~500,000 SSVs (condition (c) in Figure 4.5A and B). At this point the required computer time is only ~100 seconds, and memory use has increased to ~80 MBytes. Thus, for only a 60% increase in memory, computer time drops by about *3.5 orders of magnitude*. If even more spatial partitions are added, the distance between adjacent partitions becomes smaller than the mean random walk step length $\bar{l}_r$ (Equation 4.6), and diffusing ligand molecules pass through a partition with virtually every movement. This introduces additional operations per time-step (Figure 4.2B), and hence computer time increases again slightly (by ~3-fold for the right-most point in Figure 4.5A). With very numerous partitions and SSVs, however, memory use can increase dramatically (e.g., to over 1 GByte as shown for ~30 million SSVs). Nevertheless, Figure 4.5A shows that most of the benefit obtained with spatial partitions is achieved with far fewer than are required for the shortest possible computer time. Hence, memory use and computer time can be "titrated" as needed for any model and computer system, with little reduction in overall throughput.

4.5.3 Detailed Output

In this particular model, diffusing ACh molecules cannot escape from the synaptic cleft space, and so a simple count of all free ACh molecules gives a spatial summation within the entire cleft volume (Figure 4.4G). To quantify the concentration of diffusing molecules in any particular region of space, a transparent box (or any other closed transparent mesh) can be placed in that region, and then the free ligand molecules can be counted within the box.** Figure 4.6A shows four such sampling boxes located in the primary cleft under the central synaptic vesicle indicated in Figure 4.4. In addition, another box is located just beneath the first four, i.e., at the

* This computer time figure reflects ACh release from a corner vesicle if the entire structure shown in Figure 4.4A is enclosed in an absorptive bounding box. The presence of the bounding box has essentially no effect on the amplitude of the mEPC regardless of vesicle position, but with release from a corner a significant fraction of diffusing ACh and Ch molecules are removed by absorption rather than hydrolysis and reuptake. With release from a central vesicle (e.g., as labeled in Figure 4.4) about *fourfold* more computer time is required because far fewer molecules are absorbed, and therefore far more ray marching is required before the simulation completes.

** As detailed in Figure 4.2B, each time a diffusing ligand molecule crosses a transparent ME, the event is detected. The direction of crossing is also known, and for a fully closed mesh would be either from outside to inside or vice versa. Thus, the number of molecules that enter and exit an enclosed space can be counted during each time-step, and the amount remaining within the space at the end of each time-step can be specified as output.

entrance to the secondary cleft created by the underlying junctional fold (ligand_sampling_boxes.mdl). To improve the signal-to-noise ratio of results shown in Figure 4.6, the simulation was run 2000 times with different seed values and the results were averaged (~45-fold reduction of noise, see Figure 4.6I). With one present-day single-processor workstation this requires 1–2 days of computer time, but since each simulation is an independent computation, many processors can be used simultaneously.

The time course of sampled ACh concentration is shown in Figure 4.6B, and such simulation results can be compared to increasingly sophisticated experimental estimates.[3,8,22,23] The difference between the primary and secondary cleft concentrations illustrates a steep ACh gradient across the height of the primary cleft (~50 nm), but the gradient persists only during exocytosis and the early rising phase of the mEPC (Figures 4.4F and G and 4.6E). The concentration of Ch sampled in the same volumes is far lower at all times, reaches its peak much later, and is virtually identical in both regions, reflecting relatively slow Ch production within the cleft spaces by AChE sites (despite AChE's fast catalytic turnover number, 16,000 s^{-1} in this example).[20] Figure 4.6C and E show the temporal evolution of each AChE and AChR state, and the remaining panels illustrate detailed output of reaction transition rates. Such data is presently being used to investigate pre- and postsynaptic factors that influence synaptic noise and variability, and to generate functions that can be incorporated into compartmental models of neuronal function. Snap-shots of an individual simulation are shown in Color Figures 4.1–4.3 for selected times at which different reactants are present at peak concentrations or amounts.

4.6 EXAMPLE MODEL: POTENTIAL SPATIAL AND TEMPORAL INTERACTION OF NEURONAL GLUTAMATERGIC CURRENTS

The preceding example focuses on the detailed time course and variability of quantal currents at a peripheral synapse. Here the focus is potential spatial and temporal interaction between central glutamatergic synapses on an interneuron. The model is strictly qualitative and is used as a conceptually interesting and considerably smaller scale example than the preceding.

The model surfaces consist of an inner and outer ovoid (Color Figure 4.4), with the inner representing the nerve cell body, and the outer defining a closed bounding layer of intercellular diffusion space (total of 470 MEs for the two surfaces). In reality of course, the nerve cell would have processes and the diffusion space would extend out between many different surrounding neural and glial elements. Effector sites representing glutamate receptors with long-lived desensitization states are present on the nerve cell (AMPA GluR; reaction mechanism shown in Figure 4.7A obtained from Reference 12), and reuptake sites (transporters) are present on the surrounding surface (as for preceding example, simple reversible binding followed by rate-limiting translocation and release).

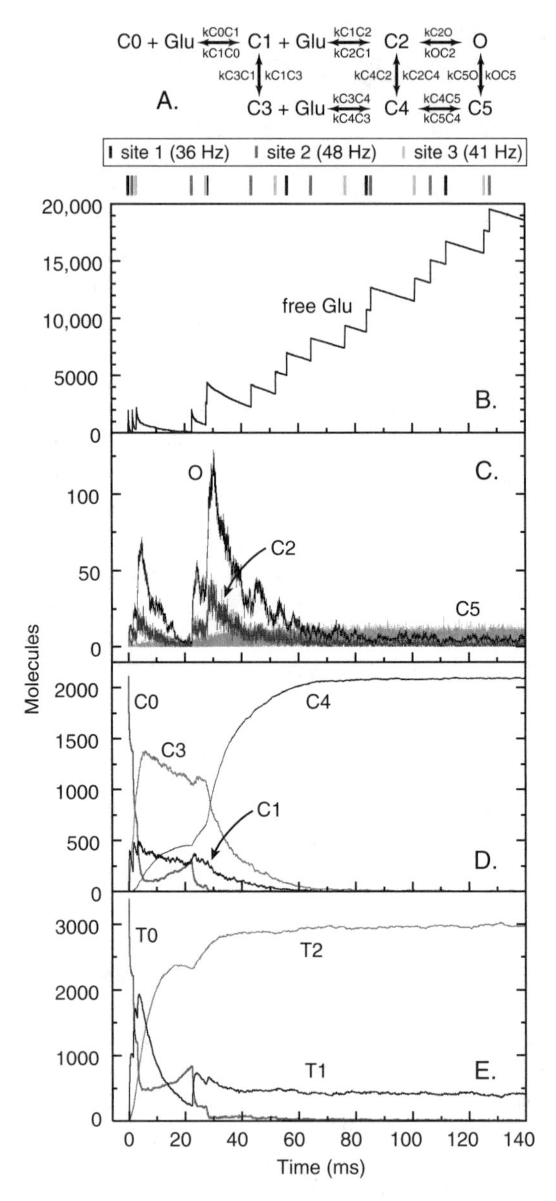

FIGURE 4.7 Glutamate Receptor Reaction Mechanism and Simulation Results. (A) Reaction mechanism with three primary closed states (C0, C1, and C2), one open state (O), and three desensitized states (C3, C4, and C5). Timing of Glu release from three different sites is indicated (2000 molecules per release). (B) Amount of free Glu in the intercellular space that surrounds the model cell. (C and D) Evolution of bound GluR states. Early release events potentiate to open channels, but at late times almost all GluRs are in the C4 desensitized state (see Figure 4.12). (E) Evolution of bound Glu reuptake (transporter) sites. T0 is unbound, T1 is reversibly bound, and T2 is the rate-limiting state that transports and releases Glu molecules. At late times, transporter sites are saturated due to the slow rate of release.

Three different sites are defined for quantal Glu release to represent different presynaptic boutons, and the MDL is also used to define a train of release events at each site (frequencies of 36, 41, or 48 Hz; for simplicity, instantaneous release is used rather than a release function). Figure 4.7A shows the composite timing of release from the three sites during the simulation, which on a present-day workstation requires about 4 hours of computer time for the conditions illustrated here (140,000 iterations for a total elapsed simulation time of 140 ms). Snap-shots are shown for two selected early times and one late time when desensitization predominates (Color Figures 4.4–4.6).

As shown in Figure 4.7B, free Glu builds up in the intercellular space over the course of subsequent release events. For the initial ~30 ms, release events potentiate to produce an increased number of open GluR channels, but thereafter the Glu reuptake sites become saturated and the GluR's are driven almost exclusively into desensitized states for the remainder of the simulation. This model is not included here to make quantitative realistic predictions, but rather to point out the important need for accurately determined 3-D input parameters, e.g., reconstructed diffusion spaces, release positions, receptor and reuptake site densities and distributions. All of these factors, together with all of the rates for transmitter release, diffusion, reuptake, and receptor activation and desensitization, will determine the physiological behavior of the system. As increasingly accurate experimental determinations of these input parameters become available, realistic 3-D MC simulations will be increasingly important to a quantitative understanding of nervous system function.

ACKNOWLEDGMENTS

We thank Erik De Schutter for his monumental patience and Ed and Mika Salpeter and Ralph Roskies for comments on Section 4.4. Supported by NIH K08NS01776 and RR-06009 and NSF IBN-9603611.

REFERENCES

1. Anglister, L., Stiles, J. R., and Salpeter, M. M., Acetylcholinesterase density and turnover number at frog neuromuscular junctions, with modeling of their role in synaptic function, *Neuron*, 12, 783, 1992.
2. Arkin, A. P., Ross, J., and McAdams, H. H., Stochastic kinetic analysis of developmental pathway bifurcation in phage λ-infected *E. coli* cells, *Genetics*, 149, 1633, 1998.
3. Barbour, B. and Hausser, M., Intersynaptic diffusion of neurotransmitter, *Trends Neurosci.*, 20, 377, 1997.
4. Barrow, G. M., *Physical Chemistry for the Life Sciences*, McGraw-Hill, New York, 1981.
5. Bartol, T. M., Jr., Land, B. R., Salpeter, E. E., and Salpeter, M. M., Monte Carlo simulation of MEPC generation in the vertebrate neuromuscular junction, *Biophys. J.*, 59, 1290, 1991.

6. Bennett, M. R., Farnell, L., and Gibson, W. G., Quantal transmission at purinergic synapses: stochastic interaction between ATP and its receptors, *J. Theor. Biol.*, 175, 397, 1995.
7. Bennett, M. R., Farnell, L., Gibson, W. G., and Lavidis, N. A., Synaptic transmission at visualized sympathetic boutons: stochastic interaction between acetylcholine and its receptors, *Biophys. J.*, 72, 1595, 1997.
8. Clements, J. D., Transmitter timecourse in the synaptic cleft: its role in central synaptic function, *Trends Neurosci.*, 19, 163, 1996.
9. Crank, J., *The Mathematics of Diffusion*, Oxford University Press, London, 1956.
10. Denes, J. and Krewski, D., An exact representation for the generating function for the Moolgavkar-Venzon–Knudson two-stage model of carcinogenesis with stochastic stem cell growth, *Math. Biosci.*, 131, 185, 1996.
11. Faber, D. S., Young, W. S., Legendre, P., and Korn, H., Intrinsic quantal variability due to stochastic properties of receptor-transmitter interactions, *Science*, 258, 1494, 1992.
12. Geiger, J. R. P., Roth, A., Taskin, B., and Jonas, P., Glutamate-mediated synaptic excitation of cortical interneurons, in *Handbook of Experimental Pharmacology*, Vol. 141, *Retinoids, Ionotropic Glutamate Receptors in the CNS*, Jonas, P. and Monyer, H., Eds., Springer-Verlag, Berlin, 363, 1999.
13. Kalos, M. H. and Whitlock, P. A., *Monte Carlo Methods*, Vol. I: *Basics*, John Wiley & Sons, New York, 1986.
14. Knuth, D., *The Art of Computer Programming*, MIT Press, Cambridge, 1969.
15. Kruk, P. J., Korn, H., and Faber, D. S., The effects of geometrical parameters on synaptic transmission: a Monte Carlo simulation study, *Biophys. J.*, 73, 2874, 1997.
16. Land, B. R., Salpeter, E. E., and Salpeter, M. M., Acetylcholine receptor site density affects the rising phase of miniature endplate currents, *Proc. Natl. Acad. Sci. U.S.A.*, 77, 3736, 1980.
17. Matthews-Bellinger, J. and Salpeter, M. M., Distribution of acetylcholine receptors at frog neuromuscular junctions with a discussion of some physiological implications, *J. Physiol.*, 279, 197, 1978.
18. McQuarrie, D. A., Jachimowski, C. J., and Russell, M. E., Kinetics of small systems II, *J. Chem. Phys.*, 40, 2914, 1964.
19. Press, W. H., Flannery, B. P., Teukolsky, S. A., and Vetterling, W. T., *Numerical Recipes. The Art of Scientific Computing*, Cambridge University Press, New York, 1986.
20. Rosenberry, T., Acetylcholinesterase, *Adv. Enzymol.*, 43, 103, 1975.
21. Rubenstein, R. Y., *Simulation and the Monte Carlo Method*, John Wiley & Sons, New York, 1981.
22. Rusakov, D. A. and Kullmann, D. M., Extrasynaptic glutamate diffusion in the hippocampus: ultrastructural constraints, uptake, and receptor activation, *J. Neurosci.*, 18, 158, 1998.
23. Rusakov, D. A., Kullmann, D. M., and Stewart, M. G., Hippocampal synapses: do they talk to their neighbors? *Trends Neurosci.*, 22, 382, 1999.
24. Salpeter, M. M., Electron microscope autoradiography as a quantitative tool in enzyme cytochemistry: the distribution of acetylcholinesterase at motor endplates of a vertebrate twitch muscle, *J. Cell Biol.*, 32, 379, 1967.
25. Salpeter, M. M., Electron microscope radioautography as a quantitative tool in enzyme cytochemistry. II. The distribution of DFP-reactive sites at motor endplates of a vertebrate twitch muscle, *J. Cell Biol.*, 42, 122, 1969.

26. Salpeter, M. M., Vertebrate neuromuscular junctions: general morphology, molecular organization, and functional consequences, in *The Vertebrate Neuromuscular Junction*, Salpeter, M. M., Ed., Alan R. Liss, New York, 1987, 1.
27. Salpeter, M. M., Smith, C. D., and Matthews-Bellinger, J. A., Acetylcholine receptor at neuromuscular junctions by EM autoradiography using mask analysis and linear sources, *J. Electron Microsc. Tech.*, 1, 63, 1984.
28. Schroeder, W., Martin, K., and Lorensen, B., *The Visualization Toolkit*, 2nd Ed., Prentice-Hall, Englewood Cliffs, NJ, 1998.
29. Smart, J. L. and McCammon, J. A., Analysis of synaptic transmission in the neuromuscular junction using a continuum finite element model, *Biophys. J.*, 75, 1679, 1998.
30. Stiles, J. R., Van Helden, D., Bartol, T. M., Salpeter, E. E., and Salpeter, M. M., Miniature endplate current rise times <100 μs from improved dual recordings can be modeled with passive acetylcholine diffusion from a synaptic vesicle, *Proc. Natl. Acad. Sci. U.S.A.*, 93, 5747, 1996.
31. Stiles, J. R., Bartol, T. M., Salpeter, E. E., and Salpeter, M. M., Monte Carlo simulation of neurotransmitter release using MCell, a general simulator of cellular physiological processes, in *Computational Neuroscience*, Bower, J. M., Ed., Plenum Press, New York, 1998, 279.
32. Stiles, J. R., Kovyazina, I.V., Salpeter, E. E., and Salpeter, M. M., The temperature sensitivity of miniature endplate currents is mostly governed by channel gating: evidence from optimized recordings and Monte Carlo simulations, *Biophys. J.*, 77, 1177, 1999.
33. Stiles, J. R., Bartol, T. M., Salpeter, M. M., Salpeter, E. E., and Sejnowski, T. J., Synaptic variability: new insights from reconstructions and Monte Carlo simulations with MCell, in *Synapses*, Cowan, W. M., Stevens, C. F., and Sudhof, T. C., Eds., Johns Hopkins University Press, Baltimore, MD, in press.
34. Tomita, M., Hashimoto, K., Takahashi, K., Shimizu, T., Matsuzaki, Y., Miyoshi, F., Saito, K., Tanida, S., Yugi, K., Venter, J. C., and Hutchison, C., E-CELL: software environment for whole cell simulation, *Bioinformatics*, 15, 72, 1999.
35. Wahl, L. M., Pouzat, C., and Stratford, K. J., Monte Carlo simulation of fast excitatory synaptic transmission at a hippocampal synapse, *J. Neurophysiol.*, 75, 597, 1996.

5 Which Formalism to Use for Modeling Voltage-Dependent Conductances?

Alain Destexhe and John Huguenard

CONTENTS

5.1 INTRODUCTION

The selective permeability of neuronal membranes to ions is the basis of various processes central to neurophysiology, such as the maintenance of a membrane potential, the genesis of neuronal excitability and the action of neurotransmitters and modulators. The rules governing ionic permeabilities were explored by Hodgkin, Huxley, Katz, and others several decades ago. It was demonstrated that the ionic permeability of the membrane can be highly dependent on the membrane potential. Hodgkin and Huxley[1] characterized these properties of voltage dependence and provided a mathematical model which proved that these properties were sufficient to account for the genesis of action potentials. The model of Hodgkin and Huxley was based on simple assumptions, reproduced well the behavior of the currents, and its parameters are easy to determine from experimental data. This explains why Hodgkin–Huxley models are still widely used today, almost 50 years later.

Hodgkin and Huxley postulated that the membrane currents result from the assembly of gating particles freely moving in the membrane. The molecular components responsible for ionic permeabilities have been later identified as being transmembrane protein complexes containing a pore permeability specific to one or several ionic species as reviewed in Reference 2. These ion channels can have their permeability modulated by various factors, such as voltage or the binding of a ligand. The sensitivity of some ion channels to voltage is a fundamental property that constitutes the core mechanism underlying the electrical excitability of membranes, and is still today an important matter of investigation.[3] Several types of voltage-dependent ion channels have been identified and are responsible for a rich repertoire of electrical behavior essential for neuronal function.[4]

The biophysical properties of ion channels have been characterized in depth following the development of single-channel recording techniques.[5] Single-channel recordings have shown that ion channels display rapid transitions between conducting and non-conducting states. It is now known that conformational changes of the channel protein give rise to opening/closing of the channel. Conformational changes of ion channels can be described by state diagrams analogous to the conformational changes underlying the action of enzymes. Markov models are based on such transition diagrams and have been used for modeling various types of ionic currents based on single-channel recordings.[5] This formalism is more accurate than Hodgkin–Huxley models, but its drawback is the greater difficulty to estimate its parameters from experimental data. On the other hand, Markov models can also be used to draw simplified representations of the current, which only capture the most salient properties of voltage-dependent or synaptic interactions, more adequate for representing currents when simulating networks involving thousands of cells.[6]

Thus, there exist various formalisms of different complexity to model ionic currents. Which formalism to adopt for modeling a given current depends on the experimental data available and its accuracy, as well as on the desired level of precision in the behavior of the model. We illustrate these aspects in this chapter by considering different types of formalisms to model processes such as the action potential and voltage-clamp recordings of the T-type calcium current in thalamic neurons. For both cases, we show the similarities and differences between the different models, how well they account for experimental data, and which is the "minimal" model required to reproduce electrophysiological behavior.

5.2 DIFFERENT FORMALISMS TO MODEL ION CHANNELS

5.2.1 THE HODGKIN–HUXLEY MODEL

The formalism of Hodgkin and Huxley was introduced in 1952 to model the ionic interactions underlying action potentials. In a remarkable series of experiments on the squid giant axon, they determined that ionic conductances can be activated or inactivated according to the membrane potential. They used the technique of voltage-clamp, introduced earlier by Cole, to record the ionic currents generated at different voltages. They identified the kinetics of two voltage-dependent currents, the fast sodium current, I_{Na}, and the delayed potassium rectifier, I_K, mediated by Na^+ and K^+, respectively. A mathematical model was necessary to establish that these properties of voltage-dependence were sufficient to explain the genesis of action potentials. The model introduced by Hodgkin and Huxley[1] incorporated the results of their voltage-clamp experiments and successfully accounted for the main properties of action potentials.

The starting point of the Hodgkin–Huxley model is the membrane equation describing three ionic currents in an isopotential compartment:

$$C_m \frac{dV}{dt} = -g_L(V - E_L) - g_{Na}(V)(V - E_{Na}) - g_K(V)(V - E_K), \tag{5.1}$$

where C_m is the membrane capacitance, V is the membrane potential, g_L, g_{Na}, and g_k are the membrane conductances for leak currents, Na^+ and K^+ currents, respectively, and E_L, E_{Na}, and E_K are their respective reversal potentials, which are given by the Nernst relation. For example, for K^+ ions:

$$E_K = \frac{RT}{ZF} \ln \frac{[K]_o}{[K]_i} \tag{5.2}$$

where R is the gas constant, T is the absolute temperature in degrees Kelvin, Z is the valence of the ion ($Z = 1$ for K^+ ions, $Z = -1$ for Cl^- ions, etc.), F is the Faraday constant, $[K]_o$ and $[K]_i$ are the concentration of K^+ ions outside and inside of the membrane, respectively (see Chapter 3).

The next step is to specify how the conductances $g_{Na}(V)$ and $g_K(V)$ depend on the membrane potential. Hodgkin and Huxley hypothesized that ionic currents result from the assembly of several independent gating particles which must occupy a given position in the membrane to allow the flow of Na^+ or K^+ ions. Each gating particle can be in either side of the membrane and bears a net electronic charge such that the membrane potential can switch its position from the inside to the outside or vice-versa. The transition from these two states is therefore voltage-dependent, according to the diagram:

$$(outside) \underset{\beta_m(V)}{\overset{\alpha_m(V)}{\rightleftharpoons}} (inside), \tag{5.3}$$

where α and β are respectively the forward and backward rate constants for the transitions from the outside to the inside position in the membrane. If m is defined as the fraction of particles in the inside position, and $(1 - m)$ as the fraction outside, one obtains the first-order kinetic equation:

$$\frac{dm}{dt} = \alpha_m(V)(1-m) - \beta_m(V)m. \tag{5.4}$$

If one assumes that particles must occupy the inside position to conduct ions, then the conductance must be proportional to some function of m. In the case of squid giant axon, Hodgkin and Huxley[1] found that the nonlinear behavior of the Na^+ and K^+ currents, their delayed activation, and their sigmoidal rising phase were best fit by assuming that the conductance is proportional to the product of several of such variables:[1]

$$g_{Na} = \bar{g}_{Na} m^3 h \tag{5.5}$$

$$g_K = \bar{g}_K n^4, \tag{5.6}$$

where $\bar{g}_{Na}$ and $\bar{g}_K$ are the maximal values of the conductances and m, h, n represent the fraction of three different types of gating particles in the inside of the membrane. This equation allowed voltage-clamp data of the currents to fit accurately, which can be interpreted to mean that the assembly of three gating particles of type m and one of type h is required for Na^+ ions to flow through the membrane, while the assembly of four gating particles of type n is necessary for the flow of K^+ ions. These particles operate independently of each other, leading to the m^3h and n^4 forms.

When it was later established that ionic currents are mediated by the opening and closing of ion channels, the gating particles were reinterpreted as *gates* inside the pore of the channel. Thus, the reinterpretation of Hodgkin and Huxley's hypothesis was that the pore of the channel is controlled by four gates, that these gates operate independently of each other, and that all four gates must be open in order for the channel to conduct ions.

The rate constants of $\alpha(V)$ and $\beta(V)$ of m and n are such that depolarization promotes opening the gate, a process called *activation*. On the other hand, the rate constants of h are such that depolarization promotes closing of the gate,* a process called *inactivation*. Thus, the experiments of Hodgkin and Huxley[1] established that three identical activation gates (m^3) and a single inactivation gate (h) are sufficient

* Therefore closing of the entire channel because all gates must be open for the channel to conduct ions.

to explain the Na^+ current's characteristics. The K^+ current does not have inactivation and can be well described by four identical activation gates (n^4).

Taking together all the steps above, one can write the following set of equations:[1]

$$\begin{aligned} C_m \frac{dV}{dt} &= -g_L(V - E_L) - \bar{g}_{Na}\, m^3 h(V - E_{Na}) - \bar{g}_K n^4 (V - E_K) \\ \frac{dm}{dt} &= \alpha_m(V)(1-m) - \beta_m(V)m \\ \frac{dh}{dt} &= \alpha_h(V)(1-h) - \beta_h(V)h \\ \frac{dn}{dt} &= \alpha_n(V)(1-n) - \beta_n(V)n. \end{aligned} \tag{5.7}$$

The rate constants (α_i and β_i) were estimated by fitting empirical functions of voltage to the experimental data.[1] These functions are given in Table 5.1.

TABLE 5.1
Rate Constants of the Hodgkin–Huxley Model

Gate	Forward Rate Constant	Backward Rate Constant
m	$\alpha_m = \dfrac{-0.1\,(V - V_r - 25)}{\exp[-(V - V_r - 25)/4] - 1}$	$\beta_m = 4\exp[-(V - V_r)/18]$
h	$\alpha_h = 0.07\exp[-(V - V_r)/20]$	$\beta_h = \dfrac{1}{1 + \exp[-(V - V_r + 30)/10]}$
n	$\alpha_n = \dfrac{-0.01\,(V - V_r + 10)}{\exp[-(V - V_r + 10)/10] - 1}$	$\beta_n = 0.125\exp[-(V - V_r)/80]$

Note: The rate constants are given for the variables m, n, h as in Equation 5.7. The rate constants are those estimated by Hodgkin and Huxley[1] in the squid giant axon at a temperature around 6°C. In the original study, the voltage axis was reversed in polarity and voltage values were given with respect to the resting membrane potential (V_r here).

Note that the Hodgkin–Huxley model is often written in a form more convenient to fit to experimental data. Equation 5.4 can be rewritten in the following form:

$$\frac{dm}{dt} = \frac{1}{\tau_m(V)}\left(m_\infty(V) - m\right), \tag{5.8}$$

where

$$m_\infty(V) = \alpha(V)/[\alpha(V)+\beta(V)] \tag{5.9}$$

$$\tau_m(V) = 1/[\alpha(V)+\beta(V)]. \tag{5.10}$$

The Hodgkin–Huxley equations then become:

$$\begin{aligned} C_m \frac{dV}{dt} &= -g_L(V-E_L) - \bar{g}_{Na}\, m^3 h (V-E_{Na}) - \bar{g}_K\, n^4 (V-E_K) \\ \frac{dm}{dt} &= (m_\infty(V)-m)/\tau_m(V) \\ \frac{dh}{dt} &= (h_\infty(V)-h)/\tau_h(V) \\ \frac{dn}{dt} &= (n_\infty(V)-n)/\tau_n(V), \end{aligned} \tag{5.11}$$

where m_∞ is the *steady-state activation* and τ_m is the *activation time constant* of the Na^+ current (n_∞ and τ_n represent the same quantities for the K^+ current). In the case of h, h_∞, and τ_h are called *steady-state inactivation* and *inactivation time constant*, respectively. These quantities are important because they can be determined easily from voltage-clamp experiments (see below).

5.2.2 Thermodynamic Models

In Hodgkin and Huxley's work, the rate constants $\alpha(V)$ and $\beta(V)$ were fit to the experimental data by using exponential functions of voltage obtained empirically. An alternative approach is to deduce the exact functional form of the voltage-dependence of the rate constants from thermodynamics. These *thermodynamic models*[7,8,9] provide a plausible physical basis to constrain and parameterize the voltage-dependence of rate constants, which are then used to fit voltage-clamp experiments.

Generally, it is assumed that the transition between two states of the channel correspond to a conformational change of the ion channel protein. Consider a transition between an initial (I) and a final (F) state, with a rate constant $r(V)$ that is voltage-dependent:

$$I \xrightarrow{r(V)} F. \tag{5.12}$$

According to the theory of reaction rates,[10,11] the rate of the transition depends exponentially on the free energy barrier between the two states:

$$r(V) = r_0\, e^{-\Delta G(V)/RT}, \tag{5.13}$$

where r_0 is a constant and $\Delta G(V)$ is the free energy barrier, which can be written as

$$\Delta G(V) = G^*(V) - G_0(V), \tag{5.14}$$

where $G^*(V)$ is the free energy of an intermediate state (activated complex) and $G_0(V)$ is the free energy of the initial state, as illustrated in Figure 5.1. The relative values of the free energy of the initial and final states (G_0 and G_1) determine the equilibrium distribution between these states, but the kinetics of the transition depend on the size of the free-energy barrier $\Delta G(V)$. Systems with a smaller energy barrier (Figure 5.1, dashed line) correspond to faster kinetics because a larger proportion of molecules will have the required energy to form the activated complex and make the transition.

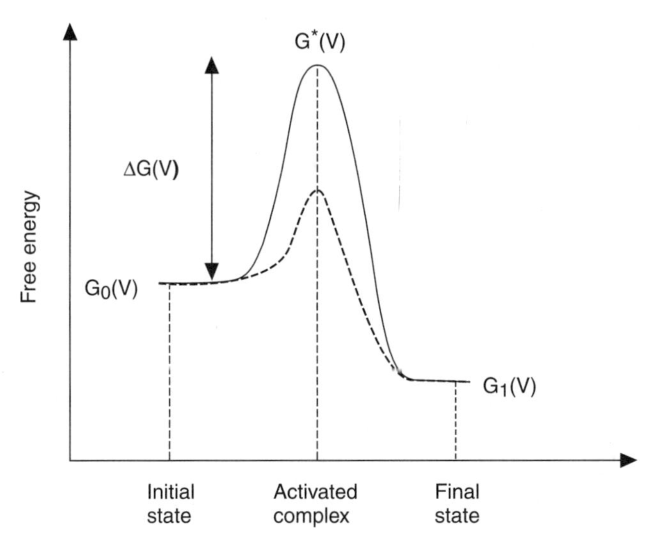

FIGURE 5.1 Schematic representation of the free energy profile of conformational changes in ion channels. The diagram represents the free energy of different states involved in a transition: the initial state, activated complex, and final state. The equilibrium distribution between initial and final states depends on the relative value of their free energy (G_0 and G_1). The rate of the transition will be governed by the *free energy barrier* ΔG, which is the free energy difference between the activated complex and the initial state. If the energy barrier is smaller (dashed line), the kinetics of the reaction are faster because a larger proportion of ion channels will have the required energy to make the transition. (Modified from Reference 15.)

In ion channels, these different states correspond to different conformations of the ion channel protein. How the transition rates between these conformational states depend on membrane potential is given by the voltage-dependence of the free energy barrier, which is in general difficult to evaluate. The effect of the electrical field on a protein will depend on the number and position of its charged amino acids, which will result in both linear and nonlinear components in the free energy. Without assumptions about the underlying molecular structure, the free energy of a given state can be written as a Taylor series expansion of the form:

$$G_i(V) = A_i + B_i V + C_i V^2 + \ldots \tag{5.15}$$

where A_i, B_i, C_i ... are constants specific to each conformational state. The constant A_i corresponds to the free energy that is independent of the electrical field; the linear term $B_i V$ refers to the interaction between electrical fields with isolated charges and rigid dipoles.[7,8,9,12] For example, linear terms in V will result if the conformations differ in their net number of charges, or if the conformational change accompanies the translation of a freely-moving charge inside the structure of the channel.[2,8] Nonlinear terms result from effects such as electronic polarization and pressure induced by V[8,9,12] or mechanical constraints in the movement of charges due to the structure of the ion channel protein.[15]

Thus, each conformational state of the ion channel protein will be associated with a given distribution of charges and will therefore be characterized by a given set of coefficients in Equation 5.15. This is also true for the activated state, which is a particular case of conformation. Applying Equations 5.13–5.15, the rate constant becomes:

$$\begin{aligned} r(V) &= r_0\, e^{-[(A^* + B^* V + C^* V^2 + \ldots) - (A_0 + B_0 V + C_0 V^2 + \ldots)]/RT}, \\ &= r_0\, e^{-(a + bV + cV^2 + \ldots)/RT}, \end{aligned} \tag{5.16}$$

where $a = A^* - A_0$, $b = B^* - B_0$, $c = C^* - C_0$, ... represent differences between the linear and nonlinear components of the free energy of the initial and activated states, according to Equation 5.15).

Considering the particular case of a reversible open-closed transition

$$C \underset{\beta(V)}{\overset{\alpha(V)}{\rightleftharpoons}} O, \tag{5.17}$$

where C and O are respectively the closed and open states, and α and β are the forward and backward rate constants. Applying Equation 5.16 to forward and backward reactions leads to the following general expression for the voltage dependence:

$$\begin{aligned} \alpha(V) &= \alpha_0\, e^{-(a_1 + b_1 V + c_1 V^2 + \ldots)/RT} \\ \beta(V) &= \beta_0\, e^{-(a_2 + b_2 V + c_2 V^2 + \ldots)/RT} \end{aligned} \tag{5.18}$$

where a_1, a_2, b_1, b_2, c_1, c_2, ... are constants specific of this transition. It is important to note that these parameters are not necessarily interrelated because the three different conformations implicated here (initial, activated, final, as in Figure 5.1) may have very different distributions of charges, resulting in different coefficients in Equation 5.15, and thus resulting in different values for $a_1 \ldots c_2$. In the following, this general functional form for the voltage dependence of rate constants will be called the *nonlinear thermodynamic model.*

In the "low field limit" (during relatively small transmembrane voltages), the contribution of the higher order terms may be negligible. Thus, a simple, commonly used voltage dependence equation results from the first-order approximation of Equation 5.18 and takes the form:

$$\begin{aligned} \alpha(V) &= \alpha_0\, e^{-(a_1+b_1V)/RT} \\ \beta(V) &= \beta_0\, e^{-(a_2+b_2V)/RT}. \end{aligned} \tag{5.19}$$

In the following, this form with simple exponential voltage dependence of rate constants will be called the *linear thermodynamic model.*

A further simplification is to consider that the conformational change consists of the movement of a gating particle with charge q.[1,13] The forward and backward rate constants then become:

$$\begin{aligned} \alpha(V) &= \alpha_0\, e^{-\gamma qFV/RT} \\ \beta(V) &= \beta_0\, e^{-(1-\gamma)qFV/RT}, \end{aligned} \tag{5.20}$$

where γ is the relative position of the energy barrier in the membrane (between 0 and 1). The constants α_0 and β_0 can be equated to a fixed constant A by introducing the half-activation voltage V_H, leading to:

$$\begin{aligned} \alpha(V) &= A\, e^{-\gamma qF(V-V_H)/RT} \\ \beta(V) &= A\, e^{(1-\gamma)qF(V-V_H)/RT}. \end{aligned} \tag{5.21}$$

This form was introduced by Borg-Graham[13] for modeling the gating of ion channels. Its parameters are convenient for fitting experimental data: V_H and q affect the steady-state activation/inactivation curves, whereas A and γ only affect the time constant with no effect on steady-state relations.

The drawback of models in which the rate functions are simple exponentials of voltage is that these functions can reach unrealistically high values, which leads to very small time constants and possibly aberrant behavior. One way to solve this problem is to force an artificial saturation of the rate constants[14] or impose a minimum value to the time constant.[13]

Another possibility is not to limit the approximation of Equation 5.18 to linear terms, but include higher-order terms in the voltage dependence of the free energy.[15] For example, the quadratic expansion of Equation 5.18 can be written as:

$$\alpha(V) = A\,e^{-\left[b_1(V-V_H)+c_1(V-V_H)^2\right]/RT}$$
$$\beta(V) = A\,e^{\left[b_2(V-V_H)+c_2(V-V_H)^2\right]/RT}, \tag{5.22}$$

and similarly, its cubic expansion:

$$\alpha(V) = A\,e^{-\left[b_1(V-V_H)+c_1(V-V_H)^2+d_1(V-V_H)^3\right]/RT}$$
$$\beta(V) = A\,e^{\left[b_2(V-V_H)+c_2(V-V_H)^2+d_2(V-V_H)^3\right]/RT}, \tag{5.23}$$

where A, b_1 … d_2 are constants as defined above.

In addition to the effect of voltage on isolated charges or dipoles, described in Equation 5.19, these forms account for more sophisticated effects such as the deformation of the protein by the electrical field[8,9] or mechanical constraints on charge movement.[15] It also allows the model to capture more complicated dependence on voltage than the simple exponential functions of Equation 5.19, which may result in more realistic behavior (see below).

Finally, another way to impose a minimal value for the time constant is to consider that the gate operates via two successive transitions:

$$C_1 \underset{\beta(V)}{\overset{\alpha(V)}{\rightleftharpoons}} C_2 \underset{k_2}{\overset{k_1}{\rightleftharpoons}} O\,, \tag{5.24}$$

where C_1 and C_2 are two distinct closed states of the gate. The second transition does not depend on voltage, and therefore acts as a rate-limiting factor when α and β are large compared to k_1 and k_2. In this case, the system will be governed essentially by k_1 and k_2, which therefore impose a limit on the rate of opening/closing of the gate. On the other hand, when α and β are small compared to k_1 and k_2, the system will be dominated by the first transition, while the two states C_2 and O will be in rapid quasi-equilibrium. Although this system apparently solves the problem of having a minimal time constant while still conserving the voltage dependence of the gate, it is nevertheless unrealistic that the simple exponential representation for α and β permits the first transition to occur arbitrarily fast at some voltages.

Reaction schemes involving multiple states, such as Equation 5.24, are reminiscent of another class of models, called *Markov models*, which are described in more detail below.

5.2.3 Markov Models

Although the formalism introduced by Hodgkin and Huxley[1] was remarkably forward-looking and closely reproduced the behavior of macroscopic currents, the advent of single-channel recording techniques revealed inconsistencies with experimental data. Measurements on Na^+ channels have shown that activation and inactivation must necessarily be coupled,[16,17,18] which is in contrast with the independence of these processes in the Hodgkin–Huxley model. K^+ channels may also show an inactivation which is not voltage-dependent, as in the Hodgkin–Huxley model, but state-dependent.[19] Although the latter can be modeled with modified Hodgkin–Huxley kinetics,[20] these phenomena are best described using Markov models, a formalism more appropriate to describe single channels.

Markov models assume that the gating of a channel occurs through a series of conformational changes of the ion channel protein, and that the transition probability between conformational states depends only on the present state. The sequence of conformations involved in this process can be described by state diagrams of the form:

$$S_1 \rightleftharpoons S_2 \rightleftharpoons \ldots \rightleftharpoons S_n, \tag{5.25}$$

where $S_1 \ldots S_n$ represents distinct conformational states of the ion channel. Defining $P(S_i, t)$ as the probability of being in a state S_i at time t, and defining $P(S_i \rightarrow S_j)$ as the *transition probability* from state S_i to state S_j according to:

$$S_i \underset{P(S_j \rightarrow S_i)}{\overset{P(S_i \rightarrow S_j)}{\rightleftharpoons}} S_j, \tag{5.26}$$

leads to the following equation for the time evolution of $P(S_i, t)$:

$$\frac{dP(S_i,t)}{dt} = \sum_{j=1}^{n} P(S_j,t)P(S_j \rightarrow S_i) - \sum_{j=1}^{n} P(S_i,t)P(S_i \rightarrow S_j). \tag{5.27}$$

This equation is called the *master equation*.[9,21] The left term represents the "source" contribution of all transitions entering state S_i, and the right term represents the "sink" contribution of all transition leaving state S_i. In this equation, the time evolution depends only on the present state of the system, and is defined entirely by knowledge of the set of transition probabilities. Such systems are called *Markovian systems*.

In the limit of large numbers of identical channels, the quantities given in the master equation can be replaced by their macroscopic interpretation. The probability of being in a state S_i becomes the *fraction of channels* in state S_i, noted s_i, and the transition probabilities from state S_i to state S_j become the *rate constants*, r_{ij}, of the reactions

$$S_i \underset{r_{ji}}{\overset{r_{ij}}{\rightleftharpoons}} S_j. \tag{5.28}$$

In this case, one can rewrite the master equation as:

$$\frac{ds_i}{dt} = \sum_{j=1}^{n} s_j r_{ji} - \sum_{j=1}^{n} s_i r_{ij}, \tag{5.29}$$

which is a conventional kinetic equation for the various states of the system.

Stochastic Markov models, as in Equation 5.27, are adequate to describe the stochastic behavior of ion channels as recorded using single-channel recording techniques.[5] In other cases, where a larger area of membrane is recorded and large numbers of ion channels are involved, the macroscopic currents are continuous and more adequately described by conventional kinetic equations, as in Equation 5.29.[22] In the following, only systems of the latter type will be considered.

It is to be noted that Markov models are more general than the Hodgkin–Huxley formalism, and include it as a subclass. A Markov scheme can be written for any Hodgkin–Huxley scheme, but the translation of a system with multiple independent gates into a Markov description results in a combinatorial explosion of states. For example, the Markov model corresponding to the Hodgkin–Huxley sodium channel is:[23]

$$\begin{array}{ccccccc}
C_3 & \underset{\beta_m}{\overset{3\alpha_m}{\rightleftharpoons}} & C_2 & \underset{2\beta_m}{\overset{2\alpha_m}{\rightleftharpoons}} & C_1 & \underset{3\beta_m}{\overset{\alpha_m}{\rightleftharpoons}} & O \\
\alpha_h \downarrow\uparrow \beta_h & & \alpha_h \downarrow\uparrow \beta_h & & \alpha_h \downarrow\uparrow \beta_h & & \alpha_h \downarrow\uparrow \beta_h. \\
I_3 & \underset{\beta_m}{\overset{3\alpha_m}{\rightleftharpoons}} & I_2 & \underset{2\beta_m}{\overset{2\alpha_m}{\rightleftharpoons}} & I_1 & \underset{3\beta_m}{\overset{\alpha_m}{\rightleftharpoons}} & I
\end{array} \tag{5.30}$$

The states represent the channel with the inactivation gate in the open state (top) or closed state (bottom) and (from left to right) three, two, one, or none of the activation gates closed. To be equivalent to the m^3 formulation, the rates must have the 3:2:1 ratio in the forward direction and the 1:2:3 ratio in the backward direction. Only the O state is conducting. The squid delayed-rectifier potassium current modeled by Hodgkin and Huxley[1] with four activation gates and no inactivation can be treated analogously,[23,24] giving

$$C_4 \underset{\beta_m}{\overset{4\alpha_m}{\rightleftharpoons}} C_3 \underset{2\beta_m}{\overset{3\alpha_m}{\rightleftharpoons}} C_2 \underset{3\beta_m}{\overset{2\alpha_m}{\rightleftharpoons}} C_1 \underset{4\beta_m}{\overset{\alpha_m}{\rightleftharpoons}} O. \tag{5.31}$$

Color Figure 1.1 Principle of the density trajectory-based fitness function. (A) The evolution of the membrane potential of a Hodgkin–Huxley spiking neuron is transformed into a trajectory density plot in a (V, dV/dt) phase plane. A third dimension is added (color coded) that gives the number of times (duration) each point of the phase plane is found in the series. (B) The same transformation for a real intracellular recording of a bursting neuron. The fitness coefficient is computed between two trajectory density plots by summing the squared density difference over the entire plane (Equation 1.28).

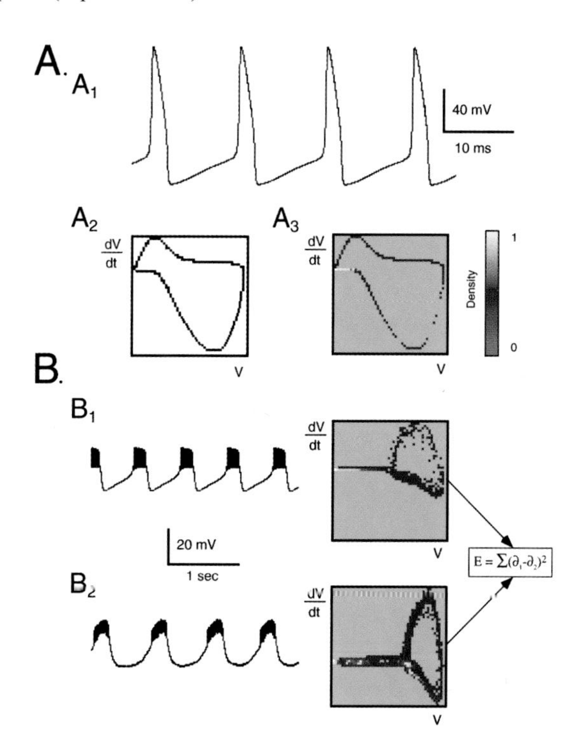

Color Figure 3.1 Representative full model calculation of a [Ca^{2+}] profile near a point source for free Ca^{2+}. Source amplitude and elapsed time are 5 pA and 1 ms, respectively. Other parameters: $[B_s]_T$ = 250 μM, $[B_m]_T$ = 50 μM, $k_s^- = k_m^- = 10^3$ s^{-1}, $k_s^+ = k_m^+ = 100$ μM^{-1} s^{-1}, D_c = 250 μm²/s, D_b = 75 μm²/s, $[Ca^{2+}]_\infty$ = 0.1 μM, Δt = 0.1 μs, Δr = 0.01 μm, R_{max} = 8 μm. (Modified from Reference 21 in Chapter 3. With permission.)

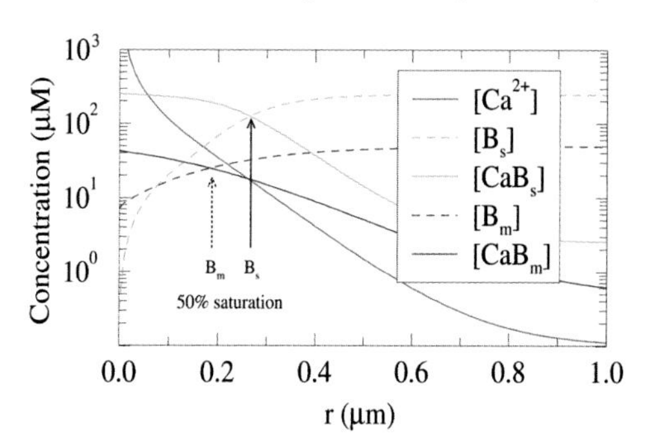

Color Figure 3.2 Simulated line-scan image of [Ca^{2+}] in the myoplasm of a cardiac myocyte with color representing [Ca^{2+}]. Space is represented on the horizontal axis and time on the vertical axis. Same model and parameters as in Figure 3.5. The reciprocal slope of the wave front gives the wave speed, $v = 67$ μm/s. (From Reference 54 in Chapter 3. With permission.)

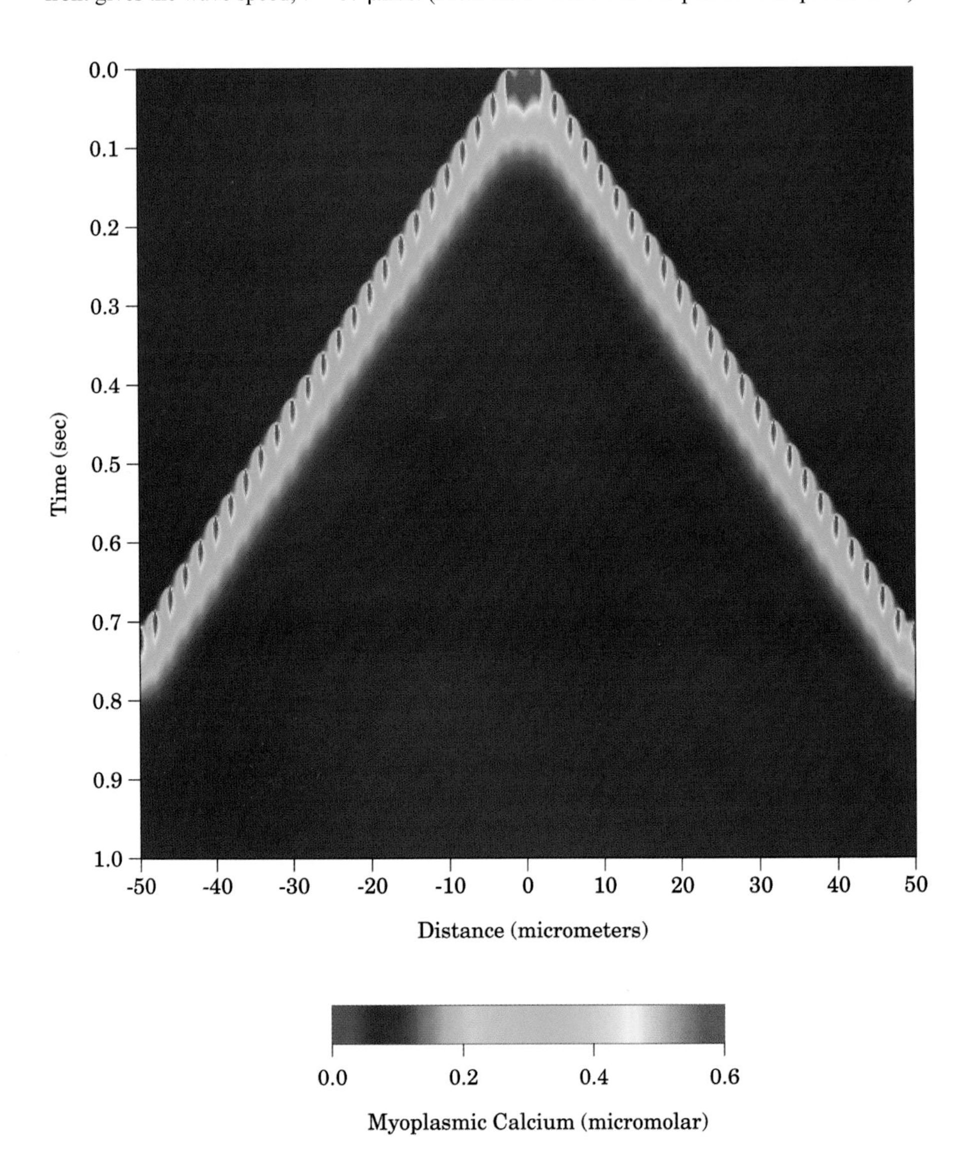

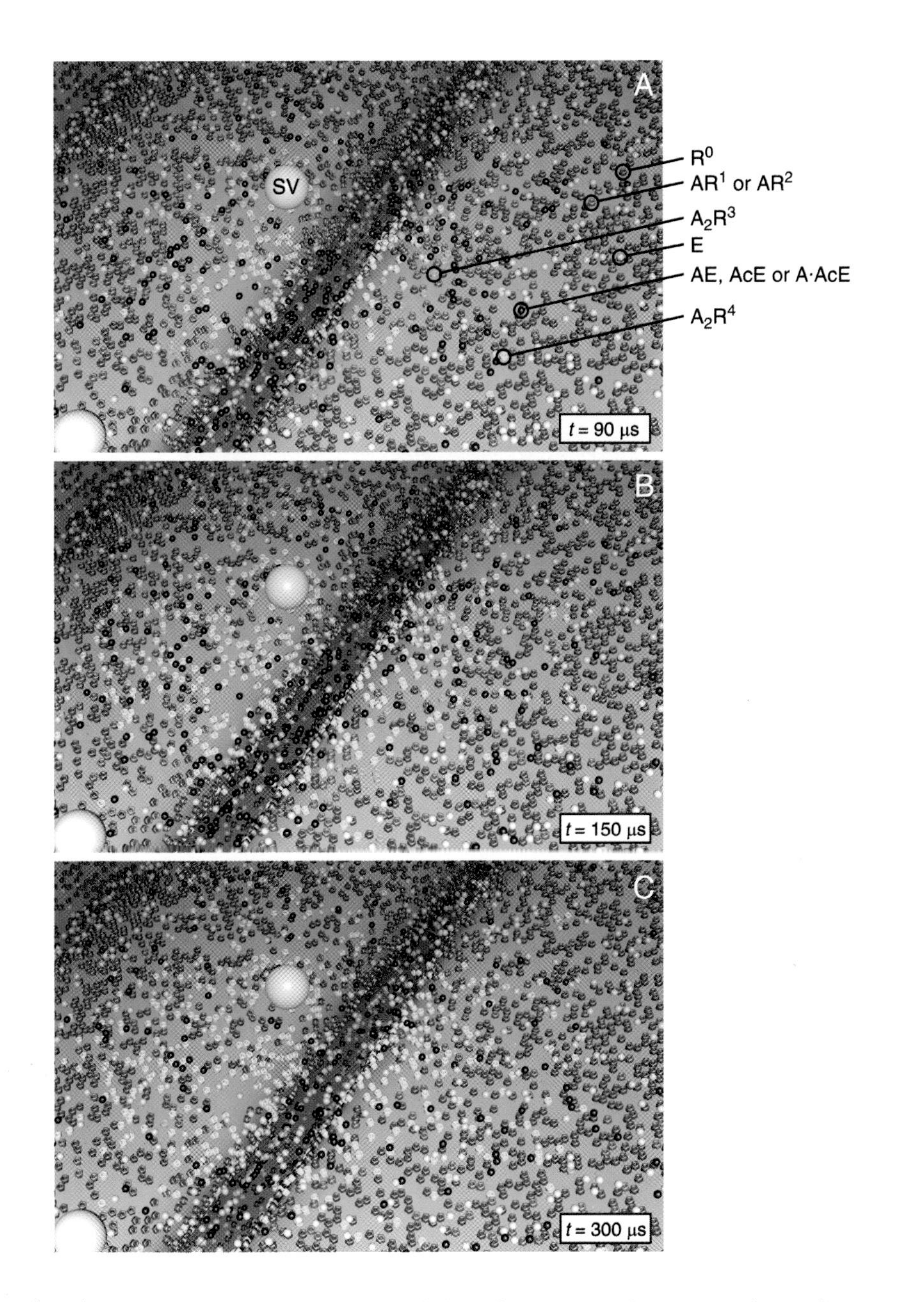

Color Figure 4.1 Spatial and temporal evolution of an mEPC. The presynaptic membrane has been removed for clarity. Postsynaptic AChR and basal lamina AChE glyphs are color-coded according to their state as indicated in A. The times for each image were chosen from Figure 4.6E to show peak numbers of A_2R^3 (A), $AR^1 + AR^2$ (B), or A_2R^4 (C). Diffusing ACh and Ch molecules are tiny (to scale) cyan and magenta spheres, respectively (see Color Figures 4.2 and 4.3). These images were rendered using Pixar RenderMan (0.5–2 hours computer time per image).

Color Figure 4.2 View of synaptic cleft at time of peak ACh concentration. The time (60 μs) was chosen from Figure 4.6B (sampling boxes not shown). ACh, Glu, AChR, and AChE molecules are color-coded as in Color Figure 4.1. The largest central synaptic vesicle (dim white) indicates the ACh release site and is seen through translucent gray nerve membrane, which contains Ch reuptake sites (blue, unbound; red, reversibly bound; green, rate-limiting intermediate that transports Ch. See text).

Color Figure 4.3 View of synaptic cleft at time of peak Ch concentration and mEPC amplitude. The time (300 μs) was chosen from Figures 4.6B and E. Molecules are color-coded as in Color Figure 4.1.

Color Figure 4.4 "Small-scale" qualitative model of neuronal glutamatergic currents. (A) Model design ($t = 0$). Inner surface (light blue) represents simplified interneuron cell body (~0.3 × ~0.5 μm), and contains ESs to simulate AMPA GluRs (blue glyphs, 4000 μm^{-2}). Translucent outer surface defines closed intercellular diffusion space, and contains Glu reuptake sites (white spheres, 4000 μm^{-2}). (B) Larger view of inner surface only, early after first release of Glu ($t = 5$ μs). Release occurred from a site at lower left. Glu molecules are small white spheres. Visible GluRs are either in the unbound state (blue, C0; see Figure 4.7) or the primary single-bound state (C1, red).

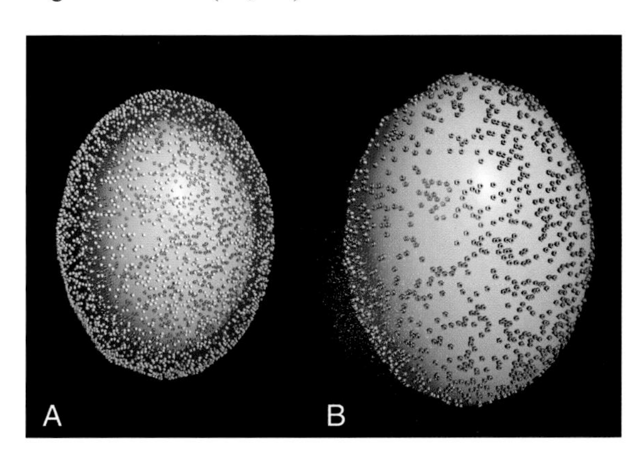

Color Figure 4.5 Second release of glutamate. Release (at $t = 1.5$ ms) preceded this image by 5 μs, and occurred from site on right side. GluRs are present in a mix of the primary closed states (C0, blue; C1, red; C2, green), the open state (O, yellow), and the initial desensitized state (C3, magenta).

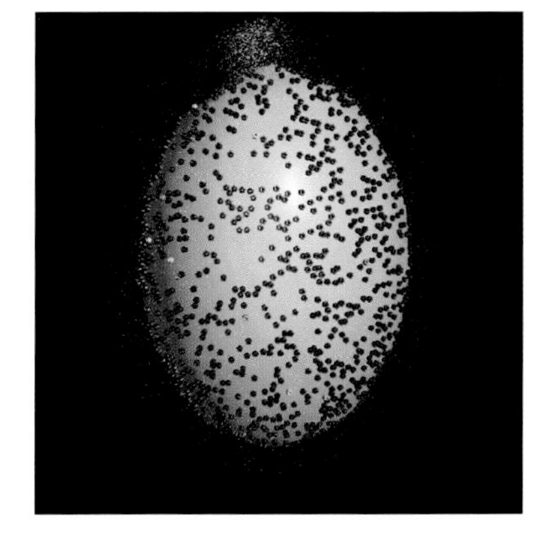

Color Figure 4.6 Late release of glutamate. Release (at $t = 76.5$ ms) preceded this image by 1 μs, and occurred from site at top of view. At this late time, Glu has accumulated in the diffusion space, and nearly all GluRs are in the longest-lived desensitized state (C4, black; cf. Figures 4.7B and D). Aside from rare open and other states, several GluRs in the short-lived C5 state (white) are also visible.

Color Figure 5.1 Fitting of different models to the T-Current in thalamic relay neurons. In each panel, the symbols show the voltage-clamp data obtained in several thalamic neurons (see Figures 5.4 and 5.5) and the continuous curves show the best fits obtained with an empirical Hodgkin–Huxley type model (blue), a linear thermodynamic model (green) and a nonlinear thermodynamic model (red). (A) Steady-state activation (m_∞^2). (B) Steady-state inactivation (h_∞). (C) Activation time constant (τ_m). (D) Inactivation time contant (τ_h). The leftmost symbols in D ≤ -80 mV) are the data from the slow recovery from inactivation of the T-current. See text for the values of the parameters. All functions were fit using a simplex method (see Chapter 1). Modified from Reference 5 in Chapter 5.

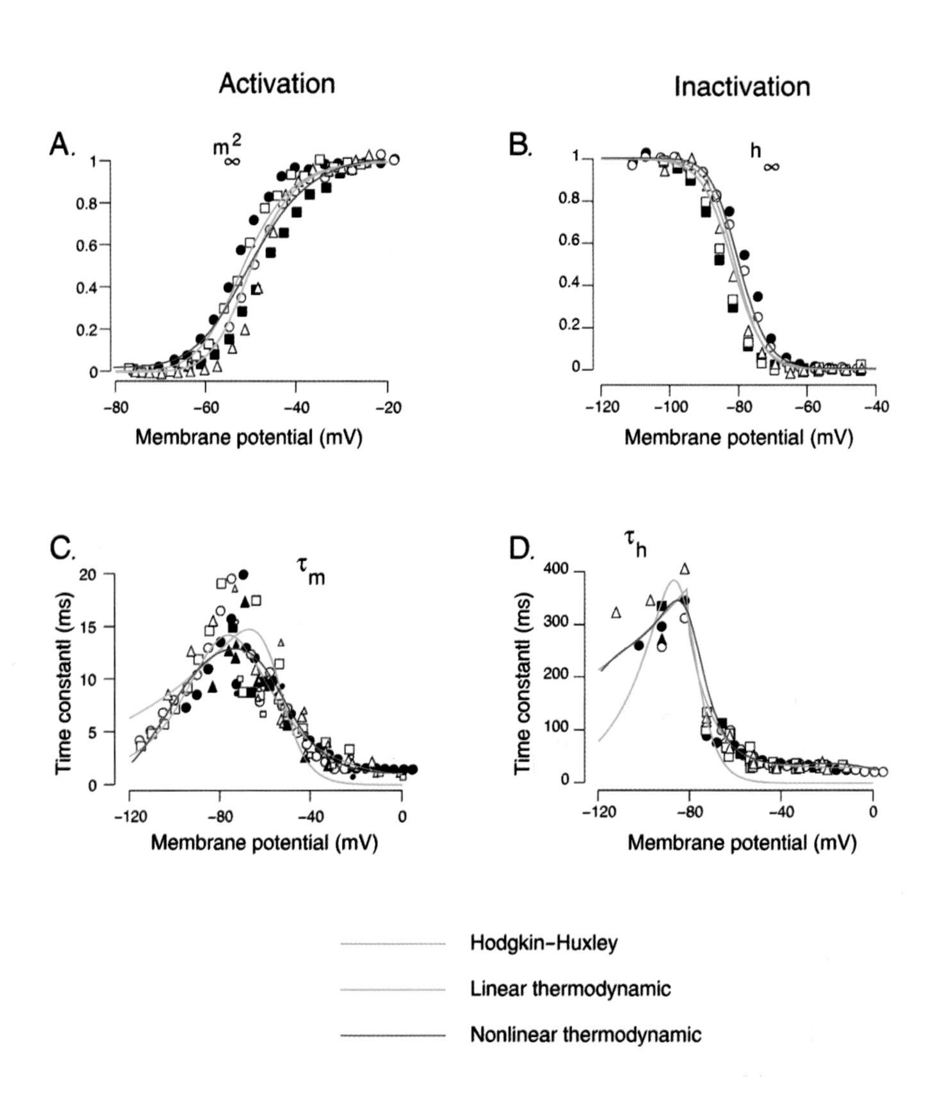

Color Figure 8.1 Semi-log plot of short (0.5 ms) current pulse responses (averages of 200) from a real CA1 pyramidal cell, somatic sharp 0.5 M KMeSO4 electrode recording, showing linear scaling, including of their undershoot ("sag"). The currents used were –1 nA (purple, red, blue), +1 nA (black, yellow, green), –0.5 nA (light green). The inclusion of several responses to the same stimulus gives an idea of the statistical fluctuations.

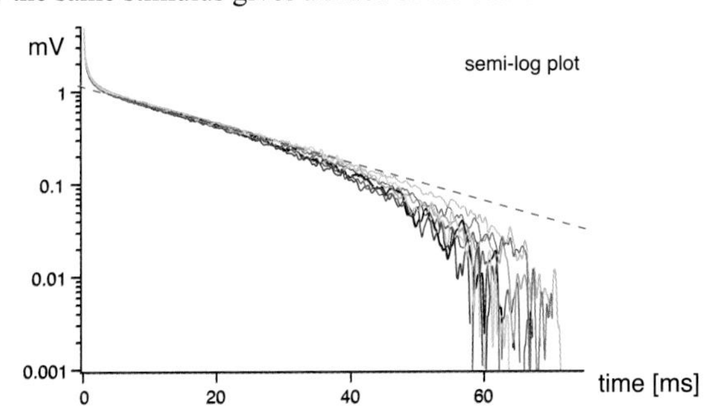

Color Figure 8.2 Short pulse (0.5 ms) responses of a cat spiny stellate cell, recorded with a sharp electrode under current clamp. Black is –1 nA response, red is +1 nA response. Notice how the apparent time constant smoothly increases as the fit interval is made progressively later. This is "creep." Notice also how the responses scale linearly.

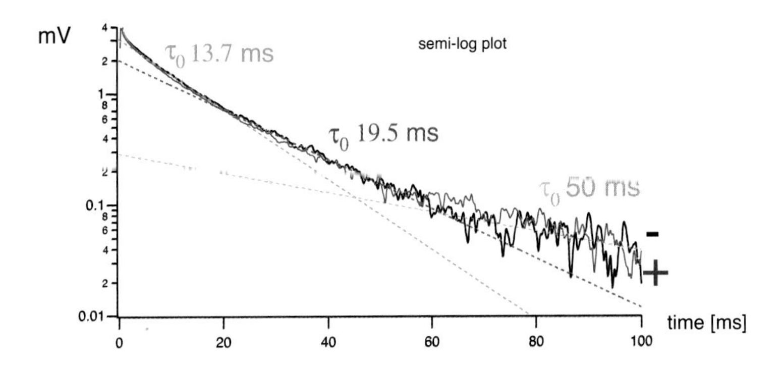

Color Figure 8.3 Schematic illustration of principles underlying direct fitting. The target waveform (noisy) is green. A simplex algorithm calling an analytical solution subroutine is given a starting (guessed) set of parameters which results in "fit 1" as the model response. Automatic adjustment of the parameters to minimize the discrepancy between the model and target waveforms results in the other smooth lines, which can be seen to progressively converge on the target. By the 100th iteration, the fit is more or less optimal.

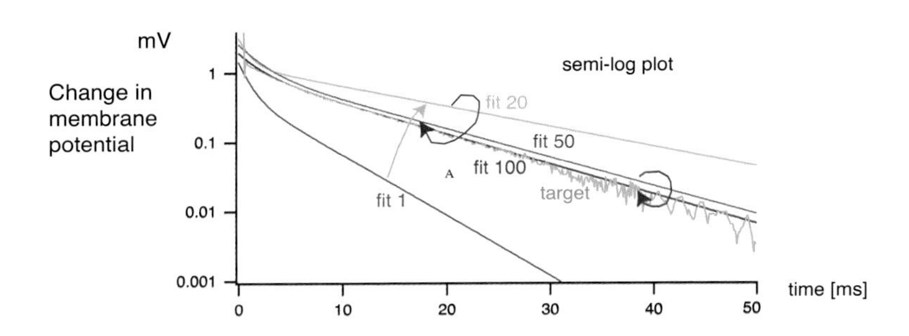

Color Figure 9.1 Simulation of the amplification of focal synchronous parallel fiber input in the Purkinje cell model. Synchronous input is applied to a distal (A,B) or a proximal branch (C, D). (A, C) False color images of membrane potential in the model (scale in mV). In the first panel the sites of input are colored white. Subsequent panels demonstrate how the dendritic depolarization spreads into nearby branches for the distal input (A), while staying local for the proximal input (C). (B, D) Averaged somatic EPSPs recorded for same synchronous inputs in a completely passive model (blue) or in model with active dendrite (yellow). In the passive model dendritic filtering attenuates the distal input much more than the proximal one (blue traces, see Chapter 8.2). In the active model both EPSPs have equal amplitudes (yellow traces, amplitude 5 mV). See Reference 59 for methods.

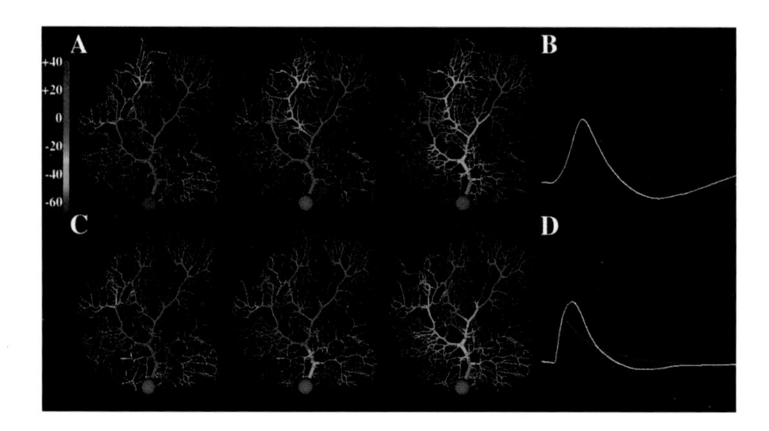

Color Figure 9.2 Effect of dendritic membrane potential on voltage-gated Ca^{2+} influx during focal parallel fiber activation on a distal branch. Calcium concentration (scale in μM) is shown for synchronous activation of zero (left panels) to 60 (right panels) parallel fibers. In the first panels the location of input is colored white. (A) Model in simulated *in vitro* condition. Increasing the strength of the stimulus leads to more Ca^{2+} influx in the region of activation only. Compare to Figure 9.4 of Reference 61. (B) The same stimuli applied during simulated *in vivo* condition (as in Color Figure 9.1) cause spreading Ca^{2+} channel activation, reflected by a high Ca^{2+} concentrations in branches surrounding the activation site. See Reference 59 for methods.

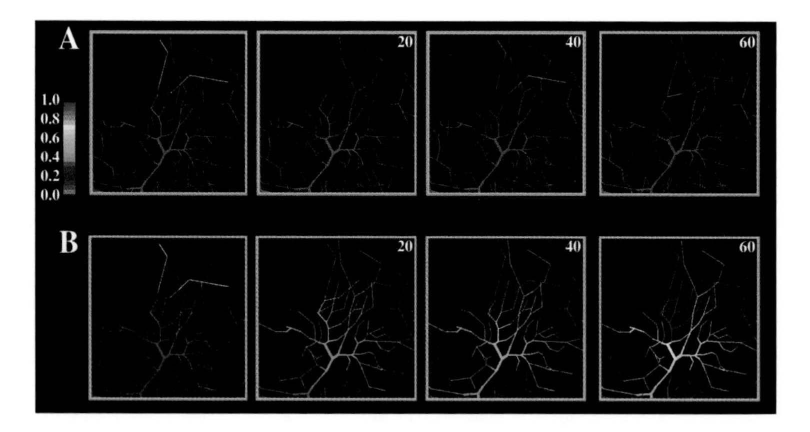

5.3 MODELS TO GENERATE ACTION POTENTIALS

We now use some of the formalisms reviewed above and compare them in similar situations. The voltage-clamp behavior of the sodium channel and the genesis of action potentials are taken as examples to illustrate the differences between these formalisms.

5.3.1 MODELS OF NA^+ AND K^+ CURRENTS UNDERLYING ACTION POTENTIALS

We will compare the model of Hodgkin and Huxley[1] with two Markov models of Na^+ channels. A nine-state Markov model was proposed by Vandenberg and Bezanilla:[25]

$$\begin{array}{ccccccccccc} C & \underset{r_6}{\overset{r_5}{\rightleftharpoons}} & C_1 & \underset{r_6}{\overset{r_5}{\rightleftharpoons}} & C_2 & \underset{r_6}{\overset{r_5}{\rightleftharpoons}} & C_3 & \underset{r_2}{\overset{r_1}{\rightleftharpoons}} & C_4 & \underset{r_4}{\overset{r_3}{\rightleftharpoons}} & O \\ & & & & & & r_{10}\downarrow\uparrow r_8 & & & & r_9\downarrow\uparrow r_7 \\ & & & & & & I_1 & \underset{r_3}{\overset{r_4}{\rightleftharpoons}} & I_2 & \underset{r_1}{\overset{r_2}{\rightleftharpoons}} & I_3 . \end{array} \tag{5.32}$$

This particular nine-state model was selected to fit not only the measurements of macroscopic ionic currents available to Hodgkin and Huxley, but also recordings of single channel events and measurements of currents resulting directly from the movement of charge during conformational changes of the protein (called *gating currents*).[2] The voltage dependence of the transition rates was assumed to be a simple exponential function of voltage, as in Equation 5.19.

To complement the sodium channel model of Vandenberg and Bezanilla, we also examined the six-state scheme for the squid delayed-rectifier channel, used by Perozo and Bezanilla:[26]

$$C \underset{r_2}{\overset{r_1}{\rightleftharpoons}} C_1 \underset{r_4}{\overset{r_3}{\rightleftharpoons}} C_2 \underset{r_4}{\overset{r_3}{\rightleftharpoons}} C_3 \underset{r_4}{\overset{r_3}{\rightleftharpoons}} C_4 \underset{r_6}{\overset{r_5}{\rightleftharpoons}} O, \tag{5.33}$$

where again rates were described by a simple exponential function of voltage (see Equation 5.19).

The third class of model considered here comprise simplified Markov models of Na^+ and K^+ currents. The model for the Na^+ channel was chosen to have the fewest possible number of states (three) and transitions (four) while still being capable of reproducing the essential behavior of the more complex models. The form of the state diagram was based on a looped three-state scheme in which some transitions were eliminated, giving an irreversible loop:[6,27]

$$\begin{array}{ccc} C & \underset{r_2(V)}{\overset{r_1(V)}{\rightleftharpoons}} & O \\ & r_4(V)\nwarrow \quad \swarrow r_3 & \\ & I & \end{array} \tag{5.34}$$

This model incorporated voltage-dependent opening, closing, and recovery from inactivation, while inactivation was voltage-independent. For simplicity, neither opening from the inactivated state nor inactivation from the closed state was permitted. Although there is clear evidence for occurrence of the latter,[28] it was unnecessary under the conditions of the present simulations. Rate constants were described by

$$r_i(V) = \frac{a_i}{1+\exp\left[-(V-c_i)/b\right]}, \tag{5.35}$$

with $c_1 = c_2$ to yield a model consisting of nine total parameters.[6]

The simplified K^+ channel model consisted of a single open or conducting state, O, and a single closed state C:

$$C \underset{r_2(V)}{\overset{r_1(V)}{\rightleftharpoons}} O. \tag{5.36}$$

Here, the rates $r_1(V)$ and $r_2(V)$ had a sigmoidal voltage dependence similar to Equation 5.35.[6]

5.3.2 Na+ Currents in Voltage-Clamp

The models reviewed above are characterized by different complexities, ranging from a two-state representation as shown in Equation 5.36 to transition diagrams involving many states, see Equation 5.32. The two-state description is adequate to fit the behavior of some channels,[6,13,29,30,31] but for most channels more complex models must be considered. To illustrate this point, we compared three different models of the fast sodium channel underlying action potentials (Figures 5.2 and 5.3).

Responses among the three sodium channel models were compared during a voltage-clamp step from rest (–75 mV) to a depolarized level of –20 mV (Figure 5.2). For all three models, the closed states were favored at hyperpolarized potentials. Upon depolarization, forward (opening) rates sharply increased while closing (backward) rates decreased, causing a migration of channels in the forward direction toward the open state. The three closed states in the

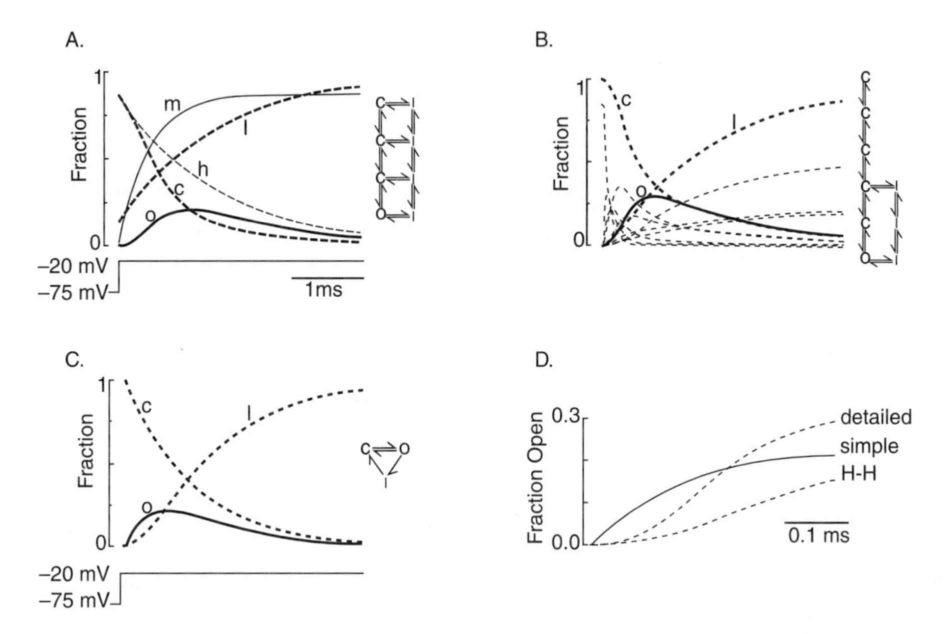

FIGURE 5.2 Three kinetic models of a squid axon sodium channel produce qualitatively similar conductance time courses. A voltage-clamp step from rest, $V = -75$ mV, to $V = -20$ mV was simulated. The fraction of channels in the open state (*O*, thick solid line), closed states (*C*, thick dashed lines), and inactivated states (*I*, thick dotted lines) are shown for the Hodgkin–Huxley model,[1] a detailed Markov model,[25] and a simple Markov model.[6] (A) Hodgkin–Huxley model of the sodium channel (Equation 5.30). The activation (m) and inactivation (h) gates were deduced from other states and are indicated by thin lines. (B) Markov model of Vandenberg and Bezanilla[25] (Equation 5.32). Individual closed and inactivated states are shown (thin lines), as well as the sum of all five closed states (C), the sum of all three inactivated states (I) and the open state (O). (C) Simplified three-state Markov model[6] (Equation 5.34). (D) Comparison of the time course of open channels for the three models on a faster time scale shows differences immediately following the voltage step. The Hodgkin–Huxley (H–H) and Vandenberg–Bezanilla (detailed) models give smooth, multiexponential rising phases, while the three-state Markov model (simple) gives a single exponential rise with a discontinuity in the slope at the beginning of the pulse. (Modified from Reference 6; parameters given in the CD-ROM.)

Hodgkin–Huxley model and the five closed states in the Vandenberg–Bezanilla model gave rise to the characteristic delayed activation and sigmoidal shape of the rising phase of the sodium current (Figure 5.2D). In contrast, the simple model, with a single closed state, produced a first-order exponential response to the voltage step and was, therefore, not sigmoidal.

These models generate different predictions about single-channel behavior. The steady-state behavior of the Hodgkin–Huxley model of the macroscopic sodium current is remarkably similar to that of the model of Vandenberg and Bezanilla,[25] but there are important differences in the relationship between activation and inactivation. First, in the Hodgkin–Huxley model, activation and inactivation are kinetically independent. This independence has been shown to be untenable on the basis

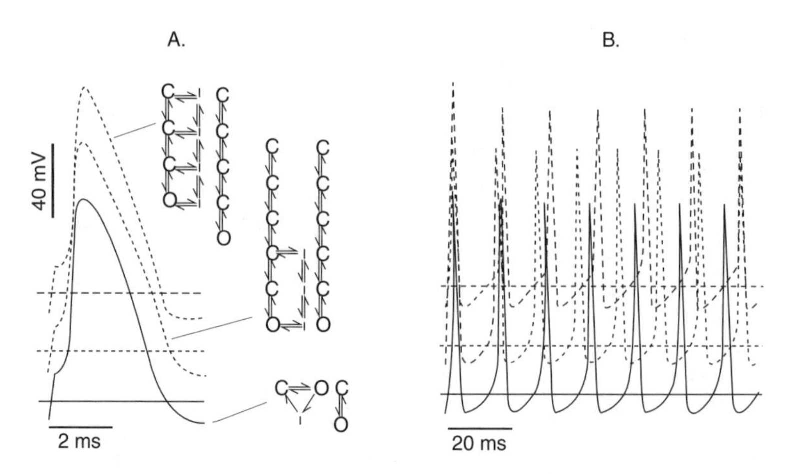

FIGURE 5.3 Similar action potentials produced using three different kinetic models of squid fast sodium and delayed-rectifier potassium channels. (A) Single action potentials in response to 0.2 ms, 2 nA current pulse are elicited at similar thresholds and produce similar wafeforms using three different pairs of kinetic models: Hodgkin–Huxley[1] (dashed line), detailed Markov models[25,26] (dotted line), and simplified kinetic models (solid line). (B) Repetitive trains of action potentials elicited in response to sustained current injection (0.2 nA) have slightly different frequencies. Sodium channels were modeled as described in Figure 5.2. The detailed Markov potassium channel model had six states[26] (Equation 5.33), and the simple model of potassium channel had two states (Equation 5.36). (Modified from Reference 6; parameters given in the CD-ROM.)

of gating and ion current measuerments in the squid giant axon.[16,17,18] Consequently, Markov models that reproduce gating currents, such as the Vandenberg–Bezanilla model examined here, require schemes with coupled activation and inactivation. Likewise, in the simple model, activation and inactivation were strongly coupled due to the unidirectional looped scheme (Equation 5.34), so that channels were required to open before inactivating and could not reopen from the inactivated state before closing.

A second difference is that, in the Hodgkin–Huxley and Vandenberg–Bezanilla models, inactivation rates are slow and activation rates fast. In the simplified Markov model, the situation was reversed, with fast inactivation and slow activation. At the macroscopic level modeled here, these two relationships gave rise to similar time courses for open channels (Figure 5.2 A–C).[12] However, the two classes of models make distinct predictions for single-channel behavior. Whereas the Hodgkin–Huxley and Vandenberg–Bezanilla models predict the latency to first-channel opening to be short and channel open times to be comparable to the time course of the macroscopic current, the simplified Markov model predicts a large portion of first-channel openings to occur after the peak of the macroscopic current and to have open times much shorter than its duration.

5.3.3 Genesis of Action Potentials

Despite the significant differences in their complexity and formulation, the three models of the sodium channel all produced comparable action potentials and repetitive firing when combined with appropriate delayed-rectifier potassium channel models (Figure 5.3) These simulations thus seem to perform similarly for fitting the macroscopic behavior of Na^+ and K^+ currents.

However, these three models generated clear differences when compared in voltage-clamp (Figure 5.2), and still larger differences would be expected at the single-channel level. Thus, which model to choose clearly depends on the scope of the model. If the detailed behavior of voltage-clamp experiments or single-channel recordings are to be reproduced, Markov models are certainly the most appropriate representation. However, if the goal is to reproduce the qualitative features of membrane excitability, action potentials, and repetitive firing, all models seem equivalent, except that simpler models are faster to compute. Thus in this case, simplified two- or three-state schemes or the Hodgkin–Huxley model would seem most appropriate.

5.4 FITTING MODELS TO VOLTAGE-CLAMP DATA

The different formalisms reviewed above are now applied to a concrete example of voltage-clamp experiments. We consider the T-type ("low-threshold") calcium current responsible for bursting behavior in thalamic neurons.[32]

5.4.1 Voltage-Clamp Characterization of the T-Current

Whole-cell voltage-clamp recordings of the T-type calcium current were obtained from thalamic relay neurons acutely dissociated from the ventrobasal thalamus of young rats (P8-P15). All voltage-clamp recordings were at a temperature of 24°C. The methods were described in detail in Reference 33.

The T-current is transient and has activation/inactivation characteristics similar to the Na^+ current, but is slower. Its voltage range for activation/inactivation typically occurs around rest. These properties are illustrated in Figure 5.4. A series of voltage steps from a hyperpolarized level (–100 mV) to various depolarized levels reveal an inward current that activates and inactivates in a voltage-dependent manner (Figure 5.4A1). Interrupting this protocol before complete inactivation generates tail currents (Figure 5.4A2), which reveal the deactivation characteristics of the current.

The fitting of these current traces was performed as follows. The most optimal template (determined by difference in residuals) included two activation gates and one inactivation gate, leading to the m^2h format in Hodgkin–Huxley equations.[33] To measure activation, the influence of inactivation must be as minimal as possible. We assumed that activation is essentially complete in 10 ms, and that there is negligible inactivation (these assumptions were checked by calculating the expected activation and inactivation at various voltages). We used the amplitude of the tail current, which reflects the number of channels open at the end of the depolarizing step, as a measure of activation (m^2). The values obtained using this procedure were

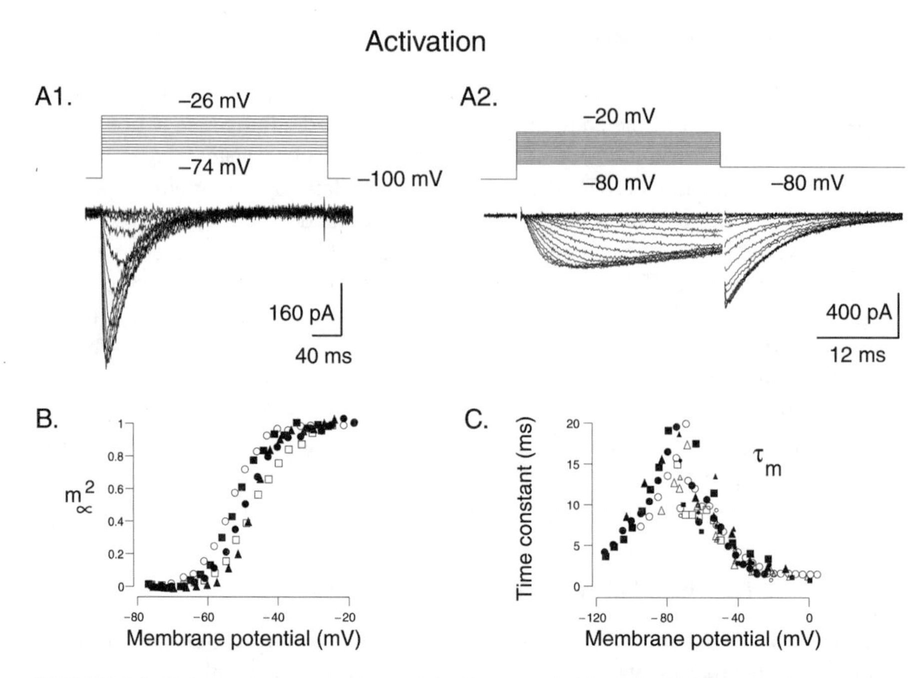

FIGURE 5.4 Voltage-clamp recordings of the T-current in dissociated thalamic relay neurons. (A) Voltage-clamp protocols for activation (A1) and deactivation (A2). Command potentials at various levels were given after the cell was maintained at a hyperpolarized holding potential, leading to the activation of the current. (B) Steady-state activation obtained from the tail currents in A2, which were fit to a m^2h template. (C) Time constants obtained using a similar procedure. Different symbols correspond to different cells. (Modified from Reference 34, where all details were given.)

very close to those obtained by fitting Hodgkin–Huxley equations to current traces.[33] The advantage of the tail current approach is that the driving force is the same for all measurements, thereby providing a direct measure of normalized conductance. This type of procedure leads to estimates of steady-state activation (Figure 5.4B). The time constants were estimated by fitting exponential templates to the current traces (Figure 5.4C).[34]

The inactivation characteristics of I_T are shown in Figure 5.5. A series of holding potentials given at various voltages, before applying a command potential at –30 mV, show different current traces that contain similar activation but different levels of inactivation (Figure 5.5A1). A particular feature of the T-current is that the recovery from inactivation is very slow (Figure 5.5A2). Estimated steady-state relations for several cells using similar protocols are shown in Figure 5.5B, and the time constants are shown in Figure 5.5C.

Thus the T-current in thalamic relay neurons has activation and inactivation characterized by relatively slow time constants, and a slow recovery from inactivation, almost an order of magnitude slower than inactivation. In the following, we examine different representations to model the behavior of this current.

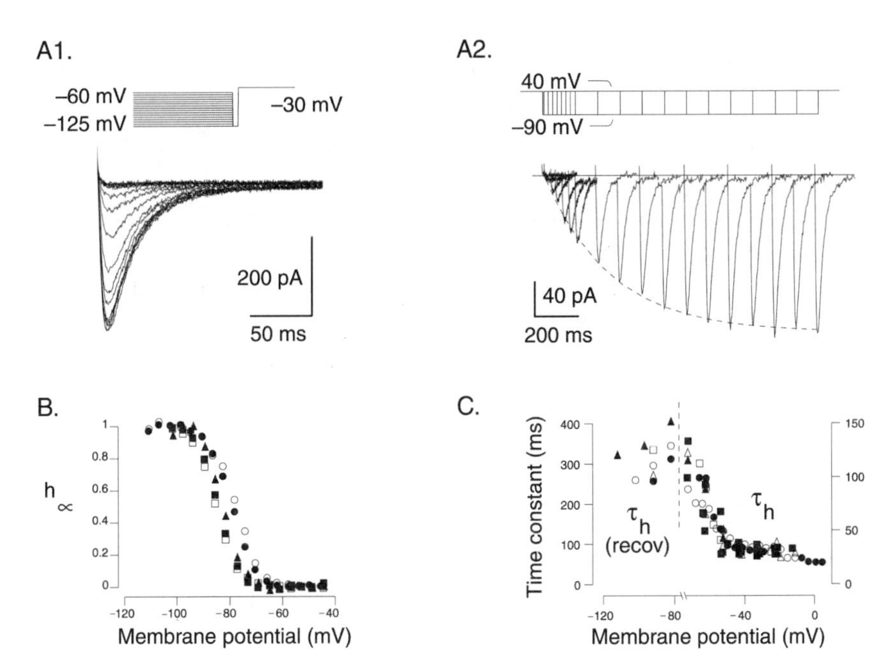

FIGURE 5.5 Voltage-clamp characterization of T-current inactivation in dissociated thalamic relay neurons. (A) Voltage-clamp protocols for inactivation (A1) and recovery from inactivation (A2). In A1, the cell was maintained at different holding potentials then stepped to –30 mV to activate the T-current with different levels of inactivation. In A2, the current is reactivated after being fully inactivated. The full recovery took about 1 second (recovery time constant of about 300 ms). (B) Steady-state inactivation calculated by the peak of currents in A1. (C) Inactivation time constants obtained by fitting a m^2h template to the data. The recovery time constants were obtained by fitting a single-exponential to the recovery experiment (dashed line in A2). Different symbols correspond to different cells. (Modified from Reference 34, where all details were given.)

5.4.2 Hodgkin–Huxley Model of the T-Current

The voltage-clamp behavior shown above was first modeled by a Hodgkin–Huxley type representation in which rate constants were fit to experimental data using empirical functions of voltage.[34] Due to the nonlinear behavior of calcium currents (the internal and external Ca^{2+} concentration differ by about four orders of magnitude), they were represented using the constant-field equations, also known as the Goldman–Hodgkin–Katz equations.[35]

$$I_T = \bar{P}_{Ca}\, m^2 h\, G(V, Ca_o, Ca_i) \tag{5.37}$$

where $\bar{P}_{Ca}$ (in cm/s) is the maximum permeability of the membrane to Ca^{2+} ions, and $G(V, Ca_2, Ca_i)$ is a nonlinear function of voltage and ionic concentrations:

$$G(V, Ca_o, Ca_i) = Z^2F^2V/RT\frac{Ca_i - Ca_o\exp(-ZFV/RT)}{1-\exp(-ZFV/RT)}, \tag{5.38}$$

where $Z = 2$ is the valence of calcium ions. Ca_i and Ca_o are the intracellular and extracellular Ca^{2+} concentrations (in M), respectively.

The variables m and h represent, respectively, the activation and inactivation variables and obey first-order equations similar to Equation 5.8. Their steady-state relations were fit using Boltzmann functions (Color Figures 5.1A and 5.1B, blue curves),* leading to the following optimal functions:

$$m_\infty(V) = 1/\left(1+\exp[-(V+57)/6.2]\right)$$

$$h_\infty(V) = 1/\left(1+\exp[(V+81)/4]\right)$$

Similarly, the voltage-dependent time constants were estimated by fitting exponential functions to the values determined experimentally (Color Figures 5.1C and 5.1D blue curves), leading to the following expression for activation:

$$\tau_m(V) = 0.612 + 1/\left(\exp[-(V+132)/16.7] + \exp[(V+16.8)/18.2]\right) \tag{5.39}$$

and for inactivation:

$$\tau_h(V) = \begin{cases} 28 + \exp[-(V+22)/10.5] & \text{for } V >= -81\ mV \\ \exp[(V+467)/66.6] & \text{for } V < -81\ mV. \end{cases} \tag{5.40}$$

Here, two different functions were fit to the time constants τ_h obtained from inactivation protocols ($V >= -81$) or recovery from inactivation ($V < -81$).

The temperature dependence of these empirical functions was adjusted according to the following rule:

$$\tau' = \tau Q_{10}^{-(T-24)/10}, \tag{5.41}$$

where Q_{10} is the experimentally determined change of time constants for a 10 degree difference in temperature. For the T-current in thalamic neurons, Q_{10} was determined as equal to 5 for τ_m and 3 for τ_h.[36]

The behavior of this model is shown in Figure 5.6. The model accounted well for all protocols of Figures 5.4 and 5.5, with activation and recovery from inactivation shown in Figures 5.6A1 and 5.6A2, respectively. However, in this model, τ_m and τ_h were fit using empirical functions of voltage. Similar to the work of Hodgkin and Huxley,[1] this approach leads to a model that accounts well for the current-clamp behavior of the T-current in thalamic neurons[37] (see below).

* Color Figure 5.1, follows page 140.

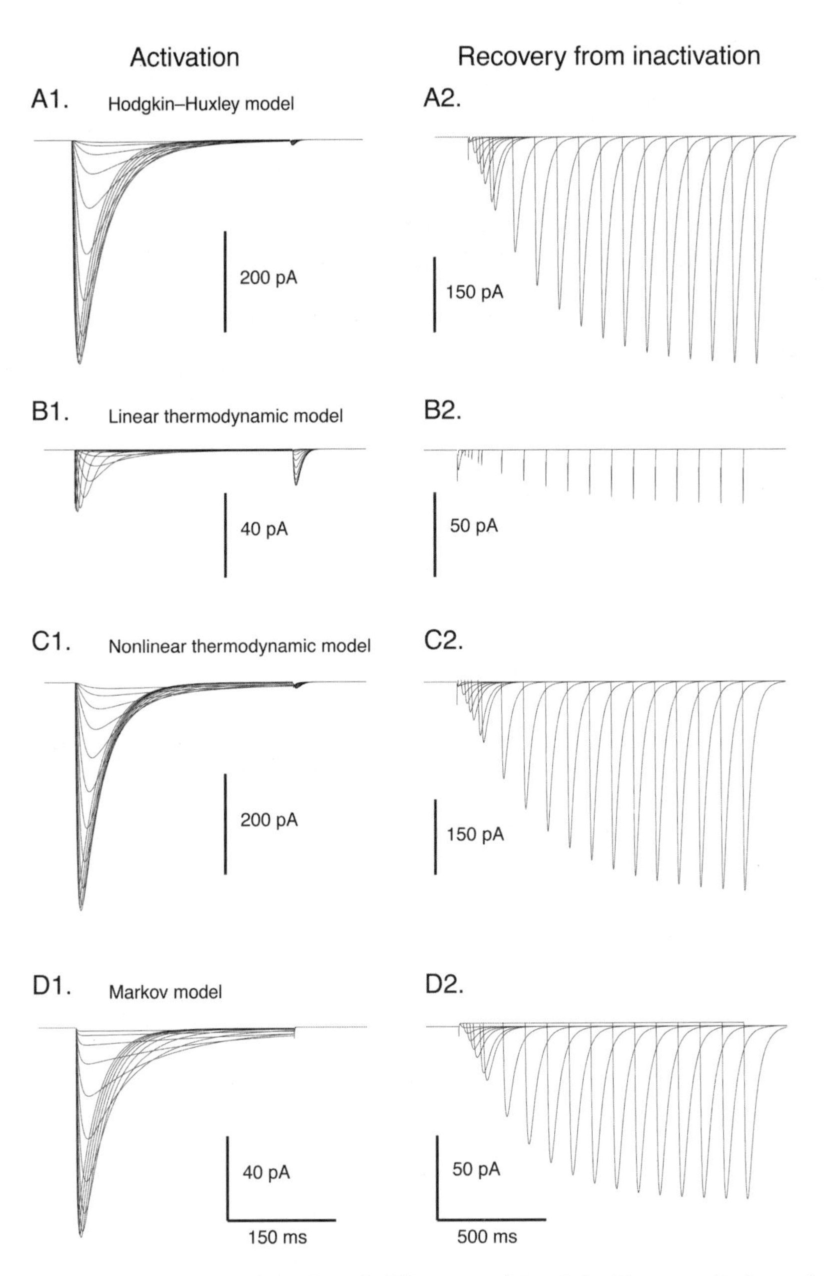

FIGURE 5.6 Voltage-clamp behavior of different models of the T-current. Left panels: activation protocol (identical to Figure 4A1); right panels: protocol for the recovery from inactivation (identical to Figure 5A2). (A) Empirical Hodgkin–Huxley type model. (B) Linear thermodynamic model. (C) Nonlinear thermodynamic model. (D) Markov model. In all cases, the same density of T-channels was used $(\bar{P}_{Ca} = 3 \times 10^{-6}$ cm/s). (Modified from Reference 15.)

5.4.3 Linear Thermodynamic Model of the T-Current

Another possibility is to deduce the functional form of rate constants from thermodynamic arguments. The first of such models is the linear approximation (Equation 5.19), which corresponds to Equation 5.37 above, but with a functional form dictated by a linear voltage-dependence of the free energy barrier. Constraining the fitting using rate constants described by Equation 5.19 (Color Figure 5.1, green curves) led to the following optimal expressions:

$$\alpha_m = 0.049 \exp\left[444\gamma_m(V + 54.6)/RT\right] \tag{5.42}$$

$$\beta_m = 0.049 \exp\left[-444(1-\gamma_m)(V + 54.6)/RT\right] \tag{5.43}$$

$$\alpha_h = 0.00148 \exp\left[-559\gamma_h(V + 81.9)/RT\right] \tag{5.44}$$

$$\beta_h = 0.00148 \exp\left[559(1-\gamma_h)(V + 81.9)/RT\right], \tag{5.45}$$

where $\gamma_m = 0.90$ and $\gamma_h = 0.25$. The steady-state relations and time constants are obtained as in Equations 5.9–5.10.

This model provided a good fit of the steady-state relations (Color Figures 5.1A and 5.1B, green curves) but the fit to time constants was poor (Color Figures 5.1 C and 5.1D, green curves). In particular, it was not possible to capture the saturation of τ_m and τ_h to constant values for depolarized membrane potentials. This poor fit had catastrophic consequences, as illustrated in Figure 5.6B. Due to the near-zero time constants at depolarized levels, the current activated and inactivated too fast and led to peak current amplitudes that were over an order of magnitude smaller than the Hodgkin–Huxley model at same channel densities (compare A and B in Figure 5.6). We conclude that linear thermodynamic models do not provide an acceptable behavior in voltage-clamp for the T-current.

A possible way to resolve this inherent limitation is to add an artificial minimum value to the time constant,[13] but this possibility was not considered here in order to stay within a physically plausible formalism. Instead, we illustrate below that this problem can be solved by including higher-order voltage-dependent contributions in the free energy barrier.[15]

5.4.4 Nonlinear Thermodynamic Model of the T-Current

Nonlinear thermodynamic models assume that the free energy barrier depends nonlinearly on voltage (see Equation 5.18) and that each conformational state involved has its own dependence on voltage, independent of other conformational states.[15] The consequence is that the coefficients $a_1 \ldots c_2$ in Equation 5.18 can take any value independently of each other. Using these nonlinear expressions to fit the voltage-clamp data of the T-current led to better fits. The quadratic expansion still provided a poor fit of the time constants, although better than linear fits (not shown). Acceptable fits were obtained for a cubic expansion of the rate constants, given by:

$$\begin{aligned}
\alpha_m(V) &= A_m \exp\left[b_{m1}(V-V_m)+c_{m1}(V-V_m)^2+d_{m1}(V-V_m)^3\right]/RT \\
\beta_m(V) &= A_m \exp\left[b_{m2}(V-V_m)+c_{m2}(V-V_m)^2+d_{m2}(V-V_m)^3\right]/RT \\
\alpha_h(V) &= A_h \exp-\left[b_{h1}(V-V_h)+c_{h1}(V-V_h)^2+d_{h1}(V-V_h)^3\right]/RT \\
\beta_h(V) &= A_h \exp\left[b_{h2}(V-V_h)+c_{h2}(V-V_h)^2+d_{h2}(V-V_h)^3\right]/RT.
\end{aligned} \tag{5.46}$$

The best fit of this nonlinear thermodynamic model is shown in Color Figure 5.1 (red curves) and was obtained with the following parameters: $A_m = 0.053$ ms^{-1}, $V_m = -56$ mV, $b_{m1} = -260$, $c_{m1} = 2.20$, $d_{m1} = 0.0052$, $b_{m2} = 64.85$, $c_{m2} = 2.02$, $d_{m2} = 0.036$, $A_h = 0.0017$ ms^{-1}, $V_h = -80$ mV, $b_{h1} = 163$, $c_{h1} = 4.96$, $d_{h1} = 0.062$, $b_{h2} = -438$, $c_{h2} = 8.73$, $d_{h2} = -0.057$. Color Figure 5.1 (red curves) shows that this model could capture the form of the voltage dependence of the time constants. In particular, it could fit the saturating values for the time constants at depolarized levels, in a manner similar to the empirical functions used for the Hodgkin–Huxley type model (Color Figure 5.1, blue curves). Nonlinear expansions of higher order provided better fits, but the difference was not qualitative (not shown).

Using these rate constants with Equation 5.37 produced acceptable voltage-clamp behavior, as shown in Figure 5.6. All protocols of activation (Figure 5.6C1), deactivation (not shown), inactivation (not shown), and recovery from inactivation (Figure 5.6C2) showed a voltage-dependent behavior similar to the experimental data.

5.4.5 Markov Model of the T-Current

To illustrate the Markov representation, we have used a model of the T-current introduced by Chen and Hess.[38] This model was obtained based on voltage-clamp recordings and single-channel recordings of the T-current in fibroblasts, and the following optimal scheme was proposed:[38]

$$\begin{array}{ccccc}
C_1 & \underset{k_d(V)}{\overset{k_a(V)}{\rightleftharpoons}} & C_2 & \underset{k_c}{\overset{k_o}{\rightleftharpoons}} & O \\
k_{-r} \downarrow\uparrow k_r & & & & k_i \downarrow\uparrow k_{-i} \\
I_1 & & \underset{k_b(V)}{\overset{k_f(V)}{\rightleftharpoons}} & & I_2.
\end{array} \tag{5.47}$$

Here, only k_a, k_d, k_f, and k_b are voltage-dependent while the other rates are constant. Thus, activation occurs through one voltage-dependent step (k_a, k_d) and one voltage-independent step (k_o, k_c), the latter being rate-limiting if k_a and k_d reach high values. Similarly, inactivation occurs first through a voltage-independent step (k_i, k_{-i}), followed by a voltage-dependent transition (k_f, k_b) and a voltage-independent return to the closed state (k_r, k_{-r}).

Fitting the parameters of Markov models to experimental data is in general difficult. It is not possible to obtain an analytic expression for both time constants and steady-state relations due to the high complexity of the model. In general, the activation and inactivation will be described by multiexponential processes with several time constants, and how to relate these multiple time constants with the time constants estimated from experimental data (Figures 5.4 and 5.5) is not trivial. Rather, the parameters of Markov models are deduced from various experimental considerations (see below). It is also possible to directly fit the Markov model to the original voltage-clamp traces by minimizing the error between the model and all experimental traces. Although in principle more accurate, this procedure is difficult to realize in practice because of the complexity of the model (11 parameters here).

The choice of these parameters was guided by the following considerations:[38] (a) the value of k_i must be close to the saturating value of the rate of inactivation at depolarized membrane potentials (Figure 5.5C), and k_{-I} must be much smaller to insure complete inactivation; (b) k_c must be close to the fastest activation time constants at negative potentials (Figure 5.4C), while k_o must be large (>1 ms^{-1}) to be compatible with the short bursts of opening in single-channel recordings;[38] (c) the sum $k_r + k_{-r}$ determines the rate of recovery from inactivation at negative membrane potentials; (d) the values of k_a and k_d were adjusted to obtain the best fit to activation/inactivation voltage-clamp recordings using a thermodynamic template with a linear dependence of the free energy on voltage:

$$k = k_0 \exp(qFV / RT), \tag{5.48}$$

where q = 3.035 is the net charge of a gating particle. As this scheme is cyclic, microscopic reversibility imposes that the clockwise product of rate constants equals the anticlockwise product, which in turn imposes that the voltage dependence of k_f and k_b must be the same as that of k_a and k_d. The optimal values of the rate constants were (all units are ms^{-1}):

$$\begin{aligned}
k_a &= 6.4 \exp[qF(V-s)/RT] \\
k_d &= 0.000502 \exp[-qF(V-s)/RT] \\
k_f &= 16 \exp[qF(V-s)/RT] \\
k_b &= 2 \times 10^{-6} \exp[-qF(V-s)/RT] \\
k_o &= 3 \\
k_c &= 0.7 \\
k_i &= 0.036 \\
k_{-i} &= 8 \times 10^{-5} \\
k_r &= 0.001 \\
k_{-r} &= 0.003.
\end{aligned} \tag{5.49}$$

Here, the parameters were adapted to recordings of the T-current in thalamic neurons. An additional parameter, $s = -5$ mV, was introduced to shift the voltage dependence to adjust the model to the thalamic T-current.

Simulation of this model was performed with the above expressions for rate constants, and the T-current was described by the following equation:

$$I_T = \bar{P}_{Ca}[O]G(V, Ca_o, Ca_i) \tag{5.50}$$

where $[O]$ is the fraction of channels in the open state. Simulated voltage-clamp experiments (Figure 5.6D) show that the Chen and Hess model reproduced well the activation characteristics of the T-current (Figure 5.6D1) as well as its slow recovery from inactivation (Figure 5.6D2). However, this model did not fit quantitatively the T-current of thalamic neurons because it was based on single-channel recordings of the T-current in fibroblasts, which is different than the "neuronal" T-current. For analysis, see Reference 38. Obtaining a better Markov representation of the thalamic T-current would require constraint of the model by single-channel recordings.

5.4.6 Comparison of the Different Models

The different models for the T-current reviewed above were compared in current-clamp. A single compartment model of the TC cell was generated, using the same parameters as in Reference 39, and containing leak currents and the T-current according to the following equation:

$$C_m \frac{dV}{dt} = -g_L(V - E_L) - I_T, \tag{5.51}$$

where $C_m = 0.88$ μF/cm² is the membrane capacitance, $g_L = 0.038$ mS/cm² and $E_L = -77$ mV are the leak conductance and its reversal potential, and I_T is the T-current as given by Equation 5.37. These parameters were obtained by matching the model to thalamic neurons recorded *in vitro*.[39]

Using this model, the genesis of low-threshold spike (LTS) was monitored through return to rest after injecting hyperpolarizing currents. The empirical Hodgkin–Huxley type model of the T-current generated LTS in a grossly all-or-none fashion (Figure 5.7A). The linear thermodynamic model (Figure 5.7B) did not generate LTS, consistent with the very small amplitude of the current evidenced above (Figure 5.6B). On the other hand, the nonlinear thermodynamic model (Figure 5.7C) and the Markov model of the T-current (Figure 5.7D) presented a behavior more consistent with the Hodgkin–Huxley type model. The peak amplitude of the LTS was compared between different models in Figure 5.7E. Although the shape of the LTS was not identical, Hodgkin–Huxley and nonlinear thermodynamic models produced remarkably similar peak amplitudes (filled circles and triangles in Figure 5.7E). Therefore, we conclude that nonlinear thermodynamic models provide fits of comparable quality to empirical Hodgkin–Huxley models, but their form is physically more plausible.

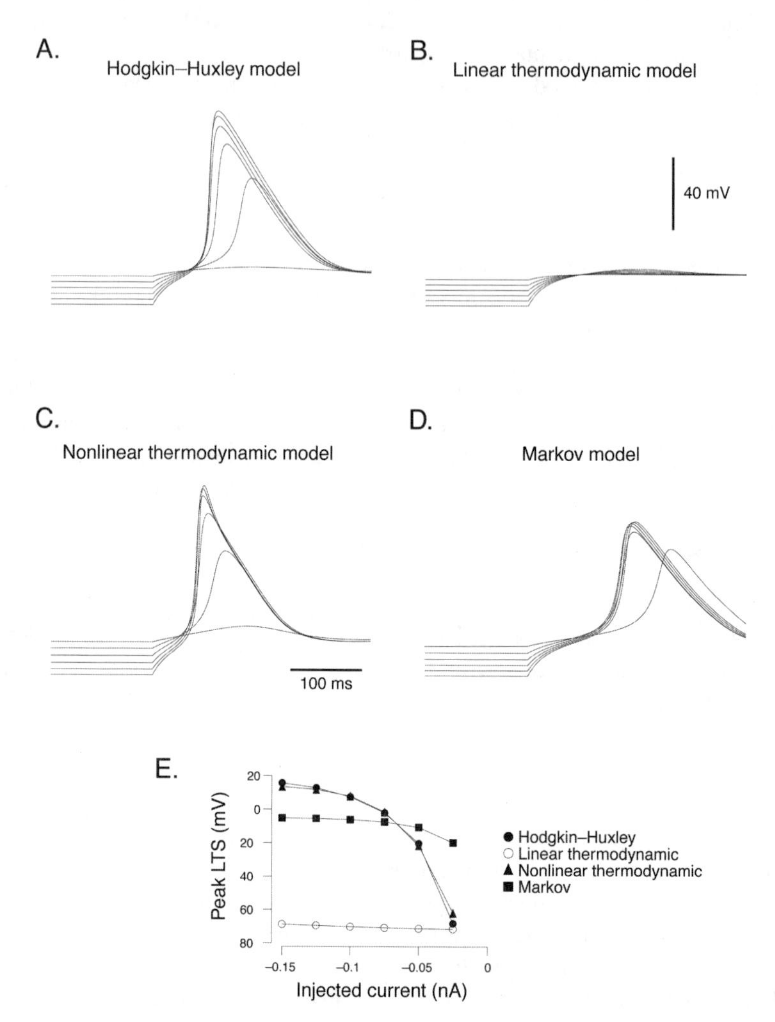

FIGURE 5.7 Low-threshold spikes generated by different models of the T-current. Comparison of the same current-clamp simulation for four different models of the T-current: an empirical Hodgkin–Huxley type model (A), a linear thermodynamic model (B), a nonlinear thermodynamic model (C), and a Markov model (D). The simulation involved injecting hyperpolarizing current pulses of various amplitudes (–0.025, –0.05, –0.075, –0.1, –0.125, and –0.15 nA) and of 1 sec duration. At the end of the pulse, the model generated a low-threshold spike upon return to rest. (E) Peak amplitude of low-threshold spikes (LTS) generated by the different models of the T-current. All simulations were done with the same single-compartment geometry, which contained leak currents in addition to the T-current (identical parameters as in Reference 39). The density of T-channels was identical in all cases ($\bar{P}_{Ca} = 5 \times 10^{-5}$ cm/s) and was in the range of densities estimated from rat ventrobasal thalamic neurons.[39] (Modified from Reference 15.)

5.5 CONCLUSION

In this chapter, we have compared different representations for modeling voltage-dependent currents and delineated some of the differences between these representations. In the case of sodium channels, models of increasing complexity, from simplified two-state representations to multistate Markov diagrams, can capture some of the features of sodium channels and of action potentials. Which model to choose depends on the type of experimental data available and its level of precision. It is clear that a two-state scheme cannot capture the features of single-channel recordings, which require Markov models of sufficient complexity to account for the data. On the other hand, we showed that even simplified two- or three-state representations can capture phenomena such as action potentials.[6] If the principal requirement is to generate action potentials, it is therefore not necessary to include all the complexity of the most sophisticated Markov diagrams of channels, and simplified representations appear sufficient. This simplistic approach may be adequate for models involving large-scale networks of thousands of cells, for which computational efficiency is a more important concern than reproducing all the microscopic features of the channels.

In the case of the T-current, we have shown that various formalisms, such as empirical Hodgkin-Huxley type models, thermodynamic models, and Markov models, can capture the behavior of the T-current in voltage-clamp and generate low-threshold spikes. In this case, Markov models are probably more accurate because they also account for single-channel recordings, while Hodgkin–Huxley type models do not. The voltage-clamp data shown here were obtained in thalamic neurons[33] and, for the particular case of these data, they were best modeled by a Hodgkin–Huxley type model in which rate constants were fit to experimental data using empirical functions of voltage. The best physically plausible approach to capture these data is to use templates taken from nonlinear thermodynamic models, which provide a fitting of comparable quality to empirical functions (compare blue and red curves in Color Figure 5.1). Therefore, we conclude that nonlinear thermodynamic models should be used to yield representations that are consistent with experimental data while having a plausible biophysical interpretation.

REFERENCES

1. Hodgkin, A. L. and Huxley, A. F., A quantitative description of membrane current and its application to conduction and excitation in nerve, *J. Physiol.*, 117, 500, 1952.
2. Hille, B., *Ionic Channels of Excitable Membranes*, Sinauer Associates Inc., Sunderland, MA, 1992.
3. Armstrong, C. M. and Hille, B., Voltage-gated ion channels and electrical excitability, *Neuron*, 20, 371, 1998.
4. Llinás, R. R., The intrinsic electrophysiological properties of mammalian neurons: a new insight into CNS function, *Science*, 242, 1654, 1988.
5. Sakmann, B. and Neher, E., Eds., *Single-Channel Recording*, 2nd ed., Plenum Press, New York, 1995.

6. Destexhe, A., Mainen, Z. F., and Sejnowski, T. J., Synthesis of models for excitable membranes, synaptic transmission and neuromodulation using a common kinetic formalism, *J. Compu. Neurosci.*, 1, 195, 1994.
7. Tsien, R. W. and Noble, D., A transition state theory approach to the kinetics of conductances in excitable membranes, *J. Membr. Biol.*, 1, 248, 1969.
8. Hill, T. L. and Chen, Y. D., On the theory of ion transport across nerve membranes. VI. Free energy and activation free energies of conformational change, *Proc. Natl. Acad. Sci. U.S.A.*, 69, 1723, 1972.
9. Stevens, C. F., Interactions between intrinsic membrane protein and electric field, *Biophys. J.*, 22, 295, 1978.
10. Eyring, H., The activated complex in chemical reactions, *J. Chem. Phys.*, 3, 107, 1935.
11. Johnson, F. H., Eyring, H., and Stover, B. J., *The Theory of Rate Processes in Biology and Medicine*, John Wiley & Sons, New York, 1974.
12. Andersen, O. and Koeppe, R. E., II, Molecular determinants of channel function, *Physiol. Rev.*, 72, S89, 1992.
13. Borg-Graham, L. J., Modeling the nonlinear conductances of excitable membranes, in *Cellular and Molecular Neurobiology: A Practical Approach*, Wheal, H. and Chad, J., Eds., Oxford University Press, New York, 1991, p. 247.
14. Willms, A. R., Baro, D. J., Harris-Warrick, R. M., and Guckenheimer, J., An improved parameter estimation method for Hodgkin-Huxley models, *J. Comput. Neurosci.*, 6, 145, 1999.
15. Destexhe, A. and Huguenard, J. R., Nonlinear thermodynamic models of voltage-dependent currents, *J. Compu. Neurosci.*, in press.
16. Armstrong, C. M., Sodium channels and gating currents, *Physiol. Rev.*, 62, 644, 1981.
17. Aldrich, R. W., Corey, D. P., and Stevens, C. F., A reinterpretation of mammalian sodium channel gating based on single channel recording, *Nature*, 306, 436, 1983.
18. Bezanilla, F., Gating of sodium and potassium channels, *J. Membr. Biol.*, 88, 97, 1985.
19. Aldrich, R. W., Inactivation of voltage-gated delayed potassium currents in molluscan neurons, *Biophys. J.*, 36, 519, 1981.
20. Marom, S. and Abbott, L. F., Modeling state-dependent inactivation of membrane currents, *Biophys. J.*, 67, 515, 1994.
21. Colquhoun, D. and Hawkes, A. G., On the stochastic properties of single ion channels, *Proc. Roy. Soc. Lond. Ser. B*, 211, 205, 1981.
22. Johnston, D. and Wu, S. M., *Foundations of Cellular Neurophysiology*, MIT Press, Cambridge, 1995.
23. Fitzhugh, R., A kinetic model of the conductance changes in nerve membrane, *J. Cell. Comp. Physiol.*, 66, 111, 1965.
24. Armstrong, C. M., Inactivation of the potassium conductance and related phenomena caused by quaternary ammonium ion injection in squid axons, *J. Gen. Physiol.*, 54, 553, 1969.
25. Vandenberg, C. A. and Bezanilla, F., A model of sodium channel gating based on single channel, macroscopic ionic, and gating currents in the squid giant axon, *Biophys. J.*, 60, 1511, 1991.
26. Perozo, E. and Bezanilla, F., Phosphorylation affects voltage gating of the delayed rectified K^+ channel by electrostatic interactions, *Neuron*, 5, 685, 1990.
27. Bush, P. and Sejnowski, T. J., Simulations of a reconstructed cerebellar Purkinje cell based on simplified channel kinetics, *Neural Comput.*, 3, 321, 1991.
28. Horn, R. J., Patlak, J., and Stevens, C. F., Sodium channels need not open before they inactivate, *Nature*, 291, 426, 1981.

29. Labarca, P., Rice, J. A., Fredkin, D. R., and Montal, M., Kinetic analysis of channel gating. Application to the cholinergic receptor channel and the chloride channel from *Torpedo Californica, Biophys. J.*, 47, 469, 1985.
30. Yamada, W. N., Koch, C., and Adams, P. R., Multiple channels and calcium dynamics, in *Methods in Neuronal Modeling*, Koch, C. and Segev, I., Eds., MIT Press, Cambridge, 1989, p. 97.
31. Destexhe, A., Mainen, Z. F., and Sejnowski, T. J., Kinetic models of synaptic transmission, in *Methods in Neuronal Modeling*, 2nd ed., Koch, C. and Segev, I., Eds., MIT Press, Cambridge, 1998, p. 1.
32. Jahnsen, H. and Llinás, R. R., Ionic basis for the electroresponsiveness and oscillatory properties of guinea-pig thalamic neurons *in vitro*, *J. Physiol.*, 349, 227, 1984.
33. Huguenard, J. R. and Prince, D. A., A novel T-type current underlies prolonged calcium-dependent burst firing in GABAergic neurons of rat thalamic reticular nucleus, *J. Neurosci.*, 12, 3804, 1992.
34. Huguenard, J. R. and McCormick, D. A., Simulation of the currents involved in rhythmic oscillations in thalamic relay neurons, *J. Neurophysiol.*, 68, 1373, 1992.
35. De Schutter, E. and Smolen, P., Calcium dynamics in large neuronal models, in *Methods in Neuronal Modeling*, 2nd ed., Koch, C. and Segev, I., Eds., MIT Press, Cambridge, 1998, p. 211.
36. Coulter, D. A., Huguenard, J. R., and Prince, D. A., Calcium currents in rat thalamocortical relay neurons: kinetic properties of the transient, low-threshold current, *J. Physiol.*, 414, 587, 1989.
37. McCormick, D. A. and Huguenard, J. R., A model of the electrophysiological properties of thalamocortical relay neurons, *J. Neurophysiol.*, 68, 1384, 1992.
38. Chen, C. and Hess, P., Mechanisms of gating of T-type calcium channels, *J. Gen. Physiol.*, 96, 603, 1990.
39. Destexhe, A., Neubig, M., Ulrich, D., and Huguenard, J. R., Dendritic low-threshold calcium currents in thalamic relay cells, *J. Neurosci.*, 18, 3574, 1998.

6 Accurate Reconstruction of Neuronal Morphology

Dieter Jaeger

CONTENTS

0-8493-2068-2/01/$0.00+$.50

6.1 INTRODUCTION

To achieve realistic simulations of biological neurons anatomically correct morphological reconstructions of the desired cell type need to be obtained first. If significant errors are made in the reconstruction, the resulting passive and active properties of the simulated neuron could substantially deviate from the real cell. For example, if the diameter of a thin dendrite was traced at 0.5 μm but the actual diameter was 0.8 μm, an error of 60% for the surface area and of 156% for the cross-sectional area would result. Of course, the surface area is proportional to the cell capacitance and total membrane conductance, and the cross-sectional area to the axial conductance, which are important parameters describing the passive structure of neural models (Chapter 8). Errors of similar magnitude easily result from omitting the surface area of dendritic spines. Other major sources of error in reconstructions include the degradation of neurons during recording, shrinkage artifacts during histological processing, the omission of faintly stained distal processes from the reconstruction, and just the limitations in the resolution of light microscopy if this technique is used without further validation. This chapter is aimed at providing a step by step description of the techniques that can be used to reduce such problems and to obtain optimal morphological reconstructions for modeling.

6.2 OVERVIEW OF TECHNIQUES

The process of obtaining a reconstructed neuron for modeling typically begins with the injection of a dye during intracellular recording. Alternatives are juxtapositional staining during extracellular recordings *in vivo*, or anatomical stains that do not require recording, such as the Golgi method. Staining during intracellular recording is preferable because the obtained membrane potential traces are very useful as templates for tuning the model later on (see Chapters 8 and 9). The stained neuron is most commonly visualized with a histological procedure following fixation of the tissue. Fluorescent dyes, however, can be used to visualize neurons before fixation or histological processing take place. Although fluorescent stains are generally not ideal to obtain a detailed reconstruction of the neuron, they can be useful to control for shrinkage artifacts occurring later on. The most widely used technique for visualizing neurons is the injection of biocytin or neurobiotin followed by coupling to avidin-HRP and some variant of a peroxidase reaction with a chromogen to result in a darkly stained neuron. The darkly stained neuron can then be traced under a microscope with a motorized stage using reconstruction software such as Neurolucida (MicroBrightField). Last, the morphology files resulting from this process need to be converted into a format that compartmental simulation software can read.

6.3 TECHNIQUES TO RECONSTRUCT NEURONS FOR SIMULATION

6.3.1 FILLING AND STAINING NEURONS WITH BIOCYTIN IN BRAIN SLICES

6.3.1.1 Slice Preparation

The slice preparation method and composition of the slice medium have to be optimized differently for different brain structures. Individual methods to optimize slice quality can not be covered within this chapter, and the literature for the respective brain structure should be consulted. To reconstruct neurons with the biocytin method it is useful to begin the slicing procedure by deeply anesthetizing the animal and then perfusing through the heart with ice-cold slice medium. This washes out erythrocytes from brain capillaries. Erythrocytes contain endogenous peroxidase activity and are therefore darkly stained during the histological protocol described below. Although a fast perfusion (<1 min after breathing stops) is often advantageous for the physiological quality of slices as well, there may be circumstances where a perfusion is not possible. In these cases, erythrocyte staining that obscures the visibility of stained neurons can be reduced by an additional step during histological processing (see below).

6.3.1.2 Injection of Biocytin

When sharp intracellular electrodes are used, the pipette solution should contain 2–4% biocytin (Sigma) by weight in 1M potassium acetate. This is about the limit in the amount of biocytin that can be dissolved in 1M potassium acetate, and slight warming may be required to get the substance fully dissolved. To fill a neuron, biocytin needs to be ejected with current pulses at the end of the recording. Positive current pulses of 200 ms duration and an amplitude of 1–4 nA applied for 5–10 minutes at 1–3 Hz are usually suitable for filling.[1,2]

With whole cell recording a concentration of 0.5–1% biocytin in the pipette is sufficient and ejection with current is not required. Bulk fluid exchange during recording usually results in a good fill within 5 minutes. Extended recording periods (>45 min) are not recommended when an optimal morphological reconstruction is desired, because the neuron tends to degrade slowly due to dialysis in the whole cell configuration. The soma may swell, and dendrites may attain a beady or puffy look after prolonged recordings. Similar symptoms can also be due to cell degradation close to the surface of the slice and to progressive slice degradation with time after slicing. The look of dendrites with such poor quality is displayed in Figure 6.1. For optimal histological reconstructions, recordings of moderate duration early in the experiment and of cells deep in the slice are likely to give the best results. A recording of 15 minutes is usually fully sufficient to give a complete dendritic fill and to stain local axon collaterals if present.

Neurobiotin (Vector Labs) can be used instead of biocytin, and is reputed to have some advantageous properties such as being more soluble and to iontophorese more easily. Background staining that sometimes occurs due to ejection of dye during

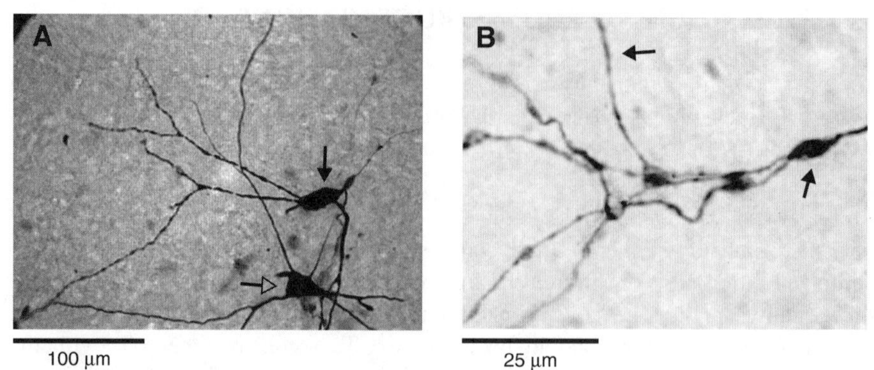

FIGURE 6.1 (A) A pair of neurons was recorded simultaneously from globus pallidus (GP) in a slice from a 16-day-old rat. Both cells were filled with biocytin during whole cell recording and processed with the Vector ABC method. The morphological quality of one neuron (open arrow) was much better than that of the second one (solid arrow), although both neurons were recorded for the same duration with an identical technique. The most likely explanation for the difference in this case is that the more degraded neuron was somewhat more damaged during the slicing procedure and had started to degrade before recording. The electrophysiological quality of this cell was also poor. (B) The dendrite of another GP neuron of poor quality is shown at increased magnification. The arrows point to typical signs of degradation, such as beads and patchy staining.

the approach of a cell with positive pressure (whole cell recording) may also be reduced with Neurobiotin. The possible disadvantages of biocytin are rarely significant in the experience of the author, however, and the higher expense of purchasing Neurobiotin may not be justified. In addition, a report that Neurobiotin may alter electrophysiological properties of neurons has been published.[3]

6.3.1.3 Fixation of Slices

When only a light-microscopic reconstruction is desired, the slice can be fixed in 4% paraformaldehyde or in a 10% formaldehyde solution. For convenience, prepared stabilized and buffered fixatives can be purchased, for example from Sigma (HT50-1). When a verification of dendritic diameters with electron microscopy is desired, a suitable EM fixative such as 2.5% paraformaldehyde, 1.25% glutaraldehyde, and 15% picric acid in buffer needs to be used.[4] Other authors use less picric acid and more aldehyde fixative.[5,6] For good EM quality it is also essential to avoid prolonged incubation of slices before fixation or extended periods of dialysis during whole cell recording. To obtain good EM ultrastructure from cells recorded in the slice remains a challenge in the best of cases.

6.3.1.4 Histological Processing of Slices

Slices may be left in the fixative for a few hours up to several weeks. Since the visualization of biocytin depends on binding with avidin inside the filled cell, access of the avidin to the cell needs to be ensured. One method is to resection the slice before histological processing.[2,6] A vibratome is suitable to resection slices that have been embedded in agar. To do so, it is important to put the slice flat on a glass slide before embedding to be able to obtain good sections from the entire slice. Overall, however, the author does not recommend resectioning slices, because as a result parts of the stained neuron may be lost and the cell likely will extend through several sections. This may cause serious problems when tracing the morphology of the neuron due to differential shrinkage and deformation of individual sections. In addition, keeping track of many small branches across sections can be challenging. An alternative to resectioning is permeabilizing membranes throughout the entire slice. This can be done by including 0.1% Triton-X-100 (Sigma #T-9284) in the incubation step of histological processing. The full sequence of steps of histological processing to stain neurons using the Vector ABC kit is listed in Box 6.1. Slices can be treated individually in small vials that need 1–2 ml of solution for each step. The Vector peroxidase substrate (Vector SK-4700) may be replaced by a standard diaminobenzidine reaction with Ni^{++} intensification. The author prefers the ease of use and low toxicity of the Vector product.

Box 6.1 Histological Processing of Slices to Stain Neurons Filled with Biocytin

1. Rinse fixed slice three times in 0.01 M potassium phosphate buffered saline (KPBS, pH 7.4).
2. ONLY IF ANIMAL WAS NOT PERFUSED: Incubate in 1% H_2O_2 + 10% Methanol + 2% albumin in 0.1 M KPBS for 30 min at room temperature while gently shaking to reduce endogenous peroxidase activity. Then rinse three times in KPBS, as above.
3. Incubate slice in 0.1 M KPBS with 2% bovine albumin (Sigma A-2153) and Vector ABC solutions (Vector Pk-4000) overnight in refrigerator while gently shaking. If slices need to be permeabilized, add 0.1% Triton-X100 (Sigma T 9284). Note that the ABC kit A and B solutions need at least 30 min to form a complex after mixing before they can be used.
4. Rinse *thoroughly* five times with KPBS.
5. Use Vector substrate kit (SK-4700) or a standard diaminobenzidene (DAB) protocol for the final peroxidase reaction. If DAB is used, prepare a solution of 0.06 % NiCl, 0.05% DAB, and 0.06% H_2O_2 for immediate use. Staining takes 10–15 minutes. It should be discontinued earlier if the non-specific background stain becomes very dark.

6.3.1.5 Mounting and Clearing of Thick Slices

A major problem in obtaining good morphological reconstructions of neurons is tissue shrinkage. Some shrinkage is unavoidable during fixation, and correction factors may need to be determined for this reason (see below). Additional severe shrinkage as well as tissue deformation occurs when thick slices are dried on microscopic slides and dehydrated with increasing alcohol concentrations before clearing with xylene (Figures 6.2 and 6.3). Although this procedure results in very clear slices and thus highly visible neurons, the accompanying tissue deformation and shrinkage is usually unacceptable. Two alternative protocols can be used. The first protocol consists of clearing floating slices in increasing concentrations of glycerol.[7,8] A one hour immersion in 50% glycerol (Sigma) in buffered water followed by one hour immersion in 90% glycerol is sufficient. Slices are then mounted and coverslipped in 100% glycerol. This method results in reasonable visibility and little additional shrinkage. Stained neurons, however, tend to fade over the next several weeks. The fading can be largely eliminated by using Immu–Mount (Shandon, Inc.) instead of glycerol. Evaporation of the mounting medium over time can remain a problem even when the slides are sealed with nail-polish. The second method makes use of osmium to harden slices before they are dehydrated. While this method uses highly toxic osmium, and care needs to be taken to obtain flat slices, it can result in stable and clearly visible stained neurons with minimal tissue shrinkage.[9] A commercially available 4% osmium tetroxide solution (Ted Pella) is diluted to 0.02–0.2% in phosphate buffer and slices are incubated for 5 minutes to 2 hours depending on slice thickness and the amount of desired tissue hardening. The more osmium is used, the darker the tissue will get, which may ultimately obscure stained neurons. After hardening during osmification, slices can be dehydrated with increasing alcohol concentrations, cleared with xylene, and coverslipped.

6.3.2 Filling and Staining Neurons with Biocytin *In Vivo*

6.3.2.1 *In Vivo* Intracellular Recordings

Such recordings are typically performed with sharp electrodes, and the same dye-ejection technique as described above for slice recordings with these electrodes can be used. A key to successful intracellular recording *in vivo* is the reduction of brain pulsations, for example by pouring wax over the exposed brain surface after inserting the electrode.[10] The wax is prepared by mixing paraffin (Electron Microscopy Sciences 19280) with light mineral oil (Fisher 0121) in a waterbath until the mixture has a melting point just above body temperature. The wax is kept melted at this temperature in a waterbath during the experiment. A syringe w/o needle can be used to pour the liquid around inserted electrodes, where it quickly hardens.

At the end of the experiment, the animal is typically perfused with buffered saline followed by fixative. The brain should be cut with sections as thick as possible (e.g., 100 μm) to prevent the need for tracing neurons across many sections. Subsequent tissue processing is identical to that of brain slices.

In vivo recordings are common in invertebrate preparations. The dye injection and histological techniques described here are generally applicable to invertebrate preparations as well. Brain sections are usually replaced by whole-mounts of a ganglion etc., and the tissue can be treated following the guidelines above.

6.3.2.2 Juxtapositional Staining

Extracellular recordings can be used to fill neurons *in vivo* with the technique of juxtapositional staining.[11] Recordings are obtained with a glass capillary pulled for intracellular recording, but broken to a tip diameter of 1.5–3.0 μm. Breaking can be

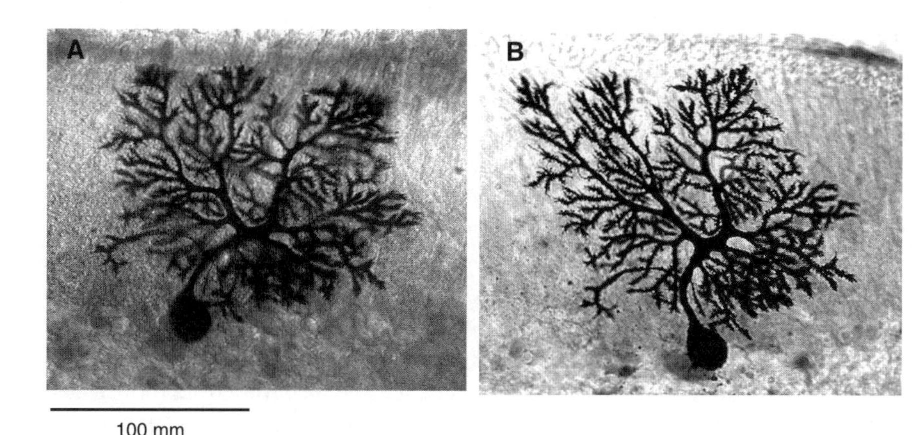

FIGURE 6.2 (A) Picture of a cerebellar cortical Purkinje cell recorded from a slice of a 14-day-old rat and filled with biocytin. This image was taken after fixation and staining, but while the slice was still in buffer solution. (B) Picture of the same cell after drying on a glass slide, dehydration, and coverslipping. It is clear that the shape of the cell has become distorted in a non-uniform way. The upper left dendrite in particular looks stretched.

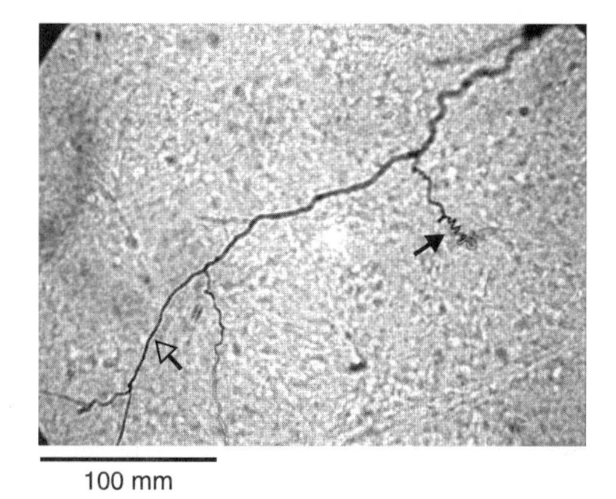

FIGURE 6.3 Picture of a globus pallidus dendrite. This cell was lightly air-dried on a glass slide, and then dehydrated using alcohol and xylene before coverslipping. The open arrow points at a straight segment of dendrite coursing in the xy plane. The solid arrow points at a segment of dendrite that is ascending in the z-axis. This segment is curled up due to the large shrinkage of the entire slice in the z-axis. Curled dendrites frequently are common when severe shrinkage of slices occurs, indicating that individual processes do not shrink at the same rate as the slice in which they are embedded.

done by mounting the pulled capillary on a micromanipulator and then advancing the tip into a vertical rough glass surface at an angle of approximately 60 degrees. The process is best done under the microscope using a 20× or 40× objective. This electrode can be filled with 0.5 M or 1.0 M NaCl and 3% biocytin. Extracellular recordings need to be obtained in close proximity to the cells, which can be confirmed by a large spike size. Following recording, biocytin is ejected with 5–20 nA positive current pulses. The action potential firing of neurons should accelerate during the applied pulses to indicate close proximity of the electrode tip to the neuron. In a not completely understood fashion the recorded neuron takes up some of the ejected dye and nicely stained cells may be obtained.[12] The histological processing is identical to the case of intracellularly injected neurons.

6.3.3 Other Methods to Obtain Neuron Morphologies

6.3.3.1 Photoconversion of Fluorescent Dyes

In some studies neurons may be filled with fluorescent dyes for particular reasons. For example, the dye may be used for imaging or for localizing dendrites during recording. Fluorescent dyes can subsequently be photoconverted to form a dark reaction product using DAB.[13] This technique is unlikely to achieve the quality of direct biocytin fills, however. In most instances it seems more advisable to inject biocytin along with the fluorescent dye for later use of the ABC protocol to obtain a dark stain (see above). In some cases injected biocytin may be conjugated to a fluorescent label (e.g., Cy3-conjugated avidin) to allow confocal imaging. If subsequently processed with the ABC protocol, a dark stain can still be obtained.[14]

6.3.3.2 Golgi Method

Clearly one of the most successful anatomical techniques ever invented, the Golgi method allows the intense full staining of individual neurons in fixed tissue. Most commonly, the rapid Golgi technique that was already known to Cajal is used. Tissue is exposed first to potassium dichromate and then to silver nitrate. For a recent reference to a detailed protocol of the rapid Golgi procedure see Reference 15. A black deposit fills a small number of neurons fully. Why particular neurons are filled while most others are not remains unknown even after 100 years of use of this technique. A large disadvantage of the Golgi method for neuronal modeling is that physiological data for the stained neurons are not available. Using this technique for modeling is thus not generally recommended. It may be the best technique in special cases, however, like generating models of neurons in human tissue. In addition, a larger sample of cells than could be obtained with intracellular recording might be desirable for the study of morphological variability.

6.3.3.3 Filling Individual Neurons in Fixed Tissue

Individual neurons in already fixed tissue can still be filled with dyes. Both fluorescent dyes such as Lucifer Yellow,[16] and dyes for a later peroxidase reaction like HRP or biocytin[17] can be used. To monitor the success of injection, the presence of a fluorescent dye is desirable. A conjugate of biocytin and tetramethylrhodamine (Molecular Probes) appears particularly useful.[18] In distinction to the Golgi technique, one can aim for particular types of neurons with this technique, or stain the detailed morphology of neurons that have previously taken up a retrograde tracer. As with the Golgi technique, however, electrophysiological data are missing, which makes the obtained morphologies less than ideal for compartmental modeling.

6.3.4 Cross-Validation of Shrinkage Artifacts and Dendritic Diameter

6.3.4.1 Rationale

From the time the brain is removed until coverslipping, neurons may undergo significant changes in morphological structure (Figure 6.2). Even just the event of slicing a brain and subsequent incubation may change the morphology significantly by triggering the growth of a sizable proportion of dendritic spines within hours.[19] Thus, it may be difficult to obtain cell morphologies that fully preserve the *in vivo* situation from slice tissue. A comparison with cells of the same type obtained from perfused brains can help judging the presence of slice artifacts such as swelling or spine growth. Both in slices and in perfused brains, neuron morphologies will get distorted to a varying degree during fixation and subsequent histological processing. Before relying on the obtained morphology, some cross-validation of this process is advisable. Of course, the need for such a validation depends on the accuracy that the modeler desires to achieve with morphological reconstructions.

6.3.4.2 Assessing Fixation Shrinkage

Fixation shrinkage of an entire slice can be assessed by measuring slice size and thickness before and after fixation. Shrinkage factors are estimated from these measurements and applied to the obtained neural reconstructions. These methods assume, however, that shrinkage was uniform throughout the slice and that individual cells shrink at the same rate as the entire slice. This assumption may not hold true in some cases. Shrinkage at the edge of a slice can be different than in the center, leading to a distortion of cells. Also, individual dendrites may curl up rather than shrink (Figure 6.3), and curled dendrites may retain their original length. To assess these possibilities it is desirable to obtain pictures of the filled cell before and after fixation. A fluorescent dye such as Lucifer Yellow or rhodamine (Molecular Probes) can be injected during recording. These dyes diffuse throughout a neuron within minutes. It should be noted that Lucifer Yellow in particular is highly phototoxic and neurons can not be exposed to ultraviolet light during electrophysiological recordings. After recording but before slices are fixed, the neuron can be visualized using the correct filter set for the dye used, and pictures can be taken. Water immersion lenses with long working distances are practical to use. For cells with fine dendrites or cells deep in the slice it may be necessary to obtain images with a confocal microscope. After taking a set of pictures of the cell, the tissue is fixed. A second picture of the cell is then taken. Comparing between the pictures before and after fixation, the amount of fixation related shrinkage and distortion of the filled cell can be determined. If precise measurements in the diameter of dendritic processes are desired, measurements should be calibrated by imaging fluorescent beads of known diameter (Molecular Probes). Following histological processing and coverslipping, a third set of pictures can be taken to assess further shrinkage and deformation.

6.3.4.3 Electron-Microscopic Methods for Detailed Analysis of Thin Processes and Dendritic Spines

The accuracy of EM is much higher, of course, than that even of confocal techniques. The strength of this technique in the process of making accurate anatomical reconstructions for modeling lies in the ability to make good estimates of the shape and diameter of fine dendrites and spines. Spines can not be traced during LM reconstructions and are often added during the modeling process either just as additional membrane surface or as stereotypical appendages of two compartments each (see Chapter 9).[22] EM measurements of spines allow for improved accuracy of the assumed spine characteristics. Since the surface area of spines can contribute more than 50% of the total membrane area of a neuron like in the case of the cerebellar Purkinje cell, a good sample of EM spine measurements may aid greatly in the modeling process. Neurons can be carried to the EM level even after preceding LM analysis.[4,5] Thus, neurons can be traced using LM reconstruction software (see below) and then particular dendritic areas can be singled out for EM analysis. To allow for the use of EM, care has to be taken to preserve tissue ultrastructure as well as possible. The fixative to be used (see above) should be made fresh. Triton needs to be avoided during incubation with avidin. To make sections more permeable, a quick freezing protocol can be used instead. The tissue is cryoprotected (e.g., with 25% sucrose, 10% glycerol in 0.05 M phosphate buffer at pH 7.4) and then rapidly frozen to –80°C and thawed again after 20 min. Slices can be mounted in glycerol for reconstruction under LM, and then washed into phosphate buffer for EM processing. To identify the same dendritic sections with EM that were measured before with LM, a careful alignment strategy is required. Landmarks such as blood vessels or dendritic junctions can be used, and additional marks can be placed by making razor nicks or pinholes in the epoxy-embedded block. The location of these marks can be identified under LM and later used to verify EM positioning.

6.3.4.4 High-Voltage EM Tomography

Three dimensional imaging of small pieces of dendrite and spines can be achieved with EM methods, when the sample is tilted at various angles under the EM microscope.[23,24] Few scientists have direct access to the required equipment, however. The use of this technique may still be possible through the National Center for Microscopy and Imaging Research under Dr. Mark Ellisman in San Diego (*http://www-ncmir.ucsd.edu/us.html*), which does allow access to its superb equipment. Software that will make the use of the Center's instruments possible across the Internet are under development. Beyond pushing the envelope of EM technology, the obtained high-resolution images of small dendrites and spines may be of particular use to modelers concerned with the intracellular spatial and temporal dynamics of calcium and second messengers or with the dimensions of synaptic contacts (see Chapter 4). Modeling detailed intracellular diffusion processes is becoming an active area of research, since these processes have important implications, for ion channel and synaptic dynamics (Chapter 3).

6.3.5 Using the Light Microscope to Trace Stained Neurons

6.3.5.1 Limitations

Light microscopy (LM) is limited by the optical resolution that can be obtained. The fundamental limit is approximately 0.6 * λ / N.A. where λ is the wavelength of the light and N.A. is the numerical aperture of the objective used.[25] For a typical wavelength of 500 nm and a numerical aperture of 1.0 the limit in resolution is thus 0.3 μm. Numerical apertures of up to 1.5 can be achieved with oil-immersion objectives, but the working distances of high N.A. lenses are relatively short, and generally do not allow focusing deep enough to visualize stained cells in thick brain slices. Water immersion lenses with long working distances and a N.A. of close to 1 are sold both by Zeiss and Olympus. A resolution limit of 0.3 μm does not mean that thinner structures are not imaged. Rather, a thin line of 0.1 μm structure will have an apparent diameter of about 0.3 μm due to light diffraction. This effect leads to an apparent increase in diameter of small dendrites that can lead to a significant error in estimates of axial resistance and membrane surface area. When cells with abundant small profiles are reconstructed for the purpose of accurate passive modeling, a calibration of processes with small diameters using EM is therefore recommended. This is particularly relevant when accurate reconstructions of axon collaterals are desired.

Another problem can arise when processes take a vertical course through the slice, especially underneath or above the soma. Such profiles may be missed altogether, or the reconstruction may suffer in accuracy. If neurons have a preferred orientation in the target tissue, a slice orientation perpendicular to this orientation should be considered. Cerebellar cortical Purkinje cells provide an extreme example of this point, since these flat cells with extensive branching in two dimensions only would appear as a single line in a horizontal or frontal slice and can not be reconstructed at this orientation at all.

6.3.5.2 Standard LM Reconstruction Technique Using Neurolucida

The cell reconstruction system Neurolucida is sold by MicroBrightField, Inc. To our knowledge, this system is the only commercially available option for 3-D computerized single cell reconstructions at the present time. The Eutectics system that is known to many investigators has been discontinued. Most major brand microscopes can be fitted with Neurolucida. A motorized xyz stage is required. It is typically acquired together with Neurolucida, since only a limited number of stage types are supported by the software. A total Neurolucida system including this hardware and computer system (without microscope) can be expected to cost around U.S. $26,000. Several labs have also developed their own reconstruction software, and may make it available in certain circumstances. Information about one of these products (TRAKA) developed in Zürich can be found at *http://www.ini.unizh.ch:80/~jcal*. The following description is based on the use of Neurolucida.

To allow tracing thick slices, a microscope objective with a long working distance and a high magnification such as a Zeiss 63x water immersion objective (N.A. 0.9) needs to be installed. If tissue sections are below 100 μm thick, the working distance of a 100x oil immersion objective is likely sufficient, and such an objective with an N.A. of greater than 1.0 will provide superior optical resolution. During the installation of Neurolucida great care has to be taken that the stage movement and optical properties of the objectives to be used are properly calibrated. Large errors can result otherwise. It should be noted in particular, that the vertical movement of the microscope stage by some number of μm does not mean that the specimen is scanned vertically by the same number of μm. Due to the different optical diffraction indexes of glass, oil, and water the movement of the stage can be very different from the distance traversed inside the optical sample. Correction factors need to be determined and set in the software to allow for accurate reconstructions along the z-axis.

To accurately trace the microscopic image of a cell it is required that this image is superimposed with the computer reconstruction. This can be done in two ways with Neurolucida. Either the Lucivid system (MicroBrightField) is used to superimpose the computer drawing onto the microscope image inside the microscope optics, or a camera is mounted on the microscope and the cell image is superimposed with the reconstruction on a computer screen. The latter method is preferred by the author, since looking at a screen is less strenuous than looking through a microscope, and an additional magnification of the image that can be provided by the camera system is useful in judging dendritic diameters.

The actual tracing of a single neuron can take between 30 minutes and several days, depending on the complexity of the neuron. To begin a reconstruction, the soma is traced as a 2-D contour around its circumference. Dendrites and axons are attached to the soma and consist of chains of tubular segments. Branch points can be inserted in the reconstruction and subbranches are traced consecutively. The diameter of each segment is chosen to match the superimposed microscope image. This match can be somewhat subjective, and a reconstruction of the same cell by several people can show instructive differences (Figures 6.4, Table 6.1). Processes with thin diameters are particularly prone to large differences when reconstructed by different people. To optimize this process, a sample of thin diameter processes should be cross-checked with other means (see above) and the personnel performing the reconstructions should be instructed to use the resulting criteria of choosing optimal process diameters. The choice of segment boundaries is another subjective factor in performing reconstructions. Obvious reasons for starting a new segment are a change in process diameter, or a bend in the process. Following a process smoothly in the z-axis requires coordinated use of the z-axis control and the tracing device (mouse or digitization tablet). If the z-axis is only moved when the process becomes quite out of focus a stair-case reconstruction of the process in the z-axis may result. Another problem with the z-axis is the occurrence of a small drift over time. Therefore, when the reconstruction returns to a major branch point several minutes or even hours after it has been placed, a mismatch in the z-axis position of the in-focus image with the previous position stored by the computer may have developed. The current in-focus z-axis position recorded by the computer should be checked against the previous one, and the current z-axis position should be adjusted. If this schedule is not adhered to, an otherwise perfectly good looking reconstruction will suddenly show large vertical excursions at branch points when it is later rotated with other software. While editing such mistakes has become possible with version 3 of Neurolucida, it is still a cumbersome process prone to failure. Because problems may develop over the course of the reconstruction, it is advisable to save backup copies of work in progress.

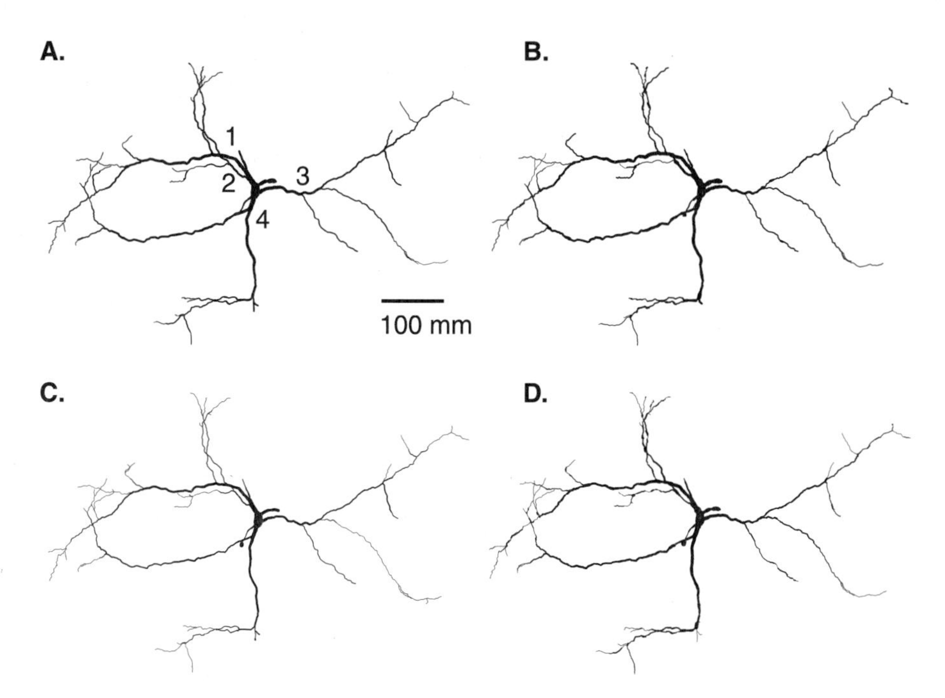

FIGURE 6.4 Four pictures of the same globus pallidus neuron reconstructed by different people. Some differences in dendritic diameters used can be discerned. The resulting large differences in cell properties and especially cell surface area are listed in Table 6.1.

6.3.5.3 Confocal Microscopy for Neurolucida Reconstructions

A confocal module can be purchased with the Neurolucida software that allows reconstructions as described above out of confocal stacks. Confocal microscopy has so far not been a good tool to reconstruct detailed morphologies of entire neurons, however. This is largely due to the fact that at the high resolution required for tracing fine branches, a full confocal representation of a neuron will literally extend over hundreds if not thousands of confocal stacks (John Miller, personal communication). In addition, photobleaching generally limits the amount of time fluorescent neurons

TABLE 6.1
Cell Statistics of Four Reconstructions of the Same GP Neuron

	Rec. A	Rec. B	Rec. C	Rec. D
Length Dendrite 1 (μm)	1160	1226.3	1378.1	1307.1
Surface Area Dendrite 1 $(\mu m)^2$	4829.09	5230.72	4963.62	4344.62
Branch Points Dendrite 1	8	8	11	15
Length Dendrite 2 (μm)	540.6	508.4	481.3	650.6
Surface Area Dendrite 2 $(\mu m)^2$	1899.23	1931.47	1057.34	2321.12
Branch Points Dendrite 2	4	3	5	6
Length Dendrite 3 (μm)	1107.5	1158.4	1133.1	1268.7
Surface Area Dendrite 3 $(\mu m)^2$	3980.27	4204.36	2641.09	4147.13
Branch Points Dendrite 3	7	7	8	9
Length Dendrite 4 (μm)	1249.6	1251.9	1242.7	1385
Surface Area Dendrite 4 $(\mu m)^2$	5352.67	5535.01	3723.33	5007.53
Branch Points Dendrite 4	11	9	11	20

Note: Pictures of the reconstructions are shown in Figure 6.4. The surface area of the cell in particular is quite variable between reconstructions. All people performing these reconstructions had previous experience in the use of Neurolucida. Specific instructions as to how to trace thin processes were not given.

can be visualized. Nevertheless, confocal microscopy can achieve approximately 1.4 times improvement in resolution compared to traditional light-microscopic images.[26] This high resolution can aid in the reconstruction of sample areas of the neuron containing thin profiles.

6.3.6 Statistical Description of Neuronal Reconstructions

When Neurolucida is used for neuronal reconstructions, the companion software Neuroexplorer (MicroBrightField) is available for a statistical description of the traced neuron. The geometry of the neuron can be drawn as a dendrogram, which is a good visual representation of the branching structure of a neuron. Tables of the number, length, surface area, and volume can be constructed for processes of any order, where order signifies the number of branch points between the soma and the respective process. If a sample of neurons has been reconstructed, the distribution of these values can be plotted to examine whether there are distinct anatomical subpopulations of neurons within the sample. A clear bimodal distribution of the mean dendritic length, distance between branch points, or surface area per dendrite likely indicates the presence of two distinct cell types in the recorded sample. Measures of total membrane surface and total dendritic length should be interpreted with care in neurons obtained from brain slices, since truncated or missing dendrites have a large impact on these measures.

6.3.7 Conversion of Morphological Description to Computer Simulations

6.3.7.1 File Conversion

The morphology files created by Neurolucida or other reconstruction programs need to be converted to a different file format before they can be used as cell description files for the most popular compartmental simulation packages, GENESIS (*http://www.bbb.caltech.edu/GENESIS*) and NEURON (*http://www.neuron.yale.edu*). The morphology file formats of all of these applications have in common that one line in the file describes the diameter, size, and absolute location of a single cylindrical compartment. Manual editing therefore is feasible to convert between the specific reconstruction and simulation file formats. Except possibly for very small cells manual conversion is not recommended, however. Automated conversion tools can be written in script languages such as Perl. For conversions from Neurolucida v3 file format to compartmental representations used by NEURON or GENESIS a free Java application named CVAPP is available on the CD-ROM. In addition to providing file conversion routines, this application is a general cell editor and display tool (Figure 6.5). These features can for example be used to cut away reconstructed axons, if axonal modeling is not intended. Note, however, that the axon can make an important contribution to the passive properties of small neurons, especially if local branching is abundant. Matching fast time constants of a neural recording to a passive model without axon may yield substantial errors in the resulting choice of passive parameters. An example of a minimal cell in Neurolucida asci file format (cell.asc) and in GENESIS morphology file format (cell.p) after conversion with CVAPP is given in Box 6.2. The resulting GENESIS file demonstrates the use of a hierarchical naming scheme for the branching structure of the cell. If a modeler needs to make a new conversion tool, this structure could be used as a model. Hierarchical naming is useful, for example, to reflect the logical position of each compartment in the dendritic tree of a neuron. In GENESIS this naming scheme allows that particular areas of the dendrite or branch orders are quickly identified in the simulation and endowed with particular properties, such as a unique ion channel distribution.

6.3.7.2 Recompartmentalizing the Electrotonic Structure of the Obtained Morphology

Neural simulations are dependent on the correct electrotonic representation of a neuron, not the exact 3-D shape such as angles at branch points. The segmentation of a neuron as it is drawn during reconstruction under the microscope is very unlikely to be a good electrotonic representation of the cell. Long straight stretches of a dendrite can be drawn as a single segment, while curvy parts may be broken into many very small parts. For optimal simulation performance all compartments should have approximately the same electrotonic length, i.e., $l \,/\, (R_m/R_i)^{1/2}$ should be constant, where l is the actual length of a compartment, and R_m and R_i are the specific

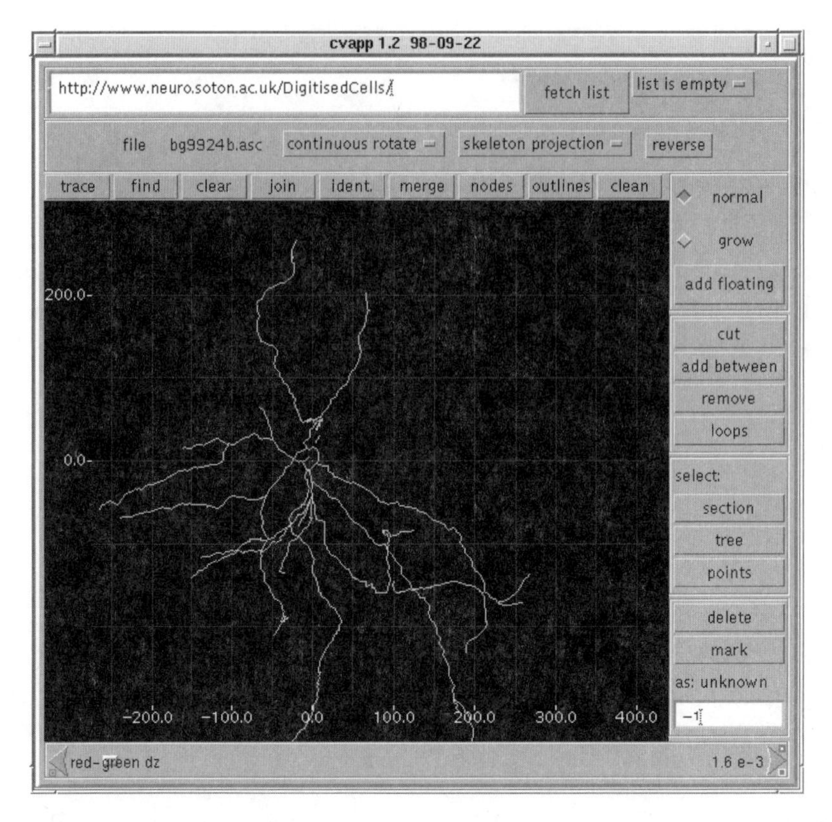

FIGURE 6.5 Image of the user interface provided by the Java cell viewer cvapp. A globus pallidus neuron is displayed in the cell window. Neurons can be rotated in 3-D, and morphology files can be written in different formats.

membrane and axial resistance, respectively. The actual electrotonic length of a compartment should never exceed 0.1 length constants, but much smaller values may be required, especially when the membrane resistance decreases radically due to the opening of active conductances. A value of 0.01 length constants may serve as a good rule of thumb. The conversion utility CVAPP mentioned above contains an algorithm that recompartmentalizes a neuron to a specific desired electrotonic length per compartment.

Box 6.2 Neurolucida and GENESIS Morphology File Formats

A: Neurolucida asci representation of a small cell. Segment endpoints are given in absolute x,y,z coordinates and diameter.

```
; V3 text file written for MicroBrightField products.
(Sections)
("CellBody"
```

(continued)

Box 6.2 (continued)

```
  (Color DarkRed)
  (CellBody)
  (  -0.93     3.57     0.00     0.12)  ;   1, 1
  (   3.14     3.57     0.00     0.12)  ;   1, 2
  (   4.42     1.04     0.00     0.12)  ;   1, 3
  (   4.30    -1.38     0.00     0.12)  ;   1, 4
  (   2.56    -2.42     0.00     0.12)  ;   1, 5
  (  -0.35    -2.30     0.00     0.12)  ;   1, 6
  (  -2.09    -1.84     0.00     0.12)  ;   1, 7
  (  -2.67     0.23     0.00     0.12)  ;   1, 8
  (  -2.33     2.88     0.00     0.12)  ;   1, 9
) ; End of contour
( (Color Yellow) ; [10,1]
  (Dendrite)
  (   0.93     3.69     0.00     1.40)  ;   Root
  (   0.93     6.22     0.00     1.40)  ;   R, 1
  (   1.28     9.45     0.00     1.40)  ;   R, 2
  (   2.09    12.56     0.00     1.40)  ;   R, 3
  (   2.67    15.90     0.00     1.40)  ;   R, 4
  (
  (  10.00    21.77     0.00     1.16)  ;   R-1, 1
  (
  (  20.35    22.12     0.00     0.93)  ;   R-1-1, 1
  Normal
 |
  (  11.39    29.37     0.00     0.70)  ;   R-1-2, 1
   Normal
 ) ; End of split
 |
  (  -5.93    27.42     0.00     1.05)  ;   R-2, 1
   Normal
 ) ; End of split
) ; End of tree
( (Color Blue) ; [10,1]
  (Dendrite)
  (   4.42    -0.12     0.00     1.63)  ;   Root
  (  22.09    -0.35     0.00     1.63)  ;   R, 1
  (
  (  27.91     9.68     0.00     1.16)  ;   R-1, 1
   Normal

  (  35.35    -7.83     0.00     0.93)  ;   R-2, 1
   Normal
) ; End of split
) ; End of tree
```

B: Same cell in genesis morphology file format after conversion with cvapp. The soma contour has been replaced by a single spherical compartment. The coordinate scheme for dendritic compartments is retained.

```
// genesis
//
```

(continued)

Box 6.2 (continued)

```
// Cell morphology file for GENESIS.
// Written by cvapp (http://www.neuro.soton.ac.uk/cells/#software).

*absolute
*asymmetric
*cartesian

// End of cvapp-generated header file.

*origin 0.706 0.756 0

*compt /library/soma
soma none 0.706 0.756 0 6.067
*compt /library/dendrite
p0[1]        soma       1.28     9.45   0   1.4
p0[2]        p0[1]      2.09    12.56   0   1.4
p0[3]        p0[2]      2.67     15.9   0   1.4
p0b1[0]      p0[3]      10      21.77   0   1.16
p0b1b1[0]    p0b1[0]    20.35   22.12   0   0.93
p0b1b2[0]    p0b1[0]    11.39   29.37   0   0.7
p0b2[0]      p0[3]      -5.93   27.42   0   1.05

p1[1]        soma       22.09   -0.35   0   1.63
p1b1[0]      p1[1]      27.91    9.68   0   1.16
p1b2[0]      p1[1]      35.35   -7.83   0   0.93
```

6.3.7.3 Databases of Neuronal Morphologies

It is highly desirable that scientists interested in compartmental simulations should be able to share reconstructed morphologies, given that it requires a lot of effort and time to obtain neural reconstructions of high quality. In this regard the on-line archive of reconstructed hippocampal neurons in Southampton[27] is a promising start. Other efforts of constructing shared neuron databases are under way, but only few labs participate in donating their data. Interesting approaches by several groups can be visited on line (*http://www.nervana.montana.edu/projects/neurosys; http://ycmi.med.yale.edu/SenseLab; http://www.bbb.caltech.edu/hbp*). Several issues likely need to be clarified before such databases become widely popular. One issue is that of quality control, for example by including an assessment of possible errors with morphologies contained in databases. A crucial issue is that of who should finance installing and maintaining such databases. Last not least, a strategy to reward authors for submitting data to common databases needs to be designed. While these issues will likely require general changes in the process of how science operates, the contribution of each investigator towards such goals may provide important stepping stones. To reach the goal of increasingly complex realistic simulations of brain structures a high degree of cooperativity between investigators is certainly in order.

6.4 LIST OF SUPPLIERS

Electron Microscopy Sciences, 321 Morris Rd., Box 251, Fort Washington, PA, 19034, (800) 523-5874, fax: (215) 646-8931, http://www.emsdiasum.com.

Fisher Scientific, Pittsburgh, PA., U.S.A., (800) 766-7000 fax: (800) 926-1166, http://www.fishersci.com.

MicroBrightField, Inc., 74 Hegeman, Ave, Colchester, VT, 05446, U.S.A., (802) 655-9360, fax: (802) 655-5245, http://www.microbrightfield.com.

Molecular Probes, Inc., 4849 Pitchford Ave., Eugene, OR, 97402-9165, U.S.A., (541) 465-8300, fax: (541) 344-6504, http://www.probes.com.

Olympus America, Inc., Two Corporate Center Dr., Melville, NY, 11727-3157, U.S.A., (631) 844-5000, http://www.olympusamerica.com.

Shandon, Inc., 171 Industry Dr., Pittsburgh, PA, 15275, U.S.A., (800) 547-7429.

Sigma-Aldrich, Inc., SIGMA, P.O. Box 14508, St. Louis, MO, 63178, U.S.A., (800) 325-3010, fax: (800) 325-5052, http://www.sigma-aldrich.com.

Ted Pella, Inc., P.O. Box 492477, Redding, CA, 96049-2477, U.S.A., (800) 237-3526, fax: (530) 243-3761, http://www.tedpella.com.

Vector Laboratories, Inc., 30 Ingold Rd., Burlingame, CA, 94010, U.S.A., (650) 697-3600, fax: (650) 697-0339, http://www.vectorlabs.com.

Carl Zeiss, U.S.A., (800) 233-2343, fax: (914) 681-7446, http://www.zeiss.com/micro. Note: Expect shipment delays of several months.

ACKNOWLEDGMENTS

The author wishes to thank Drs. Charles Wilson, John Miller, and Yoland Smith for helpful comments in the preparation of this chapter. Lisa Kreiner and Jesse Hanson have provided valuable technical assistance.

REFERENCES

1. Horikawa, K. and Armstrong, W. E., A versatile means of intracellular labeling: injection of biocytin and its detection with avidin conjugates, *J. Neurosci. Methods*, 25, 1, 1988.
2. Kawaguchi, Y., Wilson, C. J., and Emson, P. C., Intracellular recording of identified neostriatal patch and matrix spiny cells in a slice preparation preserving cortical inputs, *J. Neurophysiol.*, 62, 1052, 1989.
3. Schlösser, B., ten Bruggencate, G., and Sutor, B., The intracellular tracer neurobiotin alters electrophysiological properties of rat neostriatal neurons, *Neurosci. Lett.*, 249, 13, 1998.
4. Tamas, G., Somogyi, P., and Buhl, E. H., Differentially interconnected networks of GABAergic interneurons in the visual cortex of the cat, *J. Neurosci.*, 18, 4255, 1998.
5. Kawaguchi, Y. and Kubota, Y., Physiological and morphological identification of somatostatin- or vasoactive intestinal polypeptide-containing cells among GABAergic cell subtypes in rat frontal cortex, *J. Neurosci.*, 16, 2701, 1996.
6. Ceranik, K., Bender, R., Geiger, J. R. P., Monyer, H., Jonas, P., Frotscher M., and Lübke, J., A novel type of GABAergic interneuron connecting the input and the output regions of the hippocampus, *J. Neurosci.*, 17, 5380, 1997.
7. Thurbon, D., Field, A., and Redman, S., Electronic profiles of interneurons in stratum pyramidale of the CA1 region of rat hippocampus, *J. Neurophysiol.*, 71, 1948, 1994.
8. Chitwood, R. A., Hubbard, A., and Jaffe, D. B., Passive electrotonic properties of rat hippocampal CA3 interneurones, *J. Physiol.* (London), 515, 743, 1999.
9. Kawaguchi, Y., Physiological, morphological, and histochemical characterization of three classes of interneurons in rat neostriatum, *J. Neurosci.*, 13, 4908, 1993.
10. Wilson, C. J., Chang, H. T., and Kitai, S. T., Origins of postsynaptic potentials evoked in identified rat neostriatal neurons by stimulation in substantia nigra, *Exp. Brain Res.*, 45, 157, 1982.

11. Pinault, D., A novel single-cell staining procedure performed in vivo under electrophysiological control: morpho-functional features of juxtacellularly labeled thalamic cells and other central neurons with biocytin or neurobiotin, *J. Neurosci. Methods*, 65, 113, 1996.
12. Pinault, D. and Deschênes, M., Projection and innervation patterns of individual thalamic reticular axons in the thalamus of the adult rat: a three-dimensional, graphic, and morphometric analysis, *J. Comp. Neurol.*, 391, 180, 1998.
13. Maranto, A. R., Neuronal mapping: a photooxidation reaction makes lucifer yellow useful for electron microscopy, *Science*, 217, 953, 1982.
14. Sun, X. J., Tolbert, L. P., Hildebrand, J. G., and Meinertzhagen, I. A., A rapid method for combined laser scanning confocal microscopic and electron microscopic visualization of biocytin or neurobiotin-labeled neurons, *J. Histochem. Cytochem.*, 46, 263, 1998.
15. Sultan, F. and Bower, J. M., Quantitative Golgi study of the rat cerebellar molecular layer interneurons using principal component analysis, *J. Comp. Neurol.*, 393, 353, 1998.
16. Wouterlood, F. G., Jorritsma-Byham, B., and Goede, P. H., Combination of anterograde tracing with Phaseolus vulgaris-leucoagglutinin, retrograde fluorescent tracing and fixed-slice intracellular injection of Lucifer yellow, *J. Neurosci. Method*s, 33, 207, 1990.
17. Coleman, L.-A. and Friedlander, M. J., Intracellular injections of permanent tracers in the fixed slice: a comparison of HRP and biocytin, *J. Neurosci. Methods*, 44, 167, 1992.
18. Liu, W.-L., Behbehani, M. M., and Shipley, M. T., Intracellular filling in fixed brain slices using Miniruby, a fluorescent biocytin compound, *Brain Res.*, 608, 78, 1993.
19. Kirov, S. A., Sorra, K. E., and Harris, K. M., Slices have more synapses than perfusion-fixed hippocampus from both young and mature rats, *J. Neurosci.*, 19, 2876, 1999.
20. Mott, D. D., Turner, D. A., Okazaki, M. M., and Lewis, D. V., Interneurons of the dentate-hilus border of the rat dentate gyrus: morphological and electrophysiological heterogeneity, *J. Neurosci.*, 17, 3990, 1997.
21. Pyapali, G. K., Sik, A., Penttonen, M., Buzsaki, G., and Turner, D. A., Dendritic properties of hippocampal CA1 pyramidal neurons in the rat: intracellular staining in vivo and in vitro, J. Comp. Neurol., 391, 335, 1998.
22. De Schutter, E. and Bower, J. M., An active membrane model of the cerebellar Purkinje cell I. Simulation of current clamp in slice, *J. Neurophysiol.*, 71, 375, 1994.
23. Wilson, C. J., Morphology and synaptic connections of crossed corticostriatal neurons in the rat, *J. Comp. Neurol.*, 263, 567, 1987.
24. Wilson, C. J., Mastronarde, D. N., McEwen, B., and Frank, J., Measurement of neuronal surface area using high-voltage electron microscope tomography, *Neuroimage*, 1, 11, 1992.
25. Perlman, P., *Basic Microscope Techniques*, Chemical Publishing, New York, 1971.
26. van der Voort, H. T. M., Valkenburg, J. A. C., van Spronsen, E. A. et al., Confocal microscopy in comparison with electron and conventional light microscopy, in *Correlative Microscopy in Biology. Instrumentation and Methods*, Hayat, M. A., Ed., Academic Press, Orlando, FL, 1987, 59.
27. Cannon, R. C., Turner, D. A., Pyapali, G. K., and Wheal, H. V., An on-line archive of reconstructed hippocampal neurons, *J. Neurosci. Methods*, 84, 49, 1998.

7 Modeling Dendritic Geometry and the Development of Nerve Connections

Jaap van Pelt, Arjen van Ooyen, and Harry B.M. Uylings

CONTENTS

0-8493-2068-2/01/$0.00+$.50

INTRODUCTION

The two methods described in this chapter focus on the development of neuronal geometry and interneuronal connectivity. The first *model for dendritic geometry* is based on a stochastic description of elongation and branching during neurite outgrowth. This model allows the user to generate random trees by means of (Monte Carlo) computer simulations. Using optimized parameters for particular neuron types, the geometrical properties of these modeled trees can be made to correspond with those of the empirically observed dendrites. The second *model for the development of nerve connections* describes competition for neurotrophic factors. This model is formulated in terms of differential equations, which can be studied analytically using well-known tools for nonlinear system analysis.

7.1 MODELING DENDRITIC GEOMETRY

Interest in the geometry of dendritic branching patterns stems from a variety of reasons. Anatomists are interested in the morphological characterization and differences among neuronal classes as well as in the morphological variations within these classes. Developmental neuroscientists seek to discover the rules of development and the mechanisms by which neurons attain their final morphological appearance. Physiologists are interested in how dendritic morphology is involved in synaptic connectivity within neuronal networks, and in the integration and processing of postsynaptic potentials. Computer scientists are interested in algorithms for generating tree structures. The enormous amount of structural and functional variation with which nature confronts us is a major challenge providing strong motivation to search for "fundamental rules" or minimal parsimonious descriptions of architecture, development, and function. Modeling the geometry of dendritic branching patterns can provide answers to a variety of morphological, physiological, and developmental questions. A variety of approaches have been developed. In these modeling approaches, a distinction can be made between reconstruction and growth models.

Reconstruction models use the abstracted geometrical properties of a set of observed trees, and provide algorithms for randomly generating trees with identical statistical geometrical properties. A typical example is given by the work of Burke et al.,[1] who developed a parsimonious description on the basis of empirically obtained distribution functions for the length and diameters of dendritic branches and for the diameter relations at bifurcation points. Random dendrites are generated by a repeated process of random sampling of these distributions in order to decide whether a traced neurite should branch, and for obtaining the diameters of the daughter branches. The modeled dendrites obtained in this way conform to the original distribution functions of shape characteristics. An important assumption in this approach is that the sampled shape properties are independent from each other.

Hillman[2] emphasized the statistical correlation of segment diameters across branch points and their relation to segment lengths as fundamental parameters of neuronal shape. Tamori[3] postulated a principle of least effective volume in deriving equations for dendritic branch angles. A sophisticated implementation of the reconstruction approaches by Hillman, Tamori, and Burke has recently been developed by Ascoli et al.[4] in L-Neuron, a modeling tool for the efficient generation and parsimonious description of dendritic morphology (*http://www.krasnow.gmu.edu/L-Neuron/index.html*). This modeling tool implements local and recursive stochastic and statistical rules into the formalism of L-systems. Kliemann[5] followed a different approach by considering all the segments at a given centrifugal order as individuals of the same generation, which may or may not give rise to individuals in a subsequent generation (by producing a bifurcation point with two daughter segments). Mathematically, such a dendritic reconstruction can be described by a *Galton–Watson process*, based on empirically obtained splitting probabilities for defining per generation whether a segment will be a terminal or an intermediate one. Applications of this method to dendrites grown *in vitro* can be found in Reference 6.

Growth models, in contrast, aim at revealing rules of neuronal growth in relation to the geometrical properties of the trees emerging from these rules. The outgrowth of neurons proceeds by the dynamic behavior of growth cones—specialized structures at the tips of outgrowing neurites that mediate neuritic elongation and branching.[7] Mature neurons have attained their morphological features as result of this process. Growth models include these processes of elongation and branching. Several implementations have been studied, differing in the level of detail of the mechanisms involved. *Topological growth models* focus on the branching process only and ignore all metrical aspects of dendrites. They have shown how the probability of occurrence of topologically different tree types depends on the rules for outgrowth. These rules include, for instance, (1) growth by sequential or synchronous branching and (2) random selection of segments participating in the branching process, based on uniform or order- and type-dependent branching probabilities. Examples of such topological approaches are in References 8, 9, 10, and 11 for a review. *Metrical growth models* include rules for both branching and elongation of segments. These models allow the study of both metrical and topological properties of the generated dendrites, in relation to the growth assumptions, and have been developed in References 11, 12, 13, and 14.

The studies reviewed above concern *phenomenological approaches* in the sense that both the reconstruction and the growth models are based on simple probabilistic schemes, still aiming at the reproduction of the empirically obtained geometrical characteristics. A point of distinction is that the growth models are based on a developmental process in time and as such are able to include mechanisms that depend on time, and on the growing tree itself, such as its size. Reconstruction models do not include this dimension, and the underlying probabilistic schemes are based on empirical correlation and distribution functions of the shape parameters of particular sets of trees. Growth models should, in principle, be able to describe groups of trees, reconstructed at different time points during development, using the same set of parameters. Reconstruction models, however, may need different sets of optimized parameter values for each age group. Additional to these phenomenological approaches, models have been and are being developed which include more detailed intracellular and local environmental mechanisms and processes in dendritic growth models. Such biophysical and biochemical processes concern, for instance, the polymerization of the intracellular cytoskeleton[15] and neuritic tension and lateral inhibition.[14]

The growth model described in this chapter includes branching and elongation as stochastic processes. The stochasticity assumption is based on the notion that the actual behavior of growth cones, mediating elongation and branching, is subject to so many intracellular and extracellular mechanisms that a probabilistic description is appropriate. The stochasticity assumption, thus, does not imply that the processes involved are stochastic by themselves, but only that their outcome can be described as such. The model has a modular structure, evolved in the course of time by separately studying the branching process (a) with respect to the choices of the segments at which branching events occur, (b) with respect to the time pattern of the branching events, and (c) by finally including the elongation of the segments. Each phase was validated by comparison with specific empirical findings. The modular structure of the model facilitates the optimization of the model parameters, which will be illustrated in this chapter. The model allows the generation of random dendritic trees, and it will be shown how these trees conform in their geometrical properties to empirical observations. This will be illustrated for a set of rat cortical pyramidal cell basal dendrites and for a small set of three cerebellar Purkinje cell dendritic trees from the guinea pig.

7.1.1 Geometry of Dendritic Trees

Typical shape characteristics of dendritic trees are the *number* of terminal tips (degree) or branch points, the *lengths* and *diameters* of the segments between these branch points and tips, and the *connectivity pattern* of the segments (topological structure). A distinction is made between *terminal* segments (ending in tips) and *intermediate* segments (ending in a branch point) (Figure 7.1). Related shape properties are *path lengths* (total length of the path from the dendritic root to a branch point or terminal tip) and the number of segments at a particular centrifugal order. The *centrifugal order* of a segment is equal to the number of branch points on the path from the dendritic root up to the segment, thus indicating the position of the segment in the tree. The embedding of the segments in 3D space and their irregular shapes, although prominent features of dendritic trees, are not dealt with in this chapter. The *topological asymmetry* of a given tree α^n with n terminal segments will be quantified by means of the *tree asymmetry index* A_t defined as

$$A_t(\alpha^n) = \frac{1}{n-1}\sum_{j=1}^{n-1} A_p(r_j, s_j),$$

being the mean value of all the $n-1$ *partition asymmetries* $A_p(r_j, s_j)$ in the tree. Partition asymmetries indicate, at each of the $n-1$ bifurcation points, the relative difference in the number of bifurcation points $r_j - 1$ and $s_j - 1$ in the two subtrees emerging from the jth bifurcation point.[11] The *partition asymmetry* A_p at a bifurcation is defined as $A_p(r, s) = |r - s|/(r + s - 2)$ for $r + s > 2$, with r and s denoting the number of terminal segments in the two subtrees. $A_p(1,1) = 0$ by definition. Note that a binary tree with n terminal segments has $n - 1$ bifurcation points. The elements of a tree are further illustrated in Figure 7.1.

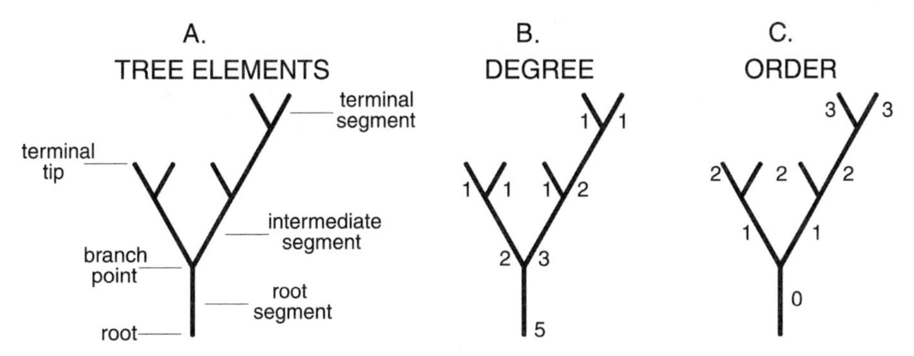

FIGURE 7.1 (A) Elements of a topological tree, with a distinction of branch points, terminal tips and root, intermediate and terminal segments. Segments are labeled according to (B) the number of tips in their subtrees (degree) and (C) their distance from the root (centrifugal order).

7.1.2 Dendritic Growth Model Assumptions

Basic actions in the growth model are elongation and branching of segments, assumed to be independent stochastic processes in time. At each branching event a bifurcation is formed at the tip of a terminal segment, from which two new daughter terminal segments emerge. Elongation is assumed to occur at terminal segments only. The branching probability of a terminal tip is assumed to depend on the momentary number of tips in the growing tree and on its position in the tree. The branching process results in a proliferation of tips, and this process fully determines the final variation among dendritic trees with respect to the number of terminal segments. The branching parameters can be derived from the shape of empirical terminal segment number distributions. The topological structure of a fully grown dendrite is determined by the sequence of particular segments at which branching occurs. The segment lengths are determined both by the elongation rates of the segments and by the elapsed time between successive branching events. Segment length distributions can therefore only be studied once the branching process has been optimized. No developmental rules have been incorporated for the diameter of segments. Rather, these diameters will be assigned to the segments once the skeleton tree has been grown.

7.1.3 Dendritic Growth Model Equations

7.1.3.1 Branching Process: Describing the Variation in the Number of Segments and the Variation in Topological Tree Types

First, we describe the branching process on a series of time bins, without specifying explicitly the duration of each bin. The branching probability of a terminal segment at time bin i is given by

$$p_i = C2^{-S\gamma} B / N n_i^E , \tag{7.1}$$

with N denoting the total number of time bins in the full period of development and n_i denoting the actual number of terminal segments or tips (degree) in the tree at time bin i.[10] The parameter B denotes the expected number of branching events at an isolated segment in the full period, while parameter E determines how strongly the branching probability of a terminal segment depends on the number of tips in the tree. Such a dependency turns out to be essential for reducing the proliferating effect of the increasing number of tips on the total branching rate of the tree.[10] Parameter γ denotes the centrifugal order of the terminal segment while

$$C = n / \sum_{j=1}^{n} 2^{-S\gamma_j}$$

is a normalization constant, with a summation over all n terminal segments. Parameter S determines how strongly the branching probability of a terminal segment depends on the proximal/distal position of the segment in the tree. For $S = 0$, all terminal segments have equal probabilities for branching, a mode of growth called random terminal growth. The frequency distribution of tree types produced by this mode of growth has an expected value for the tree asymmetry index of 0.46 for large trees.[11] For $S \neq 0$, the branching probability of a segment depends on its position in the tree, resulting in more symmetrical trees when $S > 0$ and more asymmetrical trees when $S < 0$. The number of time bins N can be chosen arbitrarily, but such that the branching probability per time bin remains much smaller than unity, making the probability of there being more than one branching event in a given time bin negligibly small.

The distribution of the number of terminal segments in dendritic trees after a period of growth can be calculated by means of the recurrent expression

$$P(n,i) = \sum_{j=0}^{n/2} P(n-j, i-1) \binom{n-j}{j} \left[p(n-j)\right]^j \left[1 - p(n-j)\right]^{n-2j}, \tag{7.2}$$

with $P(n,i)$ denoting the probability of a tree of *degree* n at time bin i with $P(1, 1) = 1$, and $p(n)$ denoting the branching probability per time bin of a terminal segment in a tree of *degree* n, with $p(n) = B/Nn^E$.[10] A tree of *degree* n at time-bin i emerges when j branching events occur at time bin $i - 1$ in a tree of *degree* $n - j$. The recurrent equation expresses the probabilities of all these possible contributions from $j = 0, \ldots, n/2$. The last two terms express the probability that, in a tree of *degree* $n - j$, j terminal segments will branch while the remaining $n - 2j$ terminal segments will not do so. The combinatorial coefficient

$$\binom{n-j}{j}$$

expresses the number of possible ways of selecting j terminal segments from the existing $n - j$ ones.

7.1.3.2 Elongation Process: Describing the Variation in Segment Lengths

Elongation was initially included in the branching model by assigning an elongation rate to growth cones at the time of their birth (i.e., after a branching event) which was randomly chosen from a predefined distribution. This implementation was successful in describing the mean and standard deviation of segment length distributions of basal dendrites of rat cortical pyramidal neurons.[10] This agreement with empirical data was obtained by additionally assuming that the elongation rate differed during two phases of dendritic development, the first with branching and elongation, the second with elongation only.

Recent studies have also focused on the shape of the segment length distributions, showing that the first implementation of the elongation process resulted in monotonically decreasing length distributions for intermediate segments. Empirical distributions of different cell types, however, consistently show intermediate segment length distributions with a modal shape.[13,16] In reconstructions of stained neurons, short intermediate segments apparently occur less frequently than was expected. Nowakowski et al. first noticed this phenomenon and suggested a transient suppression of the branching probability immediately after a branching event.[13] Such a reduction indeed resulted in a correct shape for the intermediate segment length distributions.[13] Implementing such an assumption in our approach has a drawback, however, in that it interferes with the branching process as described by the parameters B and E and consequently with the shape of the degree distribution. Therefore, we followed a different approach by giving daughter segments an initial length immediately after a branching event, and letting them further elongate at a slower rate. The elongation process is then split into a process associated with a branching event and a subsequent elongation process. Such an implementation becomes plausible by considering that a branching event is not a point process in time, but rather proceeds during a certain period of time during which a growth cone splits and the daughter branches become stabilized.

7.1.3.3 Time

Continuous time enters into the model when elongation rates are used. Thus far, the branching process (and the associated initial segment length assignments) were defined on a series of time bins, without specifying the time-bin durations. For the mapping of the time bins onto real time, the total duration of the branching and elongation period needs to be defined as well as the type of mapping, which we will assume to be linear.

It will be shown in the examples in this chapter that these assumptions result in a correct description of the segment length distribution while maintaining the proper shape of the degree distribution. The initial length, given to new daughter segments, is determined by random sampling of a predefined distribution. Both for the elongation rate and for the initial length we have chosen a gamma distribution (Box 7.1). Such a distribution is expected for distances between Poisson distributed branching events along the path of an outgrowing neurite.[17]

Box 7.1 The Gamma Distribution

A gamma distribution has the form

$$g(x;\alpha,\beta,\gamma) = \frac{1}{\beta^{\gamma}\Gamma(\gamma)} e^{\frac{x-\alpha}{\beta}} (x-\alpha)^{\gamma-1} \tag{7.3}$$

for $x > \alpha$, $\beta > 0$, $\gamma > 0$, while the gamma function $\Gamma(\gamma)$ is defined by $\Gamma(\gamma) = \int_0^\infty e^{-t} t^{\gamma-1} dt$.[18] The cumulative distribution is given by $G(x; \alpha, \beta, \gamma) = \int_\alpha^x g(x; \alpha, \beta, \gamma)dt$ with $G(\infty; \alpha, \beta, \gamma) = 1$. Parameter α indicates the start of the distribution (offset), β a scaling unit, and γ the shape of the distribution. The mean value of the distribution is given by $\bar{x} = \alpha + \beta\gamma$ and the SD is given by $\sigma_x = \beta\sqrt{\gamma}$. The modus of the distribution is at $x = \alpha + \beta(\gamma - 1)$. For a given choice of the offset α, the parameters β and γ can be estimated from the mean and SD of a distribution by

$$\beta = \frac{\sigma^2}{\bar{x} - \alpha} \quad \text{and} \quad \gamma = \frac{\sigma^2}{\beta^2} = \left(\frac{\bar{x} - \alpha}{\sigma}\right)^2. \tag{7.4}$$

For ease of interpretation, we will characterize a gamma distribution by the parameters α, $\bar{x}$, and σ_x.

7.1.3.4 Segment Diameter

Segment diameters have not been modeled as part of the growth process, but are assigned to the skeleton of the full grown tree. A power law relationship will be assumed, relating the diameters of the segments at a branch point. By the power law relation, the diameter of a parent segment (d_p) relates to the diameter of its daughter segments d_1 and d_2 as

$$d_p^e = d_1^e + d_2^e, \tag{7.5}$$

with e denoting the branch power exponent. According to this relation, the diameter of an intermediate segment d_i relates to the number n and diameter d_t of the terminal segments in its subtree as $d_i = n^{1/e} d_t$, independent of the topological structure of the subtree.

7.1.4 Dendritic Growth Model Parameters

The model includes the parameters B, E, and S defining the branching process; the parameters $\alpha_{l_{in}}$, $\overline{l_{in}}$, and $\sigma_{l_{in}}$ define the offset, mean, and SD, respectively, of the gamma distribution $g_{l_{in}}$ for the initial lengths; and the parameters α_{υ}, $\overline{\upsilon}$, and σ_{υ} define the offset, mean, and SD, respectively, of the gamma distribution for the sustained elongation rates. At a branching event, initial lengths are assigned to both newly formed daughter segments by drawing random samples from the initial length distribution $g_{l_{in}}$, and elongation rates are assigned to both daughter segments by drawing random samples from an elongation rate distribution g_{υ}. These elongation rates hold until new branching events occur at the respective segments. In addition to establishing the sustained elongation, we must specify the duration of the period of branching and elongation T_{be} and of the subsequent period of elongation only T_e.

7.1.4.1 Parameter Estimation

A summary of the parameters in the dendritic growth model is given in Table 7.1. Finding the optimal parameter values needed to describe a particular set of observed dendritic branching patterns is a multidimensional optimization task. The modular character of the model and the assumption of independent branching and elongation, however, make it possible to optimize branching and elongation processes separately. Plots of shape properties versus parameter values offer additional material for manually finding reasonable parameter estimates. This will be described below.

Parameter S — Parameter S can be estimated from the value of the topological asymmetry index. Figure 7.2A illustrates how the expected value of the asymmetry index depends on the value of parameter S and the number of terminal segments in the tree. The equations used to calculate the tree-asymmetry index are reviewed in Reference 11. Note that these equations are derived for the more general case in which intermediate segments also may branch. In the present study we assume terminal branching only. Alternatively, parameter S can be estimated from the mean centrifugal order of the tree. Figures 7.2B and C show how the mean centrifugal order depends on the parameter S and the number of terminal segments in the trees.

Parameters B and E — Parameters B and E can be estimated from Figure 7.3, illustrating the mapping from branching parameters B and E to the expected mean and SD values of the predicted degree distribution. The estimation proceeds by plotting the observed mean and SD in the figure and finding the coordinates of this point in the B, E grid.

Metrical parameters $\alpha_{l_{in}}, \overline{l_{in}}, \sigma_{l_{in}}$ for the initial length and $\alpha_{\upsilon}, \overline{\upsilon}$, and σ_{υ} for the sustained elongation rate — Once the branching process is defined, we

TABLE 7.1
Summary of Parameters Used in the Dendritic Growth Model

Parameter	Aspect of Growth	Related to
B	Basic branching parameter	Segment number
E	Size-dependency in branching	Segment number
S	Order-dependency in branching	Topological structure
$\alpha_{l_{in}}$	Initial length—offset	Segment length
$\overline{l_{in}}$ (μm)	Initial length—mean	Segment length
$\sigma_{l_{in}}$	Initial length—SD	Segment length
$\alpha_{\upsilon_{be}}$	Elongation in "branching/elongation phase"—offset	Segment length
$\overline{\upsilon_{be}}$ (μm/h)	Elongation in "branching/elongation phase"—mean rate	Segment length
α_{υ_e}	Elongation in "elongation phase"—offset	Segment length
$\overline{\upsilon_e}$ (μm/h)	Elongation in "elongation phase"—mean rate	Segment length
$c\upsilon_{\upsilon}$	Coefficient of variation in elongation rates	Segment length
$\bar{d}_t$ (μm)	Terminal segment diameter—mean	Segment diameter
σ_{d_t}	Terminal segment diameter—SD	Segment diameter
$\bar{e}$	Branch power—mean	Segment diameter
σ_e	Branch power—SD	Segment diameter

Note: Note that the segment diameter parameters are not part of the growth model, but used afterwards to assign diameter values to the skeleton tree produced by the model.

need further specification of the gamma distributions for the initial lengths $g_{l_{in}}$ and for the sustained elongation rates g_{υ}. The parameters in both $g_{l_{in}}$ and g_{υ} must be estimated from the empirical intermediate and terminal segment length distributions, and the path length distribution. Good estimates are, however, not directly obtained from these distributions but require a process of optimization. Some considerations may help in finding reasonable estimates when using a manual approach. The length of intermediate segments in the model is determined by (1) the initial length assigned after a branching event, (2) the elongation rate assigned to this segment, and (3) the time elapsing before the segment experiences the next branching event. A segment becomes a terminal segment when it fails to undergo branching before the end of the growth period.

Terminal segments are generally longer than intermediate segments, both when compared for the whole tree and when compared for a given centrifugal order within the tree, e.g., References 19, 20, 21. Many dendritic trees also show a decrease of terminal segment length with increasing centrifugal order.[19] These findings can be explained by considering that terminal segments of a given centrifugal order have had more time, on the average, to elongate than have intermediate segments of similar order, and that this time decreases for increasing order.[17] This phenomenon occurs only when segments show sustained elongation in addition to the initial length

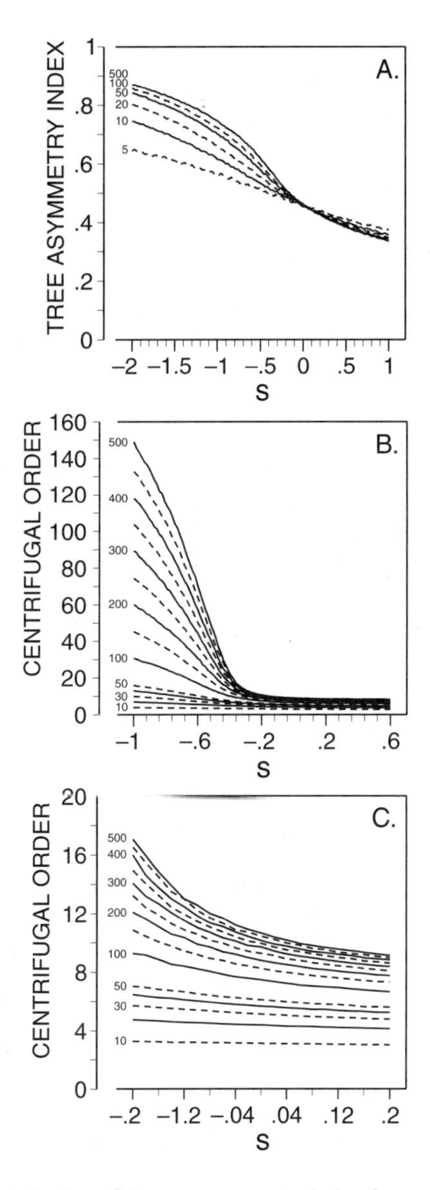

FIGURE 7.2 (A) Expected value of the tree asymmetry index for a set of trees as a function of the growth parameter S, calculated for trees of degree 5, 10, 20, 50, 100, 500, and (B) mean value of the centrifugal order of segments as a function of the growth parameter S, calculated for trees of degree 10, 20, 30, 40, 50, 100, 150, 200, 250, 300, 350, 400, 450, and 500. Panel (C) expands the area $-0.2 < S < 0.2$.

assigned to the daughter segments after branching. The length difference in terminal segments of lowest and highest order can thus be used to obtain an estimate of the sustained elongation rate during the period of branching. In the examples described in Section 7.2.6, we will see that rat neocortical pyramidal cell basal dendrites show

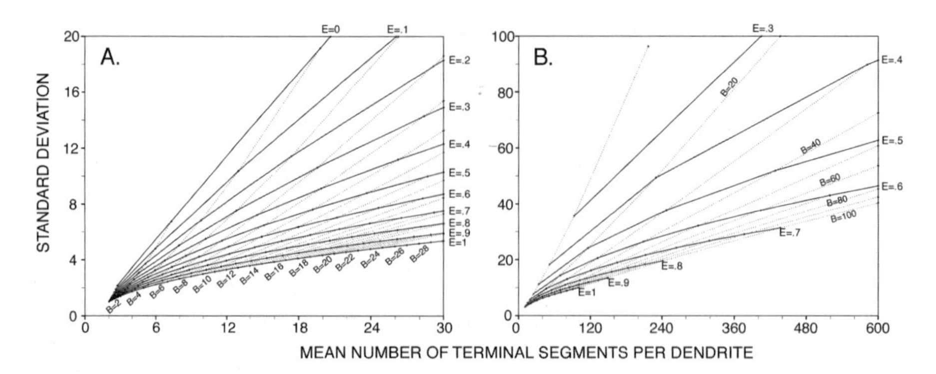

FIGURE 7.3 Mapping of the (B, E) parameter grid onto the (mean, SD)-plane. The map is obtained by calculating for many pairs of (B, E) model parameter values the mean and SD of the degree distribution, as produced by the model, which are subsequently plotted as a point in the (mean, SD)-plane. The continuous lines connect points in the (mean, SD)-plane with equal E value, the dotted lines connect points with equal B value. The map is calculated for a fine (panel A) and a course degree (panel B) scale.

such differences whereas guinea pig cerebellar Purkinje cells do not (see also Figure 7.4). Terminal segments may become much longer than intermediate ones when dendritic development includes a period of branching and elongation, followed by a period of elongation only. Additionally, the elongation rates need not be equal during these two periods.

The variation in path lengths is the final outcome of all stochasticity in elongation and branching. The SD of the path length distribution can be used to estimate the variation in the sustained elongation rates. These considerations help in estimating the parameters α_υ, $\bar{\upsilon}$, and σ_υ in the gamma distribution g_υ. The modal shape of the intermediate segment length distribution is determined by the initial length distribution $g_{l_{in}}$ and the sustained elongation rate. Estimates of the parameters in $g_{l_{in}}$ must be obtained using Equation 7.4 for a given choice for the length offset $\alpha_{l_{in}}$ and considering the choices for the parameters in g_υ.

Diameter parameters — The segment diameters in a tree have not been modeled as part of a developmental process, but have been directly assigned to the full grown skeleton tree by means of the following procedure. First, terminal segment diameters d_t are assigned by random sampling the observed diameter distribution (or a normal distribution based on the observed mean-SD values). Then, traversing the tree centripetally, at each bifurcation the diameter of the parent segment is calculated by means of Equation 7.5, using (a) the diameters of the daughter segments and (b) a branch power value e obtained by randomly sampling the observed branch power distribution.

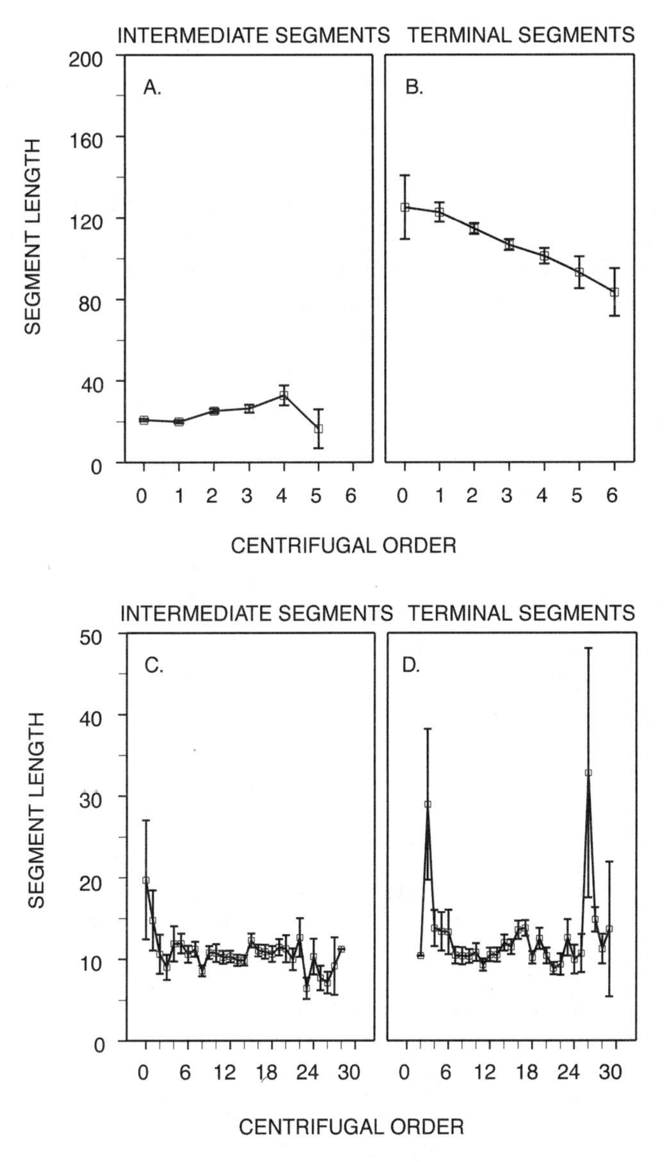

FIGURE 7.4 Length of intermediate and terminal segments plotted versus their centrifugal order, for S1-rat cortical layer 2/3 pyramidal cell basal dendrites (panels A and B) and for guinea pig cerebellar Purkinje cell dendritic trees (panels C and D).

7.1.5 Dendritic Growth Model Examples

7.1.5.1 Application to S1-Rat Cortical Layer 2/3 Pyramidal Cell Basal Dendrites

In S1-rats, the outgrowth of the layer 2/3 pyramidal cell basal dendrites starts at about 1 day after birth and continues with branching and elongation up to about day 14, followed by a period of elongation up to about day 18.[16] The geometrical properties of these dendrites are given in Table 7.2 and the segment length distributions in Figure 7.5 as hatched histograms. How segment lengths depend on centrifugal order is displayed in Figure 7.4 (A and B).

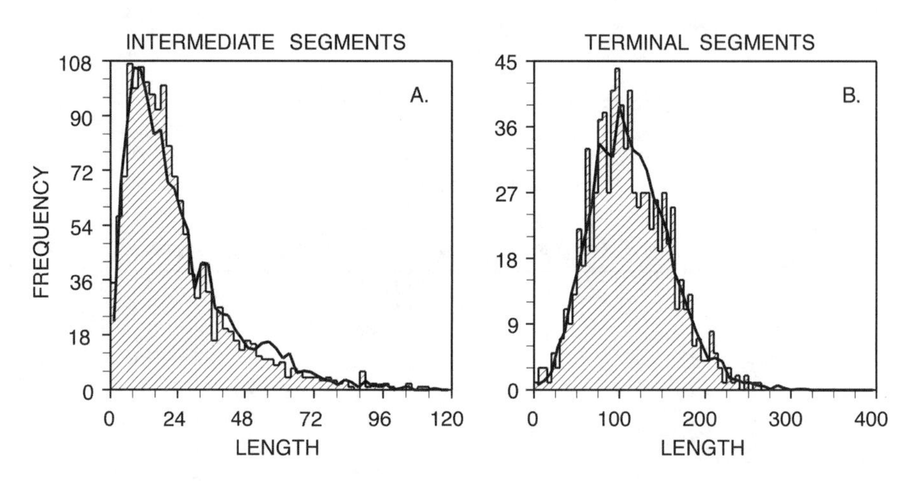

FIGURE 7.5 Comparison of (A) intermediate and (B) terminal segment length distributions of observed S1-rat cortical layer 2/3 pyramidal cell basal dendrites (hatched histograms) and model generated trees (continuous lines) for the parameter values given in Table 7.3.

Estimation of parameter S — Interpolation from the observed asymmetry index of 0.41 in Figure 7.2A results in an estimate of $S \approx 0.5$.

Estimation of parameters B and E — These parameters can be estimated from the mean and SD of the observed degree distribution when plotted as the point (4.04, 2.04) in Figure 7.3. Estimates of the corresponding coordinates in the B, E-grid are then obtained of $B = 2.52$ and $E = 0.73$.

Estimation of $g_{l_{in}}$ and g_{υ} — The observed distribution of intermediate segment lengths (Figure 7.5A) does not have a clearcut offset. We have assigned therefore a value of zero to the offset parameter $\alpha_{l_{in}}$. The difference in length between highest and lowest order terminal segments is about 50–60 μm (Figure 7.4B). Given a total duration of branching of 312 h (13 days), we obtain a rough estimate of $\overline{\upsilon_{be}} \approx 0.2$ μm/h for the sustained elongation rate in the first developmental phase of branching and elongation. Values of $\overline{l_{in}} = 6$ μm and $\sigma_{l_{in}} = 5$ μm, for the initial length distribution in combination with a sustained mean elongation rate of 0.2 μm/h during this first phase, turned out to result in a good fit of the shape of the intermediate segment

TABLE 7.2
Comparison of Shape Properties from Experimental Observations of S1-Rat Cortical Layer 2/3 Pyramidal Cell Basal Dendrites and of Model Simulated Trees

	Observed		Model Predicted	
Shape Parameter	**Mean**	**Standard Deviation**	**Mean**	**Standard Deviation**
Degree	4.04	2.04	4.05	2.02
Asymmetry index	0.41	0.24	0.4	0.23
Centrifugal order	1.86	1.2	1.85	1.19
Total dendritic length			527.6	265
Terminal length	110.7	45.2	112.62	44.8
Intermediate length	22.0	17.9	23.6	18.0
Path length	163.8	48.1	164.6	45.0

Obtained with optimized values of the growth parameters.

length distribution. The shape of the terminal segment length distribution was fitted by assuming a mean elongation rate of 0.86 μm/h during the elongation phase. A coefficient of variation of 0.47 and a zero value for α_υ were additionally assumed.

Diameter parameters — Diameter assignments can be made according to the procedure described in Section 7.1.4.1, using parameter values $\bar{e} = 1.6$, $\sigma_e = 0.2$, $\bar{d}_t = 0.6$ μm, and $\sigma_{d_t} - 0.1$.[2,21]

TABLE 7.3
Optimized Values for Growth Parameters to Match the Statistical Shape Properties of S1-Rat Cortical Layer 2/3 Pyramidal Cell Basal Dendrites

Growth Parameters										
B	E	S	α_{in}	$\overline{l_{in}}$ (μm)	$\sigma_{l_{in}}$	$\alpha_{\upsilon_{be}}$	$\overline{\upsilon_{be}}$ (μm/h)	α_{υ_e}	$\overline{\upsilon_e}$ (μm/h)	cv_υ
2.52	0.73	0.5	0	6	5	0	0.2	0	0.86	0.47

Note: Note that υ_{be} and υ_e define the sustained elongation rates during the first period of branching and elongation with a duration of 312 h (13 days), and the second period of elongation only with a duration of 96 h (4 days), respectively.

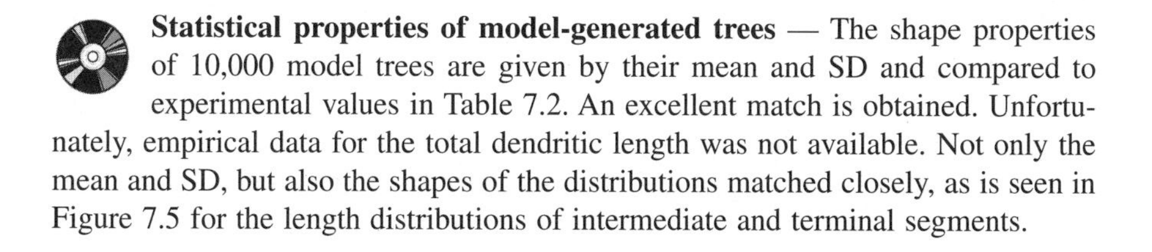

Statistical properties of model-generated trees — The shape properties of 10,000 model trees are given by their mean and SD and compared to experimental values in Table 7.2. An excellent match is obtained. Unfortunately, empirical data for the total dendritic length was not available. Not only the mean and SD, but also the shapes of the distributions matched closely, as is seen in Figure 7.5 for the length distributions of intermediate and terminal segments.

7.1.5.2 Application to Guinea Pig Cerebellar Purkinje Cell Dendritic Trees

The second example concerns the analysis of three guinea pig Purkinje cell dendritic trees, fully reconstructed by Rapp et al.[22] who analyzed in detail their physiological properties and made them available via the *http://leonardo.ls.huji.ac.il/~rapp*. The geometrical properties of these cells have been calculated from these reconstructions, and their means and SD are given in the 2nd and 3rd columns of Table 7.4.

TABLE 7.4
Comparison of Shape Properties from Experimental Observations of Guinea Pig Cerebellar Purkinje Cell Dendritic Trees and of Model Simulated Trees

	Observed Trees 1+2+3		Model Trees	
Shape Parameter	Mean	Standard Deviation	Mean	Standard Deviation
Degree	436	31.8	436	32
Asymmetry index	0.5	0.01	0.49	0.02
Centrifugal order	13.7	5.1	13.8	5.9
Total length	9577	1105	9265	683
Terminal length	11.3	8.8	10.6	7.5
Intermediate length	10.6	7.5	10.6	7.6
Path length	189.3	64.1	166	66

Obtained with optimized values of the growth parameters.

Estimation of parameter S — Interpolation from the observed asymmetry index of 0.50 in Figure 7.2A results in an estimate of $S \approx -0.15$. Interpolation from the observed mean centrifugal order of 13.7 in Figure 7.2C results in an estimate of $S \approx -0.14$.

Estimation of parameters *B* and *E* — The values for the mean and SD of the observed degree distribution (436, 31.8) form a point in the map in Figure 7.3. The *B*, *E* coordinates of this point can be obtained by reference to the *B*-*E* grid. A manual estimate of $B = 95$ and $E = 0.69$ has been used. Note that the mean and SD of the degree distribution are based on only three observations. More observations are needed in order to obtain a stable estimate for the location of the point in Figure 7.3 and, consequently, for the estimate of the corresponding *B*, *E* coordinates.

Estimation of $g_{l_{in}}$ and g_{υ} — Figure 7.4C, D shows that segment lengths do not depend on centrifugal order, and that intermediate and terminal segments have approximately equal length. Similar findings have been obtained for Purkinje cell dendritic trees in mice[8] and in rats.[19] It is therefore reasonable to assume that segments in the Purkinje cells have not (or only moderately) undergone sustained elongation, and that the observed segment length distributions (almost) fully reflect the initial lengths at the time of their origin. According to this reasoning, we can estimate the gamma distribution $g_{l_{in}}$ from the mean and SD of the intermediate segment length distribution (Table 7.4). The length offset $\alpha_{l_{in}}$ has been estimated from the observed distribution to be $\alpha_{l_{in}} = 0.7\ \mu m$.

Diameter parameters — Segment diameters can be assigned according to the procedure, described in Section 7.1.4.1 with parameter values $\bar{e} = 2.0$, $\sigma_e = 0.3$, $\bar{d_l} = 1.1\ \mu m$ and $\sigma_{d_t} = 0.1$.[2]

Examples of trees, produced with the estimates for the growth parameters (Table 7.5), are given in Figure 7.6.

TABLE 7.5
Optimized Values for Growth Parameters to Match the Statistical Shape Properties of Guinea Pig Purkinje Cell Dendritic Trees

Growth Parameters					
B	E	S	$\alpha_{\lambda_{iv}}$	$\bar{l}_{in}$	$\sigma_{\lambda_{iv}}$
95	0.69	–0.14	0.7 μm	10.63	7.53

Note: Parameters B, E, and S define the branching process, and α_{lin}, l_{in}, and σ_{lin} define the gamma distribution for the initial segment lengths.

Statistical properties of model-generated trees — Statistical properties of tree shapes, obtained by simulating 100 trees, are given in the 4th and 5th column of Table 7.4 An excellent matching is shown in both the mean and SD of the different shape parameters between the modeled and observed dendritic trees. Also the shapes of the distributions closely match, as is shown in Figure 7.7 for the length distributions of intermediate and terminal segments.

7.1.6 Discussion

The two examples discussed have shown that the model for dendritic outgrowth is able to reproduce dendritic complexity, as measured by many geometrical properties to a high degree of correspondence. The basic assumptions are (1) randomness and (2) independence in branching and elongation. The modal shape of the intermediate segment length distributions could be described by dividing the elongation process into a first phase associated with branching events, and implemented in the model by the assignment of an initial length to newly formed daughter segments, and a second phase of sustained elongation. Guinea pig Purkinje cell segment lengths turned out to be well described by the initial segment length assignments only. In contrast, rat pyramidal cell basal dendrites required, in addition to the initial segment length assignments a sustained elongation (with different rates for the first period of elongation and branching, and a second period of elongation only). The shape characteristics of the guinea pig Purkinje cells were based on dendritic reconstructions of only three cells. The empirical data for the mean and SD of the degree, asymmetry index, and total length consequently have modest stability and this could

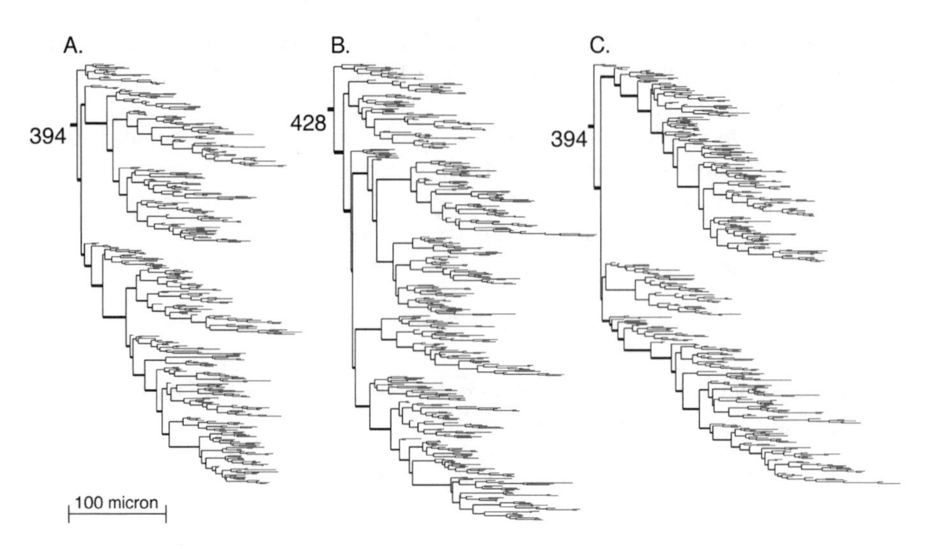

FIGURE 7.6 Examples of trees randomly produced by the growth model for parameter values, optimized for guinea pig cerebellar Purkinje cells, as given in Table 7.5. Note that the diameters of the branches are not produced by this model, but randomly assigned.

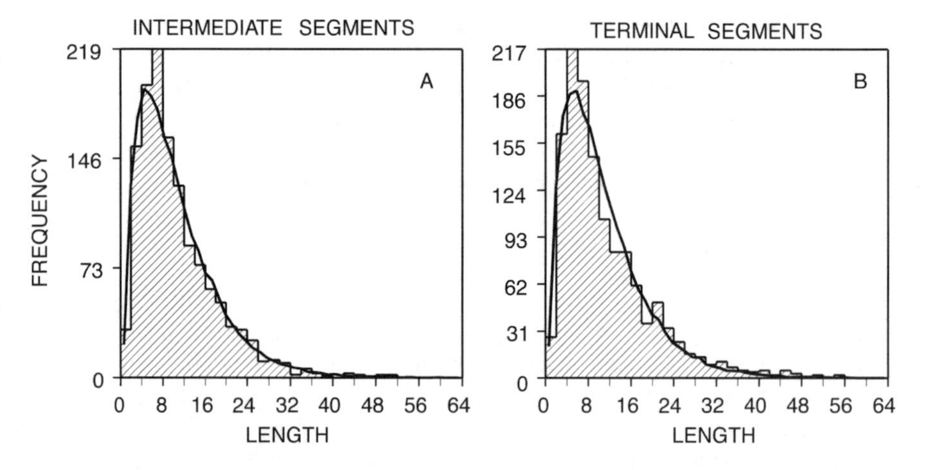

FIGURE 7.7 Comparison of (A) intermediate and (B) terminal segment length distributions of observed guinea pig Purkinje cells (hatched histograms) and model generated trees (continuous lines) for the parameter values given in Table 7.4.

be the explanation for the difference in total length SD between observed and modeled trees (Table 7.4).

The description of dendritic outgrowth as a stochastic process, defined by branching and elongation probabilities, is the reflection of a complex of molecular, biochemical, and cellular processes. It is therefore surprising that a limited set of growth rules and parameters (especially in the case of the Purkinje cells) suffices to describe dendritic complexity with such a high level of accuracy. The phenomenological approach, along with the quantified probability functions presented here, are first steps toward a further quantification of these processes underlying neurite outgrowth and neuronal morphogenesis.

The model is useful, since it can produce any number of dendritic trees with realistic variations in the number of segments, topological structure, and intermediate and terminal segment lengths. Segment diameters are assigned using a branch power rule. These model dendrites can then be used in neural simulators for studying structure–function relationships in dendrites (see Chapters 8 and 9).

7.2 COMPETITION FOR NEUROTROPHIC FACTOR IN THE DEVELOPMENT OF NERVE CONNECTIONS

The development of connections between neurons and their target cells often involves an initial stage of hyperinnervation followed by elimination of axons.[23] In some cases, elimination continues until the target is innervated by just a single axon, whereas in most other cases, several innervating axons remain. An example of single innervation is the innervation of skeletal muscle fibers.[24] The cells that act as targets for the innervating axons appear to release limited amounts of neurotrophic factors, which are taken up by the axons via specific receptors at their terminals and which affect the growth and branching of the axons.[25,26] An important class of neurotrophic factors is the neurotrophin family, with NGF (nerve growth factor) as its best characterized member.

Competition among innervating axons for neurotrophic factors is thought to be involved in axonal elimination and the generation of different patterns of innervation.[23] There is, however, little understanding of the nature of the competitive process and the underlying mechanisms. Computational models of activity-dependent development of nerve connections (e.g., of the formation of ocular dominance columns) typically enforce competition rather than model it explicitly.[27] The first way in which this can be done is to enforce synaptic normalization. Consider n synapses, with efficacies s_i, impinging upon a given postsynaptic cell. Then, synaptic normalization is the constraint that

$$\sum_i^n s_i^p = K,$$

where K is some constant and p is usually taken to be 1 or 2. Following a phase of Hebbian learning, which changes the values of s_i, the new efficacies are forced to satisfy the normalization constraint.

A second approach is that of Bienenstock et al., which does not impose synaptic normalization.[28] Here, a modified Hebb rule is used, which has the effect that inputs driving a postsynaptic cell below/above a certain threshold firing level cause a decrease/increase in synaptic efficacy. The threshold itself is a time-averaged function of the activity of the postsynaptic cell. This modified Hebb rule results in temporal competition between input patterns, rather than spatial competition between different sets of synapses.

In most existing models[29–33] of the development of nerve connections that try to explicitly model the putative underlying mechanism, competition is based on a fixed amount of neurotrophin that becomes partitioned among the individual synapses or axons, i.e., there is no production, decay, and consumption of the neurotrophin. This assumption is biologically not very realistic. Our approach, similar to that of Jeanprêtre et al.[34] in a model for the development of single innervation, considers the production and consumption of neurotrophin. By formulating a model that incorporates the dynamics of neurotrophic signalling (such as release of neurotrophin, binding kinetics of neurotrophin to receptor, and degradation processes) and the effects of neurotrophins on axonal growth and branching, competitive interactions emerge naturally. Our approach has similarities to that of Elliott and Shadbolt, although they do not model all the processes involved in a dynamic fashion (e.g., neurotrophin release and binding kinetics).[35]

7.2.1 The Model

The simplest situation in which we can study axonal competition is a single target at which there are a number of innervating axons each from a different neuron. Each axon has a number of terminals, on which the neurotrophin receptors are located (Figure 7.8). In order to model competition, we break it down into a number of subprocesses. First, neurotrophin needs to be released by the target into the extracellular space. From there it will be removed partly by degradation and diffusion, and partly by binding to the neurotrophin receptors at the terminals of the innervating axons. The binding of neurotrophin to its receptor is a reversible reaction: the forward reaction produces the neurotrophin-receptor complex, and the backward reaction dissociates the complex back into neurotrophin and unoccupied receptor. The neurotrophin-receptor complex is then taken up by the axons and is also subject to degradation. Receptor as well as neurotrophin are thereby removed. Therefore, we also need to consider the insertion of new receptors into the axon terminals, as well as turnover of unoccupied receptors. Finally, the growth and branching of each axon is affected by the amount of bound neurotrophin (neurotrophin-receptor complex) the axon has across its terminals.

7.2.1.1 Release and Removal of Neurotrophin

Because the binding of neurotrophin to receptor is what triggers the biological response, we describe, for each axon i, the time-dependent change of the axon's total amount of neurotrophin-receptor complex. The total amount of neurotrophin-receptor complex an axon has over all its terminals, C_i for axon i, increases by binding of neurotrophin to receptor, and decreases by dissociation and degradation. Thus, for the rate of change of C_i, we can formulate the following differential equation:

$$\frac{dC_i}{dt} = \left(k_{a,i} L R_i - k_{d,i} C_i\right) - \rho_i C_i , \tag{7.6}$$

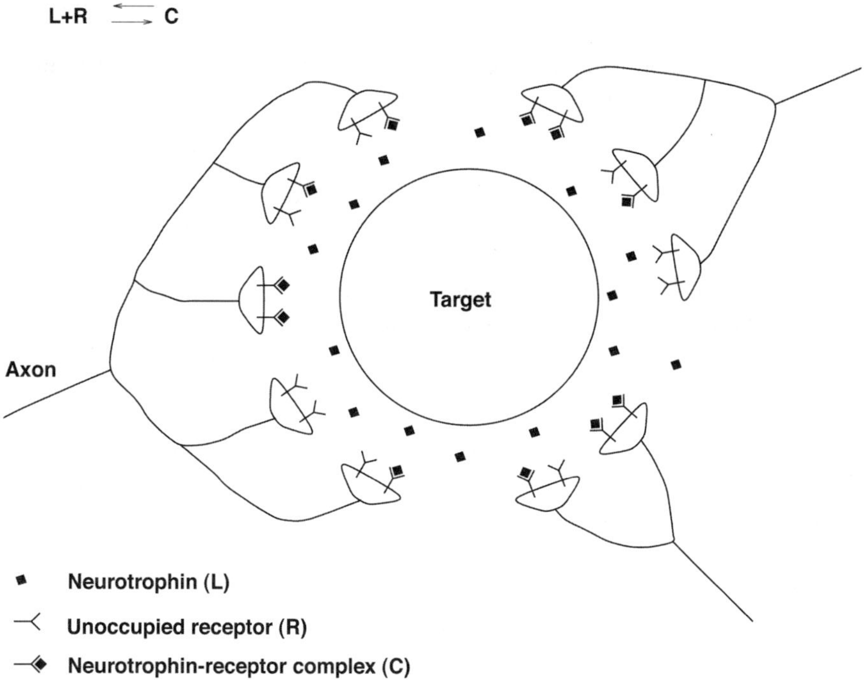

L+R ⇄ C

Axon

Target

■ Neurotrophin (L)

Unoccupied receptor (R)

Neurotrophin-receptor complex (C)

FIGURE 7.8 Single target with three innervating axons. The target releases neurotrophin that is bound by neurotrophin receptors at the axon terminals (From Reference 46. With permission.)

where L is the extracellular concentration of neurotrophin, R_i is the total number of unoccupied receptors that axon i has over all its terminals, $k_{a,i}$ and $k_{d,i}$ are the respective association and dissociation constants of the reversible binding of neurotrophin to receptor, and ρ_i is the rate constant for degradation of the complex.

The total number of unoccupied receptors and the concentration of neurotrophin in the extracellular space are not constants, but rather change in time. The total number of unoccupied receptors that an axon has over all its terminals, R_i for axon i, increases by the insertion of new receptors into the terminals as well as by dissociation of the neurotrophin-receptor complex; it decreases by the binding of neurotrophin to receptor and by receptor turnover. Thus,

$$\frac{dR_i}{dt} = \phi_i - \gamma_i R_i - \left(k_{a,i} L R_i - k_{d,i} C_i\right), \tag{7.7}$$

where ϕ_i is the rate of insertion of new receptors and γ_i is the rate constant for turnover.

The concentration of neurotrophin in the extracellular space, L, increases by the release of neurotrophin from the target and by the dissociation of neurotrophin-complex into neurotrophin and receptor; it decreases by the binding of neurotrophin to receptor and by degradation. Thus,

$$\frac{dL}{dt} = \sigma - \delta L - \sum_{i=1}^{n} \left(k_{a,i} L R_i - k_{d,i} C_i \right) / \upsilon, \tag{7.8}$$

where σ is the rate of release of neurotrophin, δ is the rate constant for degradation, n is the total number of innervating axons, and υ is the volume of the extracellular space (L is a concentration, while R_i and C_i are defined as amounts). The rate of release of neurotrophin, σ, could depend on the level of electrical activity in the target.

Equations 7.6 and 7.7 are similar to the ones used in experimental studies for analyzing the cellular binding, internalization, and degradation of polypeptide ligands such as neurotrophins.[36]

7.2.1.2 Axonal Growth

The binding of neurotrophin to receptor triggers the biological response. Many studies have shown that neurotrophins locally increase the arborization of axons, which will consequently cause an increase in the number of axon terminals.[37] It is reasonable to assume that increasing the number of axon terminals, on which the neurotrophin receptors are located, will increase the axon's total number of neurotrophin receptors. Other effects induced by neurotrophins that are likely to increase the total number of axonal neurotrophin receptors are (i) increasing the size of axon terminals[38] and (ii) upregulating the density of neurotrophin receptors.[39]

In order for the total number of receptors to be able to increase in response to neurotrophins, the total number of unoccupied receptors that is inserted into the axon per unit time, ϕ_i, must increase in response to bound neurotrophin. We assume that the larger the amount of bound neurotrophin, C_i, the larger ϕ_i will be. That is, ϕ_i is an increasing function, $f_i(C_i)$, of the amount of bound neurotrophin, C_i. We call function $f_i(C_i)$ the growth function. Compared to the dynamics of the other processes involved, axonal growth takes place on a relatively slow time scale. To account for this, ϕ_i must lag behind its target value given by $f_i(C_i)$. This lag can be modeled by the following differential equation:

$$\tau \frac{d\phi_i}{dt} = f_i(C_i) - \phi_i, \tag{7.9}$$

where the time constant τ of growth is of the order of days. The value of ϕ_i will follow changes in $f_i(C_i)$ (as a result of changes in C_i) with a lag; at steady-state, $\phi_i = f_i(C_i)$.

The precise form of the growth function, $f_i(C_i)$, is not known; we therefore use a general increasing function that can admit a range of different forms depending on its parameters. The effects of the form of the growth function on competition can then be studied. We use the general growth function

$$f_i(C_i) = \frac{\alpha_i C_i^m}{K_i^m + C_i^m}. \tag{7.10}$$

This is an increasing function that saturates towards a maximum, α_i. Parameter K_i is the value of C_i at which the response is half its maximum. Using this general growth function, we can distinguish a number of different classes of growth functions (Figure 7.9).

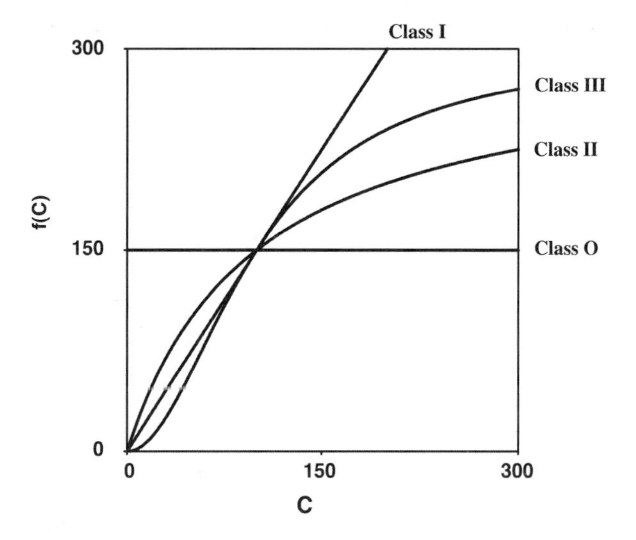

FIGURE 7.9 Growth function $f(C) = \alpha C^m/(K^m + C^m)$ for the different classes described in the text. For class O, $\alpha = 300$; for class I, $\alpha/K = 1.5$; for classes II and III, $\alpha = 300$ and $K = 100$ (From Reference 46. With permission.)

Class O: for $m = 0$, $f_i(C_i)$ is a constant ($f_i(C_i) = \alpha_i/2$) and independent of the level of bound neurotrophin, C_i.

Class I: for $m = 1$ and large K_i ($K_i \gg C_i$), growth is linear over a large range of C_i ($f_i(C_i) \approx \alpha_i C_i/K_i$).

Class II: for $m = 1$ and smaller values of K_i ($K_i \not\gg C_i$), the growth function is a Michaelis–Menten function ($f_i(C_i) = \alpha_i C_i/(K_i + C_i)$) (see Section 2.2.1).

Class III: for $m = 2$ the growth function is a Hill function

$$\left(f_i(C_i) = \alpha_i C_i^2 / \left(K_i^2 + C_i^2\right)\right),$$

which is sigmoidal.

Within each class of growth function, the specific values of the parameters (α_i and K_i), as well as those of the other parameters, may differ among axons. Various factors in the innervating axon, some dependent on and some independent of its electrical activity, may influence the values of these parameters. For example, the finding that increased presynaptic electrical activity increases the number of neurotrophin receptors[40] implies that increased electrical activity affects growth (i.e., higher α_i or lower K_i) or neurotrophic signalling (e.g., lower γ_i) or both. As the level of electrical activity and other factors can vary among innervating axons, there will be variations in parameter values among axons.

The whole model thus consists of three differential equations for each axon i, Equations 7.6, 7.7, and 7.9, and one equation for the neurotrophin concentration, Equation 7.8. By means of numerical simulations and mathematical analysis, we can examine the outcome of the competitive process. Axons that at the end of the competitive process have no neurotrophin ($C_i = 0$; equivalent to $\phi_i = 0$) are assumed to have withdrawn or died, while axons that do have neurotrophin ($C_i > 0$; equivalent to $\phi_i > 0$) are regarded as having survived.

7.2.2 Units and Parameter Values

All parameters in the model have a clear biological interpretation. For the numerical simulations, the parameter values were taken from the data available for NGF. Because the high affinity NGF receptor mediates the biological response, the association and dissociation constants of this receptor were taken: $k_a = 4.8 \times 10^7$ [M^{-1} s^{-1}], $k_d = 1.0 \times 10^{-3}$ [s^{-1}].[41] The rate constant for the turnover of receptor, γ, was calculated from the receptor half-life;[42] $\gamma = 2.7 \times 10^{-5}$ [s^{-1}]. The rate constant for the degradation of neurotrophin-receptor complex, ρ, was calculated from the half-life of complex;[43] $\rho \approx 2.0 \times 10^{-5}$ [s^{-1}]. The rate constant for degradation of neurotrophin in the extracellular space, δ, was estimated using data on neurotrophin concentration changes following blockade of axonal transport;[34] $\delta \approx 1.0 \times 10^{-5}$ [s^{-1}]. The standard value used in the model for the rate of release of neurotrophin was set at $\sigma \approx 2.0 \times 10^{-16}$ [M s^{-1}], which is well within the range of values given in References 34 and 44. Based on data on the time course of the growth of the number of receptors, τ was set at 2 days.[45] Parameter υ, the volume of the extracellular space around the target cell in which neurotrophin is released, acts as a scale parameter and was set at 1.7×10^{-11} v[l].[1]

The values of R_i, C_i, and K_i are in number of molecules; the value of L in [M] (= [mol l^{-1}]). The values of α_i and ϕ_i are expressed in [number of molecules h^{-1}]. Time is in hours [h]. Only the value of α_i varies among axons. Unless otherwise indicated, the initial value of all ϕ_i is 10.0 [molecules h^{-1}]. The initial values of R_i, C_i, and L are such that when keeping all ϕ_i at their initial value, the system is in equilibrium.

7.2.3 Examples of Results and Predictions

For an extensive overview of the results of the model, see Reference 46. Here we restrict ourselves to a few examples. The model (with growth functions of classes II and III) accounts for the experimental finding that increasing the amount of neurotrophin increases the number of surviving axons.[47,48] In the model, elimination of axons takes place until either one or several axons survive, depending on (among other parameters) the rate of release of neurotrophin, σ: the larger σ, the more axons survive (Figure 7.10a, b).

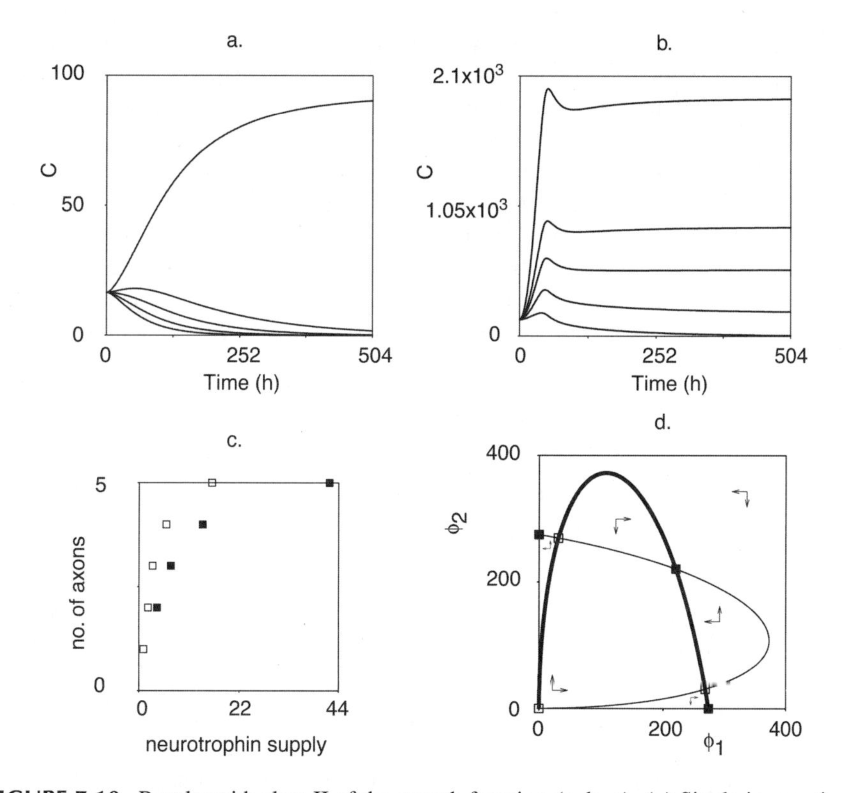

FIGURE 7.10 Results with class II of the growth function (a, b, c). (a) Single innervation. The axon with the highest value of α_i among the initial five axons survives. $\alpha_1 = 700$, $\alpha_2 = 400$, $\alpha_3 = 300$, $\alpha_4 = 200$, $\alpha_5 = 100$, and $K = 500$. (b) Multiple innervation with a rate of release of neurotrophin, σ, that is 35 times higher than the standard value. Other parameter values are the same as in **a**. (c) Relationship between the rate of release of neurotrophin (in units of the standard value) and the number of axons with $C_i > 10$ at $t = 504$, for $K = 500$ (filled squares), and $K = 150$ (open squares). Other parameter values are the same as in **a**. (d) Coexistence of equilibrium points of single and multiple innervation, in a system of two innervating axons ($n = 2$), with a class III growth function. The variables (R_i, C_i, $i = 1.2$) and L are at quasi-steady state. The bold line depicts the solutions of the equation $d\phi_1/dt = 0$ and the light line those of $d\phi_2/dt = 0$. (The lines $\phi_1 = 0$ and $\phi_2 = 0$ are also solutions of $d\phi_1/dt = 0$ and $d\phi_2/dt = 0$, respectively, but are not drawn.) The intersection points of these nullclines are the equilibrium points of the system. Vectors indicate direction of change. Filled square indicates stable equilibrium point, and open square unstable equilibrium point. Note that $\phi_i > 0 \Leftrightarrow C_i > 0$ (axon i present) and $\phi_i = 0 \Leftrightarrow C_i = 0$ (axon i eliminated). The stable equilibrium point at ($\phi_1 = 0$, $\phi_2 = 0$) is not indicated as it is too close to another, unstable point. Which of the stable equilibria will be reached depends on the initial values of ϕ_i, and the sizes of the basins of attraction of the equilibria, which are sensitive to the values of the competitive strengths, β_i. Parameters: $\alpha_1 = \alpha_2 = 300$, $K = 30$.

The axons having a survival advantage are the ones with the highest value for the quantity β_i defined as $(k_{a,i}(\alpha_i/K_i - \rho_i))/(\gamma_i(k_{d,i} + \rho_i))$, which we interpret as the axon's competitive strength. Because β_i contains parameters that may be affected by the axon's level of electrical activity (e.g., α_i), the axons having a survival advantage will be the most active ones provided that variations due to other factors do not predominate.

In agreement with the model, in skeletal muscle, stable states of single and multiple innervation can coexist, as with class III of the growth function (Figure 7.10d). Persistent multiple innervation is found in partial denervation experiments after reinnervation and recovery from prolonged nerve conduction block.[49] In terms of the model, conduction block changes the competitive strengths of the axons, which changes the sizes of the basins of attractions of the different equilibria. This can cause the system to go to an equilibrium of multiple innervation, while under normal conditions single innervation develops. When the conduction block is removed, the system will remain in the basin of attraction of the multiple innervation equilibrium, i.e., multiple innervation persists.

Our analyses suggest that of the many axonal features that change during growth in response to neurotrophin (degree of arborization and, consequently, number of axon terminals; size of terminals; and density of receptors) the consequent change in the axon's total number of neurotrophin receptors, changing its capacity for removing neurotrophin, is what drives the competition. The model predicts that axons that are in the process of being eliminated will have a relatively small number of neurotrophin receptors.

The type of dose-response relationship between neurotrophin and total number of neurotrophin receptors (i.e., the growth function), which is crucial in our model for determining what patterns of innervation can develop, can be determined experimentally *in vitro* by measuring, for different concentrations of neurotrophin in the medium, the total number of terminals of an axon or, more specifically, the axon's total number of neurotrophin receptors.[46] The model predicts that the type of growth function will determine the relationship between the concentration of neurotrophin and the number of surviving axons. For example, the smaller the value of K_i, the lower the concentration of neurotrophin needed to rescue more axons (Figure 7.10c).

7.2.4 Conclusions

Our model of competition links the formation of nerve connections with the underlying actions and biochemistry of neurotrophins. The model accounts for the development of single and multiple innervation, as well as several other experimental findings, and makes testable predictions. Although the parameter values were taken from the data available for NGF, mathematical analysis shows that our results are general and do not depend on specific choices of the parameter values.[46]

The model can be extended in several ways. In reality, axons can have more than one target. In the model, the rate of insertion of receptors could then be different for branches innervating different targets. This will also cause competition within axons between different branches in addition to competition among axons.[30,31]

In the present model, we have assumed that the concentration of neurotrophin is homogeneous in the extracellular space that surrounds the target; in other words, all innervating axons "sense" the same concentration. This assumption may not be realistic, especially if the target is large (e.g., a neuron with a large dendritic tree, onto which the axons impinge). This can be taken into account in our model by modeling the extracellular space as a collection of "compartments," into which neurotrophin is released locally from the target. Some of the compartments will have an innervating axon, which removes neurotrophin molecules locally. In addition, there will be diffusion of neurotrophin between compartments (see Chapter 3). For preliminary results of such a model, see Reference 50.

ACKNOWLEDGMENTS

We are very grateful to Dr. M. A. Corner for his critical comments and improvements of the manuscript.

REFERENCES

1. Burke, R. E., Marks, W. B., and Ulfhake, B., A parsimonious description of motoneuron dendritic morphology using computer simulation, *J. Neurosci.*, 12, 2403, 1992.
2. Hillman, D. E., Parameters of dendritic shape and substructure: intrinsic and extrinsic determination?, in *Intrinsic Determinants of Neuronal Form and Function*, Lasek, R. J., and Black, M. M., Eds., Alan R. Liss, New York, 1988, 83.
3. Tamori, Y., Theory of dendritic morphology, *Phys. Rev.*, E48, 3124, 1993.
4. Ascoli, G. and Krichmar, J. L., L-Neuron: a modeling tool for the efficient generation and parsimonious description of dendritic morphology, *Neurocomputing*, 32, 1003, 2000.
5. Kliemann, W. A., Stochastic dynamic model for the characterization of the geometrical structure of dendritic processes, *Bull. Math. Biol.*, 49, 135, 1987.
6. Uemura, E., Carriquiry, A., Kliemann, W., and Goodwin, J., Mathematical modeling of dendritic growth in vitro, *Brain Res.*, 671, 187, 1995.
7. Letourneau, P. C., Kater, S. B., and Macagno, E. R., Eds., *The Nerve Growth Cone*, Raven Press, New York, 1991.
8. Sadler, M. and Berry, M., Morphometric study of the development of Purkinje cell dendritic trees in the mouse using vertex analysis, *J. Microsci.*, 131, 341, 1983.
9. Horsfield, K., Woldenberg, M. J., and Bowes, C. L., Sequential and synchronous growth models related to vertex analysis and branching ratios, *Bull. Math. Biol.*, 49, 413, 1987.
10. Van Pelt, J., Dityatev, A. E., and Uylings, H. B. M., Natural variability in the number of dendritic segments: model-based inferences about branching during neurite outgrowth, *J. Comp. Neurol.*, 387, 325, 1997.
11. Van Pelt, J. and Uylings, H. B. M., Natural variability in the geometry of dendritic branching patterns, in *Modeling in the Neurosciences: From Ionic Channels to Neural Networks*, Poznanski, R. R., Ed., Harwood Academic Publishers, Amsterdam, 1999, 79.

12. Ireland, W., Heidel, J., and Uemura, E., A mathematical model for the growth of dendritic trees, *Neurosci. Lett.*, 54, 243, 1985.
13. Nowakowski, R. S., Hayes, N. L., and Egger, M. D., Competitive interactions during dendritic growth: a simple stochastic growth algorithm, *Brain Res.*, 576, 152, 1992.
14. Li, G.-H. and Qin, C.-D., A model for neurite growth and neuronal morphogenesis, *Math. Biosc.*, 132, 97, 1996.
15. Van Veen, M. P. and Van Pelt, J., Neuritic growth rate described by modeling microtubule dynamics, *Bull. Math. Biol.*, 56, 249, 1994.
16. Uylings, H. B. M., Van Pelt, J., Parnavelas, J. G., and Ruiz-Marcos, A., Geometrical and topological characteristics in the dendritic development of cortical pyramidal and nonpyramidal neurons, in *Progress in Brain Research, Vol. 102, The Self-Organizing Brain: From Growth Cones to Functional Networks*, Van Pelt, J., Corner, M. A., Uylings, H. B. M., and Lopes da Silva, F. H., Eds., Elsevier, Amsterdam, 1994, 109.
17. Van Veen, M. P. and Van Pelt, J., Terminal and intermediate segment lengths in neuronal trees with finite length, *Bull. Math. Biol.*, 55, 277, 1993.
18. Abramowitz, M. and Stegun, I. A., *Handbook of Mathematical Functions*, Dover, New York, 1970.
19. Uylings, H. B. M., Ruiz-Marcos, A., and Van Pelt, J., The metric analysis of three-dimensional dendritic tree patterns: a methodological review, *J. Neurosci. Meth.*, 18, 127, 1986.
20. Schierwagen, A. K. and Grantyn, R., Quantitative morphological analysis of deep superior colliculus neurons stained intracellularly with HRP in the cat, *J. Hirnforsch*, 27, 611, 1986.
21. Larkman, A. U., Dendritic morphology of pyramidal neurons of the visual cortex of the rat: I. Branching patterns, *J. Comp. Neurol.*, 306, 307, 1991.
22. Rapp, M., Segev, I., and Yarom, Y., Physiology, morphology and detailed passive models of guinea-pig cerebellar Purkinje cells, *J. Physiol.*, 474, 101, 1994.
23. Purves, D. and Lichtman, J. W., *Principles of Neural Development*, Sinauer Associates, Sunderland, MA, 1985.
24. Sanes, J. R. and Lichtman, J. W., Development of the vertebrate neuromuscular junction, *Annu. Rev. Neurosci.*, 22, 389, 1999.
25. Bothwell, M., Functional interactions of neurotrophins and neurotrophin receptors, *Ann. Rev. Neurosci.*, 18, 223, 1995.
26. McAllister, A. K., Katz, L. C., and Lo, D. C., Neurotrophins and synaptic plasticity, *Annu. Rev. Neurosci.*, 22, 295, 1999.
27. Miller, K. D., Synaptic economics: competition and cooperation in correlation-based synaptic plasticity, *Neuron*, 17, 371, 1996.
28. Bienenstock, E. L., Cooper, L. N., and Munro, P. W., Theory for the development of neuron selectivity: orientation specificity and binocular interaction in visual cortex, *J. Neurosci.*, 2, 32, 1982.
29. Bennett, M. R. and Robinson, J., Growth and elimination of nerve terminals at synaptic sites during polyneuronal innervation of muscle cells: a trophic hypothesis, *Proc. R. Soc. Lond. B*, 235, 299, 1989.
30. Rasmussen, C. E. and Willshaw, D. J., Presynaptic and postsynaptic competition in models for the development of neuromuscular connections, *Biol. Cybern.*, 68, 409, 1993.
31. Van Ooyen, A. and Willshaw, D. J., Poly- and mononeuronal innervation in a model for the development of neuromuscular connections, *J. Theor. Biol.*, 196, 495, 1999.

32. Elliott, T. and Shadbolt, N. R., A mathematical model of activity-dependent anatomical segregation induced by competition for neurotrophic support, *Biol. Cybern.*, 75, 463, 1996.
33. Harris, A. E., Ermentrout, G. B., and Small, S. L., A model of ocular dominance development by competition for trophic factor, *Proc. Natl. Acad. Sci. U.S.A.*, 94, 9944, 1997.
34. Jeanprêtre, N., Clarke, P. G. H., and Gabriel, J.-P., Competitive exclusion between axons dependent on a single trophic substance: a mathematical analysis, *Math. Biosci.*, 133, 23, 1996.
35. Elliott, T. and Shadbolt, N. R., Competition for neurotrophic factors: mathematical analysis, *Neural Computation*, 10, 1939, 1998.
36. Wiley, H. S. and Cunningham, D. D., A steady state model for analyzing the cellular binding, internalization and degradation of polypeptide ligands, *Cell*, 25, 433, 1981.
37. Cohen-Cory, S. and Fraser, S. E., Effects of brain-derived neurotrophic factor on optic axon branching and remodeling in *in vivo*, *Nature*, 378, 192, 1995.
38. Garofalo, L., Ribeiro-da-Silva, A., and Cuello, C., Nerve growth factor-induced synaptogenesis and hypertrophy of cortical cholinergic terminals, *Proc. Natl. Acad. Sci. U.S.A.*, 89, 2639, 1992.
39. Holtzman, D. M., Li, Y., Parada. L. F., Kinsman, S., Chen, C.-K., Valletta, J. S., Zhou, J., Long, J. B., and Mobley, W. C., p140^{trk} mRNA marks NGF-responsive forebrain neurons: evidence that *trk* gene expression is induced by NGF, *Neuron*, 9, 465, 1992.
40. Birren, S. J., Verdi, J. M., and Anderson, D. J., Membrane depolarization induces p140^{trk} and NGF responsiveness, but not p75^{LNGFR}, in MAH cell, *Science*, 257, 395, 1992.
41. Sutter, A., Riopelle, R. J., Harris-Warrick, R. M., and Shooter, E. M., Nerve growth factor receptors. Characterization of two distinct classes of binding sites on chick embryo sensory ganglia cells, *J. Biol. Chem.*, 254, 5972, 1979.
42. Zupan, A. A. and Johnson, E. M., Jr., Evidence for endocytosis-dependent proteolysis in the generation of soluble truncated nerve growth factor receptors by A875 human melanoma cells, *J. Biol. Chem.*, 266, 15384, 1991.
43. Layer, P. G. and Shooter, E. M., Binding and degradation of nerve growth factor by PC12 pheochromocytoma cells, *J. Biol. Chem.*, 258, 3012, 1983.
44. Blöchel, A. and Thoenen, H., Characterization of nerve growth factor (NGF) release from hippocampal neurons: evidence for a constitutive and an unconventional sodium-dependent regulated pathway, *Eur. J. Neurosci.*, 7, 1220, 1995.
45. Bernd, P. and Greene, L. A., Association of ^{125}-I-nerve growth factor with PC12 pheochromoctytoma cells. Evidence for internalization via high affinity receptors only and for long-term regulation by nerve growth factor of both high- and low-affinity receptors, *J. Biol. Chem.*, 259, 15509, 1984.
46. Van Ooyen, A. and Willshaw, D. J., Competition for neurotrophic factor in the development of nerve connections, *Proc. R. Soc. Lond. B*, 266, 883, 1999.
47. Albers, K. M., Wright, D. E., and Davies, B. M., Overexpression of nerve growth factor in epidermis of transgenic mice causes hypertrophy of the peripheral nervous system, *J. Neurosci.*, 14, 1422, 1994.
48. Nguyen, Q. T., Parsadanian, A. S., Snider, W. D., and Lichtman, J. W., Hyperinnervation of neuromuscular junctions caused by GDNF overexpression in muscle, *Science*, 279, 1725, 1998.
49. Barry, J. A. and Ribchester, R. R., Persistent polyneuronal innervation in partially denervated rat muscle after reinnervation and recovery from prolonged nerve conduction block, *J. Neurosci.*, 15, 6327, 1995.

50. Van Ooyen, A. and Willshaw, D. J., Influence of dendritic morphology on axonal competition, in *Artificial Neural Networks* - ICANN'99, Institution of Electrical Engineers, London, 9th International Conference on Artificial Neural Networks, Edinburgh, September, 1999, 1000.

8 Passive Cable Modeling — A Practical Introduction

Guy Major

CONTENTS

0-8493-2068-2/01/$0.00+$.50

8.1 INTRODUCTION

This chapter is not intended to be a rigorous review of passive cable modeling theory, nor is it a comprehensive review of the relevant literature. There are already many publications covering the first task.[1–5] Instead, I shall focus here on the practical details of passive cable modeling that often get swept under the carpet, and I shall try to offer some guidance to would-be passive cable modellers, based on my own experience.

8.2 WHAT IS PASSIVE CABLE MODELING?

Neurons have treelike morphologies. Their cell membranes are lipid bilayers with proteins floating around in them, some of which are transmembrane channels. The lipid part of the membrane has an extremely low conductance per unit area. Many of the channels are voltage- or calcium-dependent, and are capable of changes in their conductance in the millisecond to second time scale. Others have an effectively constant conductance. As a result, each small patch of membrane is electrically equivalent to a variable resistor (in series with a battery) in parallel with a capacitor, connecting the interior of the cell with the extracellular space (earth). The *passive* membrane conductance and capacitance is that part of the membrane conductance and capacitance which is effectively constant during a response, and can be approximated by a fixed resistor and capacitor, respectively (Figure 8.1).

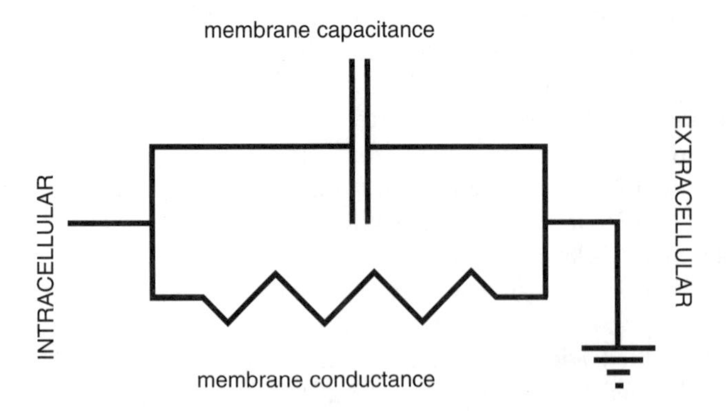

FIGURE 8.1 Electrical representation of an isopotential patch of passive membrane. The membrane is assumed to behave like a resistor and capacitor in parallel, discharging to earth.

The cytoplasm within a process of a neuron is composed of water, electrolytes, charged proteins, cytoskeleton and other obstructions such as organelles and endoplasmic reticulum (among other things). The latter ingredients probably raise cytoplasmic resistivity above the 50–70 Ωcm (in the case of mammals) that would be predicted from the composition of its electrolytes alone. Each short tube of cytoplasm can be thought of as being electrically equivalent to a resistor, which connects one short cylindrical patch of cell membrane to the next. A passive dendrite or axon can therefore be represented as shown in Figure 8.2, by parallel R–C pairs to earth coupled on the intracellular side by axial resistances. This approximation is valid provided every part of a given patch experiences approximately the same voltage and if radial (as opposed to axial) current flows in the cytoplasm are negligible. Cable theory, a related analytical theory, has generated a set of differential equations resting on the same key assumptions:[1–3,6]

A. Each cylindrical neural process can be simplified into a 1-dimensional cable, with current flowing both across the membrane and down the core.
B. The extracellular space is isopotential.

Different parts of the neuron have different structures, and may therefore have different membrane capacitance and conductances per unit area, as well as different cytoplasmic resistivities ($R_{i.}$). It is well-established now that different parts of neurons have different channel densities, and electron microscopic evidence and other considerations suggests that thinner dendrites may have higher cytoplasmic resistivities than fatter ones. Total membrane capacitance per unit area has a voltage and frequency dependent component caused in part by the gating currents in transmembrane channels, and so will also vary from branch to branch. Passive membrane capacitance per unit area (C_m) is unlikely to vary so much, although in principle the lipids and proteins and other membrane components may be nonuniform. The passive membrane conductance per unit area (G_m) and its reciprocal R_m are quite likely to be nonuniform, indeed there is mounting experimental evidence that this can be so.

C_m, R_m, and R_i are frequently referred to as the specific passive parameters, because they correspond to unit areas and length respectively.

A neuron's morphology can be brought alive in a computer simulation by converting it into an electrically equivalent structure, and then giving that structure various inputs and measuring its responses. The equivalent structure is essentially a branching tree of axial resistors, mimicking the cytoplasm, all connected together in a way that mirrors the morphology. At the connection between each pair of axial resistors are a capacitor and variable resistor in parallel to earth, mimicking the membrane. This is a *compartmental model*, a discrete numerical approximation to the branched *cable* that is electrically equivalent to the cell. A passive compartmental model is one lacking components that are rapidly modulated by voltage or simulated chemical concentration changes. Analytical solutions exist for many kinds of inputs into passive branching cable structures,[1–4] but these are all essentially numerical approximations too, albeit in many cases more efficient and accurate computationally than the equivalent compartmental simulations.

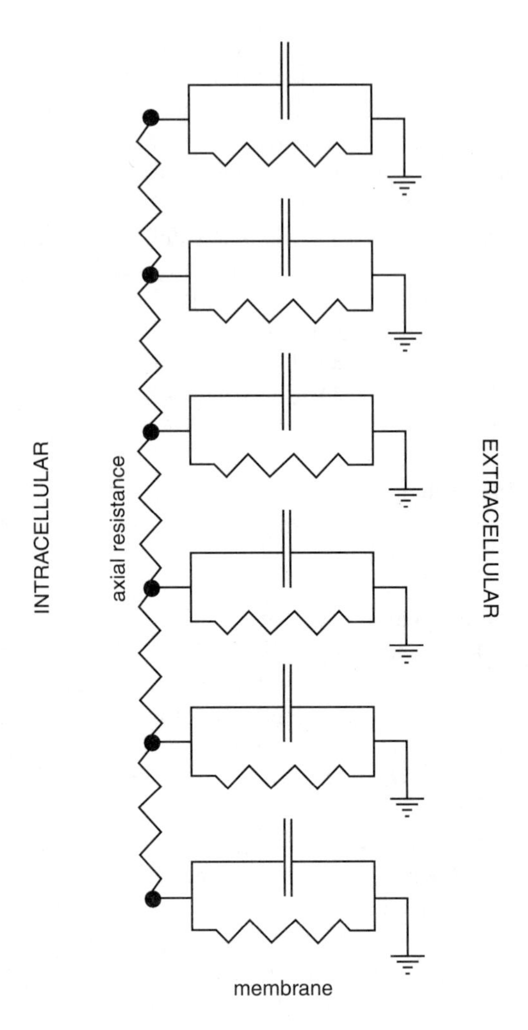

FIGURE 8.2 Compartmental electrical representation of a segment of passive cable. Several isopotential membrane patches are connected in series via axial resistances. The extracellular space is assumed to be isopotential; radial current flows (perpendicular to axial resistances) within the cytoplasm of each compartment are ignored. This is a discrete numerical approximation to a continuous cable.

The most important property of passive electrical cables is that different points experience different voltages, and the cables acts as a low pass filters, attenuating and smoothing voltage transients as they travel along the branches. A simple schematized example is shown in Figure 8.3. Charge is injected suddenly into one point of a single cylindrical passive cable. The corresponding voltage transients at either end are traces A and B (semi-log plot). Generally speaking, realistic axial resistances are much smaller than membrane resistances, so the charge travels rapidly down the cable, equalising (roughly speaking) along its length. This corresponds to the rapidly falling early components of trace A, the transient at the input site, and the rising

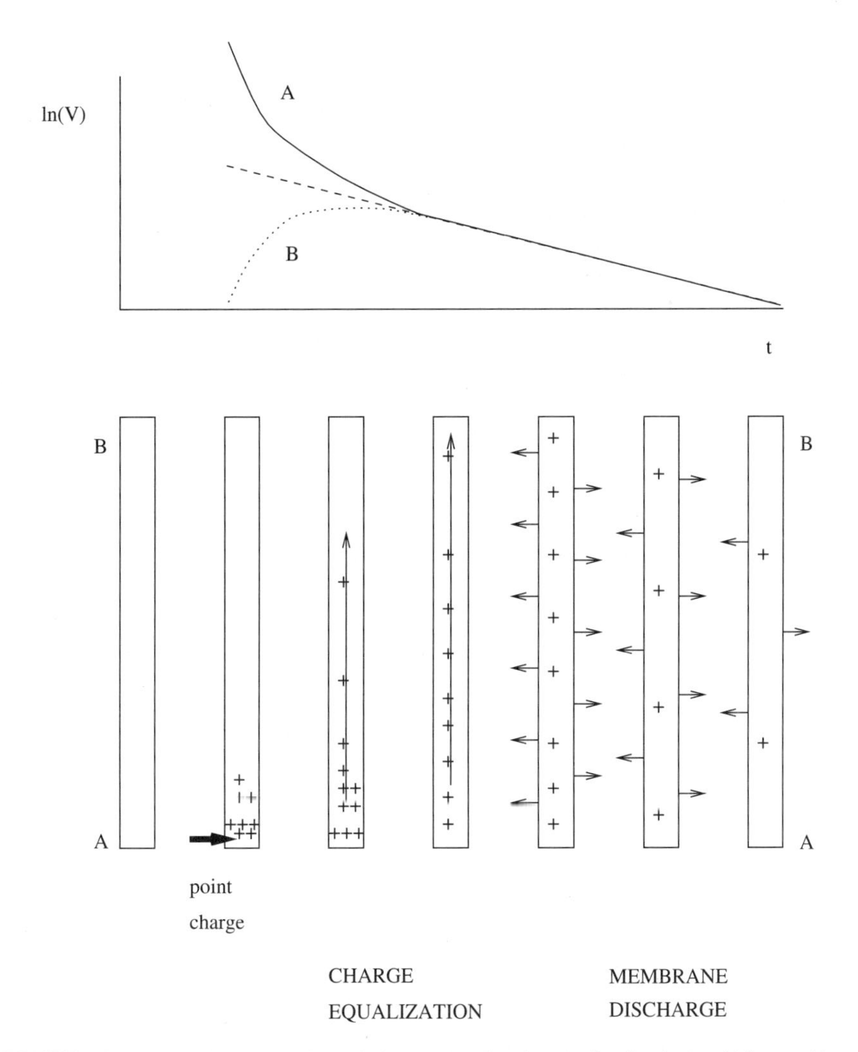

FIGURE 8.3 Schematic illustration of charge equalization (redistribution) within a uniform passive cable (sealed ends). An impulse of current is injected at A. The charge rapidly spreads axially, then discharges more slowly through the membrane. Upper panel shows semi-log plots of the responses at the two (sealed) ends of the cable. As charge redistributes, response at injection site (*A*) decays rapidly while response at far end (*B*) builds up. At later times, both responses superimpose (charge has equalized), and decay as membrane discharges more or less uniformly to earth.

phase of B, the transient at the far end. As the charge equalizes, so the voltages at the two ends (and points between) converge. At later times, the membrane discharges slowly to earth, via its R-C components. This corresponds to the single exponential decay seen on the semi-log plot (the straight line part).

Local *shunts* (see below) or nonuniformities in the specific membrane parameters complicate the picture. Instead of equalizing, charge redistributes, but never reaches an approximately uniform distribution, because some parts of the cable are discharging to earth faster than others. The direction of axial charge flow can even reverse during a response. Different points in the cable continue to experience significantly different voltages throughout the response.

Side branches off a cable can drain off charge (load effects), leading to more attenuation of signal but sealed ends can bottle charge up (end effects), leading to less attenuation of signals. In a heavily branching tree, large, fast signals in one branch lead to much smaller slower transients at distant locations.

The theory of passive cable modeling tells us that responses to transient current and voltage inputs are made up of a number of exponentially decaying components,[3] with different time constants τ_i and amplitudes A_i, where i indexes the components. The time constants hold for the entire cell, but the amplitudes vary with the input and recording positions, and the time course of the input. If the geometry is complex, there may be a large number of components contributing to response waveforms over the time interval of interest.[6] The index of the slowest component is 0 by convention, and τ_0 is commonly called the membrane time constant. If the membrane parameters are uniform, τ_0 equals $R_m C_m$.

8.3 WHAT IS THE POINT OF PASSIVE CABLE MODELING?

There are three main points:

1. To produce an *electrical skeleton* onto which active (time-varying) conductances can be grafted, as part of trying to build up a realistic computational model of a cell. If you get the passive skeleton wrong, you could run into all sorts of problems once you go active.
2. To *reconstruct signals* that originated in one part of a passive neuronal tree, but which have been recorded at another point, perhaps because of technical constraints. The most common example is the reconstruction of synaptic currents or conductances which have been heavily filtered (slowed and attenuated) by intervening dendritic cables before being recorded at the cell body.[7]
3. As a toy for neuroscientists to build up their intuitions and understanding of neuronal cables via play. One should interact with simpler passive cables and compartmental models before graduating to active compartmental models, where the number of parameters can rapidly get out of hand.

8.4 SHUNTS

A shunt is a point conductance in parallel to the membrane conductance, either to earth, or with a reversal potential. Pathological shunts can be caused by localized damage, for example, there may be a physical gap between a sharp electrode and the membrane. Such a shunt is expected to have a reversal potential near zero. In addition, even tiny gaps can let in calcium or sodium ions which will switch on calcium- or sodium-activated potassium conductances. The high ionic concentration near the tip of a hyperosmotic sharp electrode may also cause the effective local membrane conductance to increase. The latter two kinds of shunt might be expected to have negative reversal potentials — in theory close to the resting membrane potential or the potassium reversal potential. Natural shunts can be caused by high local densities of intrinsic membrane conductances or synaptic conductances. Different kinds of shunt can coexist at a given location, and their net reversal potential will be a weighted average of the individual reversal potentials. The shunt conductance at a given point is commonly abbreviated to g_{shunt}. The most common location is the soma. Shunts can exacerbate problems of model nonuniqueness (see below), by introducing extra model parameters.

8.5 SPACE CONSTANTS AND ELECTROTONIC LENGTH

Suppose a constant current is injected at one point in a cylindrical cable which extends to infinity. The *space constant* λ is the distance along the cable over which the steady state voltage decays to 1/e of its value at the injection site. This is given by ($\sqrt{R_m d/4R_i}$), where d is the diameter of the cable.[3] Another ubiquitous parameter in the field is L or electrotonic length, namely the physical length of a finite segment of cable divided by the space constant of its infinite extension. See below for common misconceptions about L.

8.6 IS PASSIVE CABLE MODELING TRIVIAL? COMMON ERRORS AND PITFALLS

Unfortunately passive cable modeling is not as trivial as it may seem at first sight and is all too easy to slip up, which has resulted in a somewhat confused picture in the literature. Even obsessive studies may have made potentially serious errors.[8]

8.6.1 Fitting a Passive Cable Model to Active Data

8.6.1.1 Linearity

A passive cable structure with current or voltage inputs is a *linear system*, that is, if you double the input, the output should also double, if you multiply the input by –1, you should get –1 times the output. The response to two stimuli is given by the sum of the responses to the individual stimuli. Linear scaling tests[8] should always be performed before using electrophysiological data for passive cable modeling, but rarely are. Many experimenters believe it is sufficient to operate in a roughly linear part of a cell's steady-state I-V curve, but this risks missing rapid voltage-dependent conductances which turn off at later times, while distorting responses at early times.

Conductance inputs however do not behave linearly in a passive system, a fact often forgotten by inexperienced modellers. If you simulate a synaptic conductance, then double it, you won't get twice the response, because of a reduced driving voltage, and because of the local increase in net membrane conductance. The responses to very small conductances, however, can scale approximately linearly, if these two effects are negligible.

8.6.1.2 Reciprocity

If you inject an input into one point on a passive cable tree, and record from another point, then swap input and recording sites, you end up with the same response (Box 8.1). This feature of passive cables allows an extremely powerful check on whether a real neuron is behaving approximately passively, if it possible to record simultaneously from several points on the cell. This is routinely possible in a number of cell classes now.

Box 8.1 An Intuitive Explanation for Reciprocity

This is based on the physical and mathematical similarity between diffusion and the spread of charge along cables. First, convert each cylindrical segment of cable into an electrically equivalent bundle of shorter, thinner cylindrical cables, all identical and connected together in parallel at both ends. The bundle of short, thin cables must have the same total surface area and axial conductance as the original segment. Choose a "lowest common denominator" or "universal" diameter, such that all segments of the original tree can be converted into an equivalent parallel bundle of thin cables with the universal diameter. Fatter original cables will have more universal cables in their equivalent bundle, thinner original cables will have fewer.

Suppose some charge is injected instantaneously at one point (A) in the original tree (spread evenly between its equivalent bundle). Each ion injected does a random walk along the cables. The voltage at another location B at a particular time t after the injection is proportional to the number of ions at B at time t. This is proportional to the number of possible routes or trips starting at A and ending at B, with a journey time t, divided by the total number of possible routes of trip duration t from A to anywhere in the tree.

To a first approximation, the total number of possible trips of duration t originating from B and ending anywhere is the same as the total number originating from A and ending anywhere — once the trip length is significantly longer than the distance between A and B, the starting point is more or less irrelevant. The number of trips from B to A having trip duration t is exactly the same as the number of trips from A to B of the same length, therefore, the direction is irrelevant.

Since the voltage at at time t in response to the charge being injected at B is proportional to the number of trips of duration t starting at B and ending at A, divided by the total number of possible trips of duration t starting from B and ending anywhere, it can be seen that the $A \rightarrow B$ response has to be the same as the $B \rightarrow A$ response — i.e., there is reciprocity — except at extremely early times. The situation is complicated by leakage of charge through the membrane, but a similar argument can still be followed. For a more rigorous discussion of trip-based algorithms and Feynmann path integrals see Reference 9.

8.6.1.3 Sag

Short pulses or impulses of current or voltage are often used to probe real neurons' passive cable properties. Unfortunately, simulations and experiments have shown that cells with the h-conductance, or hyperpolarization-activated conductance, are capable of generating impulse responses that scale linearly with the input current but show a strong undershoot at late times (40–200 ms), which also scales approximately linearly (deviations from linear scaling cannot be picked out from experimental noise). This is clearly incompatible with passive cables. Simulations of this effect are shown in Figure 8.4,[10] which shows semi-log plots of short pulse responses of a simplified cortical layer 2/3 pyramidal neuron with a uniform h-conductance density. As the soma is held at progressively more hyperpolarized voltages, the decays speed up, and show stronger negative deviations at late times (sag). In fact the responses undershoot, although this cannot be seen on the semi-log plots. What is frightening though is the fact that in the presence of realistic amounts of noise on the traces it would appear that the responses were scaling linearly with the input. An actual experimental example is shown in Color Figure 8.1*: scaled and superimposed short pulse (+0.5, –0.5, +1, –1 nA × 0.5 ms) responses from a real CA1 cell superimpose very well, even at late times when the responses are clearly sagging below the blue dashed line.

Linear scaling is clearly a necessary, but not a sufficient condition for classifying responses as passive. It is obvious from their long pulse responses that neurons like those in Color Figure 8.1 are not behaving passively: the responses sag back toward baseline at late times, and undershoot once the current is turned off (Figure 8.5). There is also clear inward rectification, input resistance increases with depolarization, and equal increments of current lead to smaller increments in voltage in the hyperpolarizing direction.

At least sag or undershoot can be picked up easily in relatively clean data (averages of large numbers of responses, taken in presence of neurotransmitter blockers). Fitting passive models to responses of models including sag can lead to potentially serious errors in the optimal passive parameters. R_i and C_m tend to be roughly right, but R_m and somatic shunt (if included) can be underestimated.[10] If transfer responses (inject at one point, record at another) are being fitted, the problems can get worse.[11] Nevertheless, if it is thought that the specific passive parameters are roughly uniform (e.g., no somatic shunt), and if fits are restricted to early parts of the responses (say less than 25 ms), well before sag appears, then the optimal fit parameters can be roughly correct.

8.6.1.4 Creep

This is the opposite of sag or undershoot, and is much harder to pick up. Instead of sagging back towards baseline, the voltage response to a long pulse of current creeps slowly away from rest, taking a long time to reach a steady state. Short pulse responses scale linearly with the input — and so appear to be passive — but do not demonstrate clear exponential decay at late times.

* Color Figure 8.1 follows page 140.

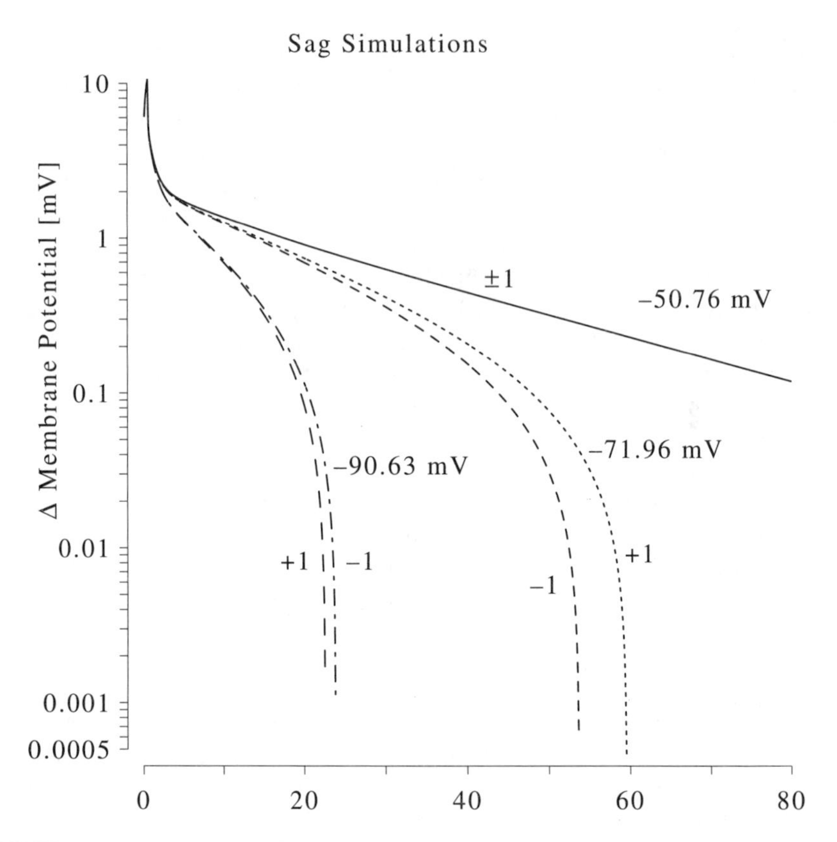

FIGURE 8.4 Semi-log plot of somatic responses of a model incorporating h-current. The morphology was a simplified representation of a cortical layer 2/3 pyramidal cell, with kinetic parameters slightly modified from Reference 24, details in Reference 10. The soma was held at three different membrane potentials and ±1 nA, 0.5 ms pulses were injected into the soma.

An example from a cat spiny stellate cell is shown in Color Figure 8.2.* The effective decay time constant of the short pulse responses appears to get steadily slower at progressively later times. There is no portion of the response showing clean single exponential decay. These cells have extremely simple geometries, with all the dendrites having roughly the same lengths and diameters. In fact, the cells can be well approximated by equivalent cylinder representations.[3] Intermediate time constant response components with large amplitudes can generate slow bends in responses, but are hard to generate from such simple cell geometries.[10,12] When fits are attempted to responses such as those in Color Figure 8.2, the results are very poor, often with nonsensically high R_is.

Creep can be caused by persistent sodium currents, and perhaps by other kinds of conductance: the only requirement is a slowly depolarization-activated conductance with a reversal potential above rest, or a slowly hyperpolarization-activated

* Color Figure 8.2 follows page 140.

conductance with a reversal potential below rest (i.e., the opposite of the h-conductance). Creep should be apparent from long pulse responses: they will not scale linearly. An example from the spiny stellate cell above is shown in Figure 8.6. The scaled responses do not superimpose and the charging and discharging curves have different time courses. The cell shows strong inward rectification, and the apparent time constant increases with depolarization.

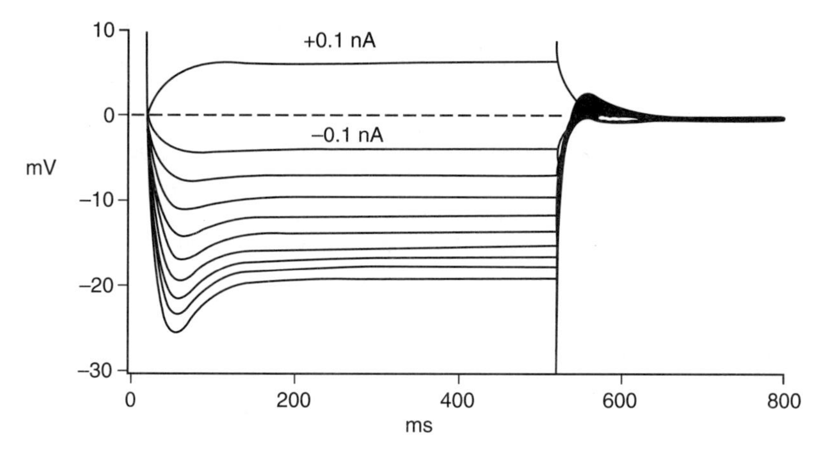

FIGURE 8.5 Long pulse responses in current clamp of same CA1 cell as in Color Figure 8.1 show pronounced sag and rebound overshoot. Increments of 0.1 nA. Notice how sag becomes more pronounced at more hyperpolarized voltages, resulting in inward rectification of steady-state responses. Responses scaled by 1/current would not superimpose.

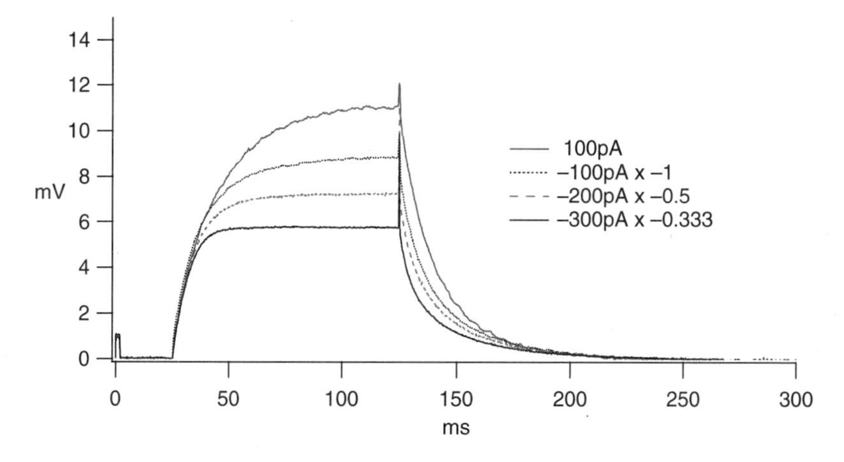

FIGURE 8.6 Long pulse responses in current clamp from same spiny stellate cell as in Color Figure 8.2, scaled by 1/current. Notice how responses show strong inward rectification: time constants and amplitudes increase at more depolarized potentials.

In short, it is important to test that a cell's responses appear to be passive at both short time scales and long time scales — using both impulse (short pulse) response linear scaling and long pulse response linear scaling. The long pulse on charging transient (approach to steady-state) should also superimpose with the inverted off discharging transient (approach to rest). Major et al.[8] made the potential mistake of not testing long pulse linear scaling, which may have compromised their results. Fitting active data to passive models can lead to horrendous errors and misleading conclusions, most famously perhaps that layer five pyramidal cells in cortex are effectively isopotential. This is now known to be completely wrong, thanks to multi-pipette recordings from single cells.[11]

The golden rule is — *if it moves, block it.* Pharmacological agents should be used to render a cell passive before taking data for passive cable modeling purposes. We now know a lot about which conductances are partially activated around rest, and all of these should be blocked, if possible. A wide range of agents are available, the most useful including ZD7288 (Tocris Cookson)[13] for blocking the h-conductance (10–20 mM) and TTX (for blocking persistent and other sodium conductances which might cause creep). Useful blockers for A-currents include 4-AP[14] and for T-type calcium currents include amiloride[15] and cadmium (150 μM — nonspecific for all calcium channels). Intracellular QX-314 and cesium block a wide range of sodium and potassium channels as well as h-channels, but a worry remains about use of these agents, since they are associated with use-dependent and voltage-dependent block and relief of block, which can artifactually distort response waveforms, as well as causing nonstationarity. Short and long pulse response linear scaling and stationarity tests should pick up this problem.

8.6.2 Fitting a Stationary Model to Nonstationary Data

Data averaging is commonly performed in order to improve signal-to-noise ratios for the small signals commonly used for passive cable modeling. If the recorded cell's physiological properties are changing over time, because of wash-out of channels, or because of some other pharmacological effect, or worse still, because of instability in the recording, it is easy to average together responses which correspond to quite different electrical systems. The worst case might be if the seal of a patch pipette to the soma, for example, was unstable, and kept springing a leak for a few hundred milliseconds at a time. One might not spot this, and average together the responses of two quite different electrical systems — one with a large shunt at the soma, and one without. This can distort the response waveforms and lead to bizarre fit parameters, for example an excessively high R_i. For this reason it is essential to inspect data by eye, and to ensure that any data that is averaged together is behaving in a stationary manner, namely, the underlying electrical behavior of the cell is constant over the averaging period. If voltage clamp is being used, nonstationarity in the series resistance is a particular problem and can mess up cable modeling data. This is one reason not to use voltage clamp transients for cable modeling.

8.6.3 Using Long Pulses Alone, or Ignoring Series Resistance

Impulse responses are the time derivative of long pulse (step) responses and therefore are more sensitive to the cable properties of the cell in question. There is more information (for a given signal-to-noise ratio) in an impulse response than in a long pulse response, because the early components, which are most sensitive to R_i and C_m, are larger for impulses.

Another problem with long pulses is that it is very hard to estimate series resistance (bridge balance) in a distributed cable system, because the cable generates fast electrical response components which blend into those of the electrode. As a result, there is no clear breakpoint between the electrode's response and the cell's response, and bridge balance is subjective guess-work, whether it is performed in current clamp or voltage clamp mode. Basically, just as with voltage clamp, using current clamp long pulse data introduces series resistance as another model parameter. For current clamp data, the series resistance parameter can be eliminated in two equivalent ways: (a) fit the charging curve relative to the steady-state voltage level achieved during the pulse, ignoring the baseline, and the initial fast change in voltage which could be due to the electrode or (b) fit the discharging curve relative to the voltage baseline (there is no voltage drop across the electrode once the current is switched off), ignoring the steady-state level and initial fast drop in voltage. In other words, choose a fit interval that steers well clear of the capacitative/bridge imbalance artifacts.

8.6.4 Ignoring Nonuniqueness

Because it seems quite probable that some of the specific passive electrical parameters, namely C_m, R_i and R_m may be nonuniform, it is quite likely that many different models of the cell may fit the experimental data equally well, while leading to significantly different predictions. This danger has to be taken on board at the start of any passive modeling study, particularly when sharp electrodes have been used and may have caused local shunts at impalement sites (see above).

8.6.5 Morphology-Related Errors

These days, the most credible passive cable models of individual cells are those that combine high-quality morphology (see also Chapter 6) with high quality physiological data from the same cell. Many morphologically related errors are possible. The first, and most worrying is swelling. It is important to preserve the osmotic balance within the cell, particularly if one is using multiple channel blockers to passify it. The main cause of this is the fact that the bulk of the intracellular anions are actually large relatively fixed negatively charged protein molecules, which cannot diffuse out of the cell and into the recording electrode or pipette. The pipette solution on the other hand contains free anions (e.g., chloride, methylsulphate or gluconate). Common symptoms of osmotic stress are unstable recordings, disappearance (loss of contrast) of the soma and apical trunk when viewed under D.I.C. or dark field microscopy, and collapse of dendritic spines into shafts, or even beading of dendrites (Chapter 6, Figure 6.1) and leak of indicator out of the cells. Loss of spines and beading may only be apparent after the event, when filled cells are processed histologically. The solution is straightforward; reduce the osmolarity of the filling solution. 0.5 M K-Methylsulphate is recommended for sharp electrodes (resistances around 100 MΩ) or 270 mOsm patch filling solutions.

Swelling is particularly insidious, because it may reduce the early components of the impulse responses, and may lead to nonstationarity in the data. Worse still, once the electrode or patch pipette is withdrawn, the cell may recover to some extent, so if it is fixed more than half an hour later, its final morphology may be very different to that actually responsible for the impulse responses recorded.

Other morphological errors occur when tissue is processed, most commonly shrinkage — relatively straight dendrites develop wiggles (Chapter 6, Figure 6.3) in the case of HRP and biocytin fills, when the tissue is being dehydrated. Wiggle is relatively easy to measure, and the measured lengths of dendrites can be compensated accordingly.

More insidious is the fact that dendrites often appear fatter at the surfaces of processed sub-sections than in the centres, because of differential diffusion of reagents. This is a particular problem with biocytin fills. Triton-processed tissue can develop small puffs of reaction product which look like tiny dendritic spines that can easily lead to spuriously high spine counts. The majority of dendrites have diameters around 0.5–0.8 μm in most vertebrate central neurons, but the resolution of the light microscope is only around 0.2 μm, making diameter measures doubly prone to error. Single spine areas have only been estimated for a comparatively small number of spines using EM, and densities can be unreliable because of occlusion. In other words, most morphological measurements, while better than nothing, need to be taken with a pinch of salt.

What is particularly troubling is the simple fact that axial resistance is proportional to $1/(\text{diameter})^2$, so a –0.2 μm error in a 0.6 μm diameter dendrite can lead to the axial resistance being out by a factor of between about 0.5 and 2, a fourfold range of uncertainty in the worst-case scenario ($0.8^2 = 0.64$, $0.6^2 = 0.36$, $0.4^2 = 0.16$). Diameters can be measured with more precision using high voltage EM, or serial EM, but the effort involved is prodigious, and the tissue reaction and shrinkage problems remain.

Luckily, one can still produce an electrically correct model of a cell, even if the morphology has some kind of uniform systematic error like all the diameter are out by a given factor, providing one directly fits the responses of the model to the responses of the corresponding real neuron. If the diameters are twice what they should be, the optimal fit R_i will end up being four times what it should be, but the axial resistances themselves will be correct: the specific passive parameters can trade off with morphological errors, so the battle is not lost. However, transplanting parameters from one cell in one study with one set of morphological errors into another cell from another study with another set of morphological errors could lead to potentially misleading models and conclusions.

8.6.6 Fit Rejection, Under-Fitting and Over-Fitting

The noise in real neurons is badly behaved for two reasons. First, it is dominated by changes in membrane potential that last for similar lengths of time to impulse responses, and second it is non-Gaussian. It is not like nice, well-behaved white noise. Even if one blocks all known neurotransmitters (which one should do, to abolish spontaneous PSPs which contribute much of the noise, as well as lowering R_m in some preparations), substantial noise remains. This is probably caused in part by a combination of single channel noise,[14] seal instabilities, and maybe spontaneous transient osmotic rupturing of dendrites or spines.

When one compares the response of a model with the real data, it is not trivial to decide when the model produces a good fit. One can inspect the two waveforms by eye, to see if they superimpose well. However, one can cheat, by making sure the data is quite noisy, allowing plenty of slope for poor fits to appear OK (under-fitting). If the data is slightly distorted by slow noise, one can also end up over-fitting — the optimal fit will not actually be the model closest to the real cell. To be rigorous, one should explore a range of models, and devise some objective statistical test to reject some and not others. One straight forward (if tedious) test is to compare fit residuals with similar stretches of baseline noise, obtained using the same length of baseline voltage (to estimate holding voltage), same number averaged. One needs to get an idea of the 95% confidence limits of the maximum and average deviation from zero of such noise traces, by comparing several of them, and then apply these confidence limits to the fit residuals. Any residual escaping from the 95% confidence band is likely to be as a result of a misfit, and can be rejected.[0] However, one should not get too obsessed by these random errors, in the light of the huge number of possible systematic errors already discussed.

8.6.7 Incorrect or Inadequate Capacitance Compensation

If the capacitance compensation dial is set incorrectly, this can distort the early components, and in bad cases, all components of the impulse or long pulse responses. For example, undersetting capacitance compensation can lead to waveforms with boosted early components which require artefactually high R_is to fit them. The best way round this problem is to keep solution levels low both inside and outside pipette, use sylgard on pipettes if necessary, optimally set "cap. comp.," record data, and then record some data with cap. comp. deliberately under-set, just to get an idea of how safe the responses are at early times.

8.6.8 Misunderstanding the Relationship Between Electrotonic Length and Cable Filtering

Having arrived at a model of a cell, there are still many mistakes one can make. For example, the concept of electrotonic length is much misunderstood. In a finite cable with a sealed end (the most usual assumption), the steady-state voltage does *not* fall to $1/e$ of its starting value after one space constant: the sealed end causes the decay to be more gradual than in the situation where the cable extends to infinity. More importantly, however, *transients* are attenuated much more strongly by cables than are steady-state signals. The space constant is, in effect, frequency dependent. If a sinusoidal signal is injected into a cable, its attenuation by the cable increases as the frequency rises. Put another way, the effective space constant decreases with frequency.[16,17] The early parts of fast voltage transients (e.g., most postsynaptic potentials) are much more dependent on R_i and C_m than they are on R_m, which principally influences the decaying phase at later times (see Parameter Dependencies section below). As mentioned, side loads cause additional attenuation. As a result, L is a dangerous guide as to what filtering effect a cable has on signals. It is far better to model the situation explicitly, exploring an appropriate range of assumed parameters.

The commonly assumed rule of thumb for compartmental modellers, that one needs about 10 or 20 compartments per space constant for reasonable accuracy, is another version of the misconception that electrotonic distance is somehow equivalent to filtering effect or attenuation. The number of compartments required for a given degree of accuracy should increase with the speed of the transient — the safest course of action is to explicitly check the accuracy of simulations by double checking with different numbers of compartments and different time steps.

8.7 RECOMMENDED PASSIVE MODELING RECIPES

The previous section highlights some of the many possible pitfalls in practical passive cable modeling. There are no doubt many others as the topic is a minefield for the unwary. The following is the recommended recipe for how to do passive cable modeling of a given cell type. This section is by no means the last word. Good cooks start with the recipe, then try and improve on it and adapt it to their own tastes.

8.7.1 Step 1: Know the Rough Morphology

Fill some cells you are interested in modeling, with biocytin or neurobiotin or HRP, or some fluorescent indicator, then ask are the dendrites all of similar lengths and diameters? If not, see if length/√(diameter) is similar between dendrites. If it is (which is rare), you can get very excited, and start using much of the equivalent cylinder cable theory developed by Rall and others,[1,3] but do not forget the warnings above. Also, beware of exponential fitting and exponential peeling routines as they are error-prone and sensitive to fit intervals. We will assume in what follows, that your cell has some long dendrites and some short dendrites, and that its morphology *cannot* be well approximated by an equivalent electrical cylinder.

8.7.2 Step 2: Impulse Responses and Long Pulse Response Behavior

Next, record from single cells, preferably with patch pipettes, but "dilute" sharp electrodes will do (e.g., 0.5 M potassium methylsulphate), and record long pulse responses over a range of membrane potentials close to resting, say between –80 and –60 mV. Add neurotransmitter blockers (e.g., 100 μM APV, 20 μM CNQX, 100 μM picrotoxin, 10 μM CGP55845A,[18] and/or anything else that seems relevant). Put in pulses of both polarities, to explore all of this 20 mV range. For pyramidal neurons, steps of 0.05 nA are useful, duration 200–400 ms.

Scale the on and off responses by 1/(current injected), invert the off responses, and superimpose. Are there any signs of non-linearity? Look for inward rectification (bigger responses in depolarizing direction), sag, creep and any other active behavior. Then attempt to block these with pharmacological agents: 10–20 mM extracellular ZD7288 for sag/undershoot.[13] Intracellular gluconate can reduce sag too.[19] Attack inward rectification with TTX, and if necessary intracellular QX-314 and cesium, although the latter two may have their own problems (above). Any signs of A-current or T-type calcium currents and these should be dealt with too.

Once the long pulse responses are behaving passively, check the responses to 0.5 ms short pulses, say averages of 50–100. The current should be adjusted so that the response is around 5 mV at 1–2 ms after the current turns off. +0.5 nA and –1 nA × 0.5 ms current pulses seem to work well for cortical and hippocampal pyramidal cells, but for smaller cells you will need smaller currents. Inspect the responses using both normal and semi-logarithmic axes (ln(V) vs. t). If the short pulse responses pass linear scaling tests by eye, then, if you are obsessive, you could try comparing subtracted waveforms with equivalent subtracted noise waveforms to see if there are any significant differences, as suggested above for comparing fit residuals to noise.[8]

Do not forget reciprocity tests if you are using multi-site recording and stimulation. Move the input and recording sites around, and see if you still end up with the same response. If not, something is wrong.

8.7.3 Step 3: Get the Morphology Right

Once you are sure you have developed the right pharmacological cocktail for making the cells behave passively around rest, check you are still able to get good fills, and that there are no signs of swelling or other problems. The most important thing is to get the pipette solution osmolarity right, perhaps by trial and error, but for mammalian brain slices, around 270–275 mOsm is a good starting point and significantly below what many people use. Very low series resistances are not necessarily an advantage. Aim for approximately 15–25 MΩ. The higher the series resistance, the slower the cell will swell.

8.7.4 Step 4: Obtaining Target Waveforms for Direct Fitting

Now combine pharmacology, filling and impulse responses. Put in as many impulse responses as you can, leaving plenty of time between trials for noise measures and for tails of responses to decay back to baseline. Just to double check, use at least two sizes and polarities, e.g., –1 nA and +0.5 nA. Stimulate and record from as many points on the cell as possible. If linear scaling still holds, pool the data (weighting by signal-to-noise ratios), normalize to + 1 nA, and form a grand average

(average of averages) for your target waveform. If the responses do not scale linearly, you must throw the cell away. Do not waste time reconstructing it.

8.7.5 Step 5: Reconstruction

There are various ways of reconstructing cells, some manual, some semi-automated. Choose the one you like best, but whatever you do, measure diameters and spine densities as carefully as you can, e.g., under high power oil immersion, or with a high resolution 2-photon or confocal microscope with as high a zoom as is consistent with the resolution limit. Measure lengths and do not forget wiggle factors and 3-D pythagorean corrections (manual reconstruction). Try not to miss out or misconnect dendrites as this can have disastrous consequences. For small neurons with a highly branched axon it may be necessary to reconstruct the axon also. If you suspect you may have lost a subsection then throw away the cell. If you have the resources, calibrate some of your measurements with serial EM or high-voltage EM (see Section 6.3.3.4).

In the case of spiny neurons, carry out a *spine collapse procedure*,[8] to incorporate the spine area morphologically into dendritic shafts while conserving their axial resistance and electrical geometry. Let F (the folding factor) be the ratio of total surface area (spines + shaft) to dendritic shaft area alone, for a particular cylindrical dendritic segment. Multiply the segment length by $F^{2/3}$ and the diameter by $F^{1/3}$. This simplification leads to a massive reduction in the number of compartments or segments in a model, and even bigger savings in computer time, without significantly altering the responses of noninput segments. The approximation works because spine necks most likely have resistances at least two orders of magnitude smaller than the spine head membrane. Alternatively, spines can be incorporated electrically by multiplying C_m and dividing R_m by F.[2] If this is done on a segment-by-segment basis, as it should be (because spine densities can vary substantially from one dendrite to another), this procedure has the possible disadvantage of introducing nonuniform C_m and R_m. However this is not a serious problem for most compartmental modeling packages. In addition, analytical solutions exist for branched cables with nonuniform passive parameters.[21] Spines receiving synaptic inputs (particularly large conductances) should probably be modelled explicitly, to ensure the correct voltage swings and reductions in driving force occur (see Chapter 9).

Next, produce a morphology file that can be input into a modeling package with fitting facilities. Do not forget to measure the soma.

8.7.6 Step 6: Automated Direct Fitting

Inject the same current (or 1 nA, if data normalized as suggested above) into the model as was injected into the cell, at the same point(s), and record from the same point(s) as during the real experiment. Select a fit interval, say 2 ms from the pulse end to when the response falls to 0.5 mV. (Check that the response doesn't alter within the fit interval when the cap. comp. dial is deliberately mis-set). Starting with $C_m = 1\ \mu F/cm^2$, $R_i = 200\ \Omega cm$, $R_m = 10{,}000\ \Omega cm^2$, and a uniform distribution of all these parameters, systematically explore the parameter space under control of a simplex or other fitting algorithm (see Chapter 1).

A schematic example of direct least-squares fitting is shown in Color Figure 8.3* (semi-log plots). The target data is the noisy green transient. The initial guess (fit 1) is the red line. A simplex routine crawls amoeba-like around the parameter space trying various parameter combinations. It compares the target transient with the responses from models with the parameters it has chosen, gradually homing in on a region of the parameter space which gives better fits. Fit 20 is still poor, but by fit 50 the simplex algorithm has chosen parameters which yield a response quite close to the target. By the 100th iteration, the fit has converged on an optimal set of parameters, yielding a response that matches the target waveform closely over the entire fit interval (2 to 50 ms in this case).

See if you can make the model responses superimpose with the target data. If you can, then decide whether you need to do fit rejection statistics to get a range of acceptable fits, and whether you ought to confront the possibility of nonuniform electrical parameters, such as a somatic shunt or lower somatic R_m.

If you cannot get good fits, see if nonuniform parameters can rescue you. Start with a somatic shunt conductance as another free parameter of the fitting, and see if that helps. If that fails, try various distribution functions of R_m or R_i, C_m is unlikely to vary much over the cell, but that is a possibility if all else fails. Double check your morphology, linear scaling and other tests of passive behaviour, and the absolute scaling of your data. It is easy to amplify electrical data by a different factor to what you thought. Check the calibration of your system with model cells (R-C networks), supplied with most commercial amplifiers.

8.8 OTHER TIPS AND TRICKS

8.8.1 Manual Fitting

Select a plausible set of starting values for C_m, R_m, and R_i. Ensure that $C_m R_m$ is equal to the apparent τ_0 of the target data. Obtain the model response to the current injected experimentally (normalized to 1 nA if that's what you did for your grand average). Then "peel" the target and model responses to obtain A_0 and τ_0, the *apparent* amplitudes and time constants of the slowest component of each.[3] You need to plot ln(V) against t, then –1/slope of the straight line component at late times is τ_0, and exp(intercept) is A_0. You need to be careful with the fit interval you use. Use the same interval for both target and model. Choose intervals that lead to nice straight line portions of semi-log plot for tens of ms. Exponential fitting routines can do the job, but make sure you check the fits with semi-log plots. Bad fits can look deceptively good with normal axes. Beware: exponential fitting and peeling[3] are notoriously ill-conditioned or error prone. Tiny changes in fit intervals can lead to huge changes in the exponential amplitudes and time constants, which is why direct fitting is preferable where possible.

We know that A_0 is proportional to $1/C_m$ in a passive cable tree with uniform specific electrical parameters.[6] We also know that A_0 is independent of R_m and R_i in the uniform case. So, you need to multiply C_m by (model A_0)/(target A_0).

* Color Figure 8.3 follows page 140.

You then need to multiply R_m by (target τ_0/model τ_0) × (old C_m/new C_m), to compensate for any errors in initial membrane time constant, and the change in C_m. How to adjust R_i? Theory says that the time constants of the early components are proportional to R_iC_m, therefore you need to compare target and model τ_1 and τ_2, obtained using the same fit intervals for both data and model, and adjust R_i as follows. Work out the median of (target τ_1/model τ_1) and (target τ_2/model τ_2) and any other early components you believe you can reliably extract from the data (ignore outlying ratios; the more peels, the more unreliable the result). Multiply R_i by (old C_m/ new C_m) × (median early τ ratio).

Next run the model with the new parameters, and compare with the target data. You may need to do a few more iterations until you are satisfied, or until you cannot reject the fit statistically. If the late components match well, but the early components cannot be matched up by the above procedure, try playing with R_i empirically. If the data has slower (bigger) early components than the model, try increasing R_i incrementally until a match is obtained. If the model has slower early components than the data, try reducing R_i. If you cannot obtain good fits this way, or you suspect there are nonuniform parameters, you really need to use direct fitting. If that fails, try introducing a somatic shunt for example, and allow it to be a free parameter.

8.8.2 Parameter Dependencies

Parameter dependences have been discussed in the literature at some length.[6,8] An intuitive illustration is shown in Figure 8.7 (semi-log plots).

If R_i is doubled, the late components are unaffected but the early components get slower (this is because fast time constants are proportional to R_i; charge equalization is slowed down by increasing axial resistances).

If R_m is doubled, the early components are unaffected, but the final time constant is doubled (slow time constants are proportional to R_m, discharge through the membrane is impeded by increasing membrane resistance, but axial charge equalization is not). A_0 is unaffected (the intercept of the final straight line part with voltage axis is the same). The other components' amplitudes A_i are also independent of R_m.

Doubling C_m produces more complicated changes to the waveform: the amplitudes of all components (A_i) are halved (amplitudes are inversely proportional to capacitance), for example, A_0 is halved, as can be seen by the intercept of the straight line part of the response with the V axis. Also the time constants of all components (which are proportional to C_m) are doubled and both early and late components can be seen to be slower. Charge equalization and membrane discharge are slowed. Equalization is slowed because charge flows down axial resistances onto membrane capacitances as it spreads (see Figure 8.2).

Inserting a 5 nS point shunt conductance (g_{shunt}) at the recording site causes the final decay to speed up, and decreases its amplitude A_0 (as can seen from the intercept of the straight line part of the response with the V axis). The very early components are little affected though. Notice that while an increase in C_m causes A_0 to decrease and τ_0 to increase, an increase in the recording site shunt causes *both* A_0 and τ_0 to decrease.

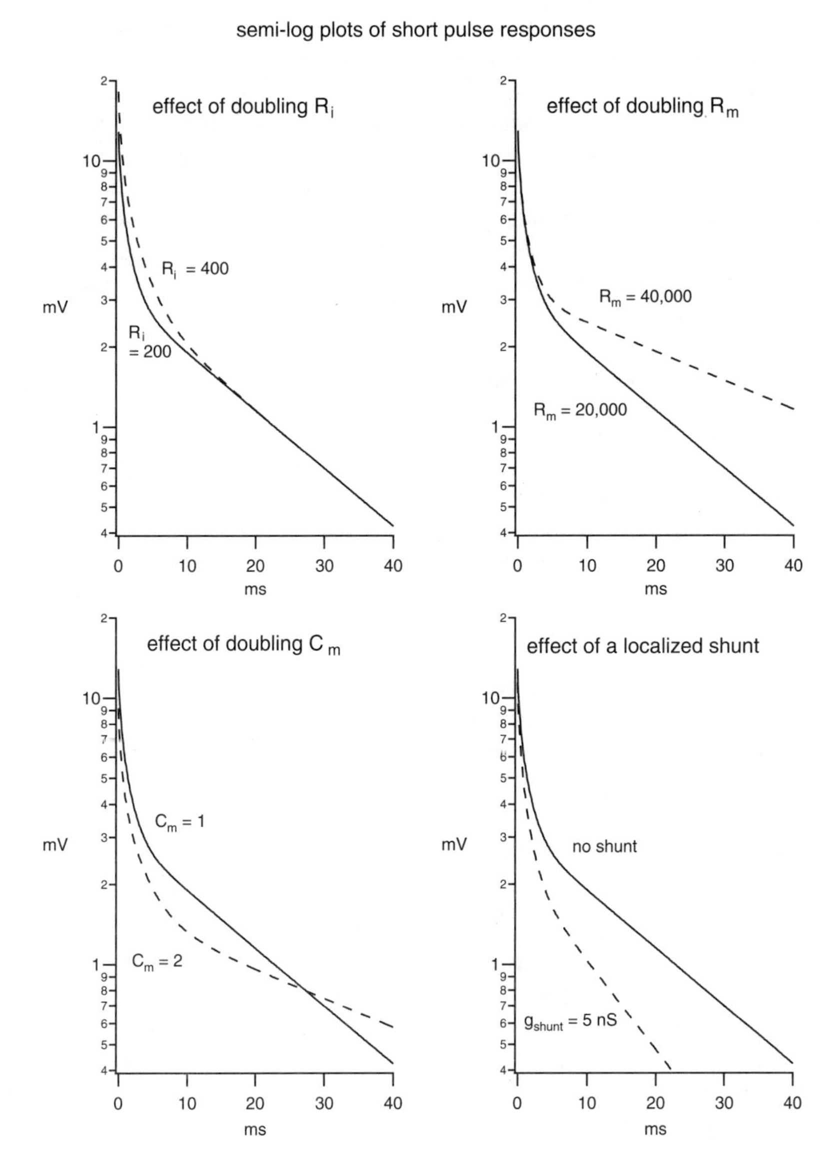

FIGURE 8.7 Illustration of parameter dependence, using a single cylinder model with default parameters (solid lines): length 1000 μm, diameter 5 μm, C_m = 1 μF/cm², R_i = 200 Ωcm, R_m = 20,000 Ωcm², no shunt; input is 1 nA × 0.5 ms.

8.8.3 Tricks for Reducing Nonuniqueness

8.8.3.1 Nucleated Patches

Clearly a sphere is easier to measure than a complex tree. Patch onto a cell of the class of interest, go whole-cell, then apply negative pressure while you withdraw the pipette.[22] If you are able to pull large approximately spherical chunk of membrane out of the soma, possibly including the nucleus, you are in a position to estimate C_m. First measure the diameter of the sphere, then inject currents of various amplitudes and measure the voltage responses. Capacitance = $1/A_0$, and resistance = $\tau_0 A_0$. Alternatively, do it all under voltage-clamp with an amplifier that can cancel and measure the cell capacitance (e.g., an Axopatch 200B). If we assume that nothing has happened to the membrane, the measurements were reasonably accurate, and the nuclear membrane has stayed disconnected from the cell membrane, then we can obtain an independent estimate of C_m. If we assume that there are no leaks from pulling off dendrites or the rest of the soma, then we can also get an estimate of somatic R_m, but this seems a little dangerous.

8.8.3.2 Raising Input Resistance

If we introduce a somatic shunt conductance, as a simple case of nonuniform R_m, the magnitude of this conductance is limited to the input conductance of the cell. The higher the input conductance, the more scope for nonuniqueness, because the shunt could be anything between zero and the input conductance. The shunt could be negative as well, if we allow dendritic R_m to be lower than somatic R_m. Anything that reduces input resistance is good, because it reduces the range over which the shunt can vary. Various tricks can be used to do this:

1. Cooling (Andrew Trevelyan, unpublished).
2. Adding further channel blockers.
3. Working on cells from younger animals.

8.8.3.3 Multi-Site Stimulation and Recording

It is now routinely possible to record from two, three, or even four points simultaneously on a single neuron,[11] using a combination of imaging and dendritic whole-cell recordings. The more passive responses one can record, the more information one has to constrain non-uniqueness, and to pin down passive electrical parameters. Fit rejection statistics should be applied, just as they are with single site responses.

However, routine recording from sub-micron diameter dendrites (the majority) is not possible with patch pipettes. This is where voltage sensitive-dyes might come to the rescue. By a bizarre twist of fate, the very weakness of the intracellularly-applied voltage-sensitive dye technique[23] could be turned into a strength from the point of view of passive cable modeling. After 50 or so optical trials lasting approximately 50 ms, phototoxicity occurs, which appears to be caused by some kind of disabling of most of the cell's ion channels. The cells turn into passive trees with intact morphology. No blebs or blow-outs occur as seen with phototoxicity caused by overexposure to more water-soluble dyes such as calcium indicators or fluorescein. Optical responses can be recorded from these passified cells. Unfortunately, only time course information, not voltage amplitude information is currently obtainable, but work is in progress to test various means of calibrating the optical signals.

8.9 CONCLUSIONS

Modellers and experimenters alike ignore passive cable modeling at their peril. It is a fundamentally important subject that all cellular and computational neuroscientists should have at least a superficial understanding of this technique. Most neurons do not behave passively but this does not mean that passive cable theory is unimportant. The opposite is true. It is just like saying that the foundations of a house do not matter just because they are buried under other more interesting structures. Get the foundations wrong, and the whole edifice collapses. The same could be true of computational neuroscience. Passive cable theory and practice are one of the foundations, but, we have not gotten it right yet. There is much more work to be done, but there are exciting new technological breakthroughs that could help us in the next few decades, or perhaps sooner.

It would help the field if the hordes of voltage-clampers currently analyzing conductance after conductance would pay serious attention to space and voltage-clamp errors[6,16,17] before trying to do their experiments or analyze their data. The active conductance literature is cluttered with pseudo-quantitative studies claiming to have measured Hodgkin–Huxley parameters (see Chapter 5) or equivalent for various conductances, but in most cases the data are severely compromised by cable problems, making the task of active compartmental modellers next to impossible.

ACKNOWLEDGMENTS

The author would like to thank Julian Jack, Andrew Trevelyan, Mike Hausser, Arnd Roth and the editor for their comments and criticisms, Neil Bannister and Kristina Tarczy-Hornoch for the cat spiny stellate cell data, Alan Larkman for his intellectual input over the years and the Wellcome Trust and Lucent Technologies for funding.

REFERENCES

1. Jack, J. J. B., Noble, D., and Tsien R. W., *Electric Current Flow in Excitable Cells*, Clarendon Press, Oxford,1975.
2. Rall, W., Core conductor theory and cable properties of neurons, in *Handbook of Physiology, The Nervous System, Cellular Biology of Neurons*, Kandel, E. R., Ed., American Physiological Society, Bethesda, MD, Sect. 1, Vol. 1 (Part 1), 39, 1977.
3. Rall, W., *The Theoretical Foundation of Dendritic Function. Selected Papers of Wilfred Rall and Commentaries*, Segev, I., Rinzel, J., and Shepherd, G. M., Eds., MIT Press, Cambridge, 1995.
4. Rall, W. and Agmon-Snir, H., Cable theory for dendritic neurons, in *Methods in Neuronal Modeling: From Ions to Networks*, 2nd ed., Koch, C. and Segev, I., Eds.), MIT Press, Cambridge, 1998, 27.
5. Koch, C., *Biophysics of Computation: Information Processing in Single Neurons*, Oxford University Press, Oxford, 1998.
6. Major, G., Evans, J. D., and Jack, J. J. B., Solutions for transients in arbitrarily branching cables: I. Voltage recording with a somatic shunt, *Biophys. J.*, 65, 423, published errata appear in *Biophys. J.*, 65, 982 and 65, 2266, 1993.

7. Jonas, P., Major, G., and Sakmann, B., Quantal components of unitary EPSCs at the mossy fibre synapse on CA3 pyramidal cells of rat hippocampus, *J. Physiol.*, 472, 615, 1993.
8. Major, G., Larkman, A. U., Jonas, P., Sakmann, B., and Jack J. J., Detailed passive cable models of whole-cell recorded CA3 pyramidal neurons in rat hippocampal slices, *J. Neurosci.*, 14, 4613, 1994.
9. Cao, B. J. and Abbot L. F., A new computational method for cable theory problems, *Biophys. J.*, 64, 303, 1993.
10. Major, G., *The physiology, morphology and modeling of cortical pyramidal neurones*, Ph.D. Thesis, Oxford University, London, 1992.
11. Stuart, G. and Spruston, N., Determinants of voltage attenuation in neocortical pyramidal neuron dendrites, *J. Neurosci.*, 18, 3501, 1998.
12. Major, G., Solutions for transients in arbitrarily branching cables: III. Voltage-clamp problems, *Biophys. J.*, 65, 469, 1993.
13. Gasparini, S. and DiFrancesco, D., Action of the hyperpolarization-activated current (Ih) blocker ZD 7288 in hippocampal CA1 neurons, *Pflugers Arch.*, 435, 99, 1977.
14. Hille, B., *Ionic Channels of Excitable Membranes*, 2nd ed., Sinauer Associates, Sunderland, MA, 1992.
15. Tang, C. M., Presser, F., and Morad, M., Amiloride selectively blocks the low threshold (T) calcium channel, *Science*, 240, 213, 1988.
16. Rall, W. and Segev, I., Space-clamp problems when voltage clamping branched neurons with intracellular microelectrodes, in *Voltage and Patch Clamping with Microelectrodes*, Smith, T. G., Jr., Lecar, H., Redman S. J., and Gage, P., Eds., American Physiological Society, Bethesda, MD, 1985, 191.
17. Spruston, N., Voltage- and space-clamp errors associated with the measurement of electrotonically remote synaptic events, *J. Neurophysiol.*, 70, 781, 1993.
18. Pozza, M. F., Manuel, N. A., Steinmann, M., Froestl, W., and Davies, C. H., Comparison of antagonist potencies at pre- and post-synaptic GABA(B) receptors at inhibitory synapses in the CA1 region of the rat hippocampus, *Br. J. Pharmacol.*, 127, 211, 1999.
19. Velumian, A. A., Zhang, L., Pennefather, P., and Carlen, P. L., Reversible inhibition of IK, IAHP, Ih and ICa currents by internally applied gluconate in rat hippocampal pyramidal neurones, *Pflugers Arch.*, 433, 343, 1997.
20. Rapp, M., Yarom, Y., and Segev, I., The impact of parallel fiber background activity on the cable properties of cerebellar Purkinje cells, *Neural Comput.*, 4, 518, 1992.
21. Major, G. and Evans, J. D., Solutions for transients in arbitrarily branching cables: IV. Nonuniform electrical parameters, *Biophys. J.*, 66, 615, 1994.
22. Martina, M. and Jonas, P., Functional differences in Na+ channel gating between fast-spiking interneurones and principal neurones of rat hippocampus, *J. Physiol.* (London), 505, 593, 1997.
23. Antic, S., Major, G., and Zecevic, D., Fast optical recordings of membrane potential changes from dendrites of pyramidal neurons, *J. Neurophysiol.*, 82,1615, 1999.
24. Spain, W. J., Schwindt, P. C., and Crill, W. E., Anomalous rectification in neurons from cat somatosensory cortex in vitro, *J. Neurophysiol.*, 57, 1555, 1987.

9 Modeling Simple and Complex Active Neurons

Erik De Schutter and Volker Steuber

CONTENTS

9.1 INTRODUCTION

Over the last decade the view of how neurons process synaptic input has shifted fundamentally. Previously the common view was that in most neurons largely passive dendrites collect synaptic charge and convey it in a non-linear but still straightforward manner to the soma.[1,2] Recently, it has become accepted that entirely passive dendrites do not exist.[3] The interaction between dendritic voltage-gated conductances and synaptic input has become a major topic of research where quantitative modeling plays an important role.[4,5] But it is not trivial to construct a detailed morphologically correct model with active membrane. Therefore this chapter is built around the question "what is the right level of complexity for a model of an active neuron?"

0-8493-2068-2/01/$0.00+$.50

9.2 POINT NEURON MODELS

The simplest spiking neuron models possible are those without any representation of morphology. In the literature such models are called point neuron models. Because of their numerical efficiency they are often used in simulations of neural networks (see Chapters 10, 11, and 12). Only a few parameters need to be tuned to replicate the spiking pattern of real neurons in great detail. The main weakness of point neuron models is the absence of dendritic synaptic integration[2] which may be essential for the interaction of neurons in a network. In the following, we will provide a brief introduction to point neuron models. Chapters 10 and 11 contain additional practical advice on how to construct and constrain such models.

Three classes of point neuron models can be distinguished: firing-rate models (described in Section 11.4.2), integrate-and-fire models and conductance based models.

9.2.1 Integrate-and-Fire Models

Integrate-and-fire models are completely phenomenological. They try to replicate the generation of spikes in response to synaptic input without representing the underlying biophysics. In its vanilla form the model computes the change in voltage $V(t)$ over a *membrane capacitance CM* which integrates the varying synaptic current $I(t)$:

$$CM\frac{dV(t)}{dt} = I(t) \tag{9.1}$$

In the absence of synaptic input, $V(t)$ equals the resting potential which is usually set to zero. Whenever $V(t)$ reaches a threshold V_{th}, an action potential is triggered (but not computed) and $V(t)$ is reset to rest. As Equation 9.1 by itself would allow for infinitely high firing rates (Figure 9.1) an absolute refractory period is added. After a spike, $V(t)$ is clamped to rest for a fixed period t_{ref}. This leads to a saturating mean firing frequency f (Figure 9.1) of:

$$f = \frac{I}{CM\,V_{th} + t_{ref}I} \tag{9.2}$$

In most models the synaptic inputs are simulated as current pulses (e.g., Reference 6) and not as conductance changes (see Section 10.2). As CM is constant each synaptic input provides for a fixed increase ΔV which may be weighted by the strength of the connection.

The standard integrate-and-fire unit integrates synaptic input over an infinite time which is not realistic. This problem can be solved by adding a leakage current to the equation:

$$CM\frac{dV(t)}{dt} = -\frac{V(t)}{RM} + I(t) \tag{9.3}$$

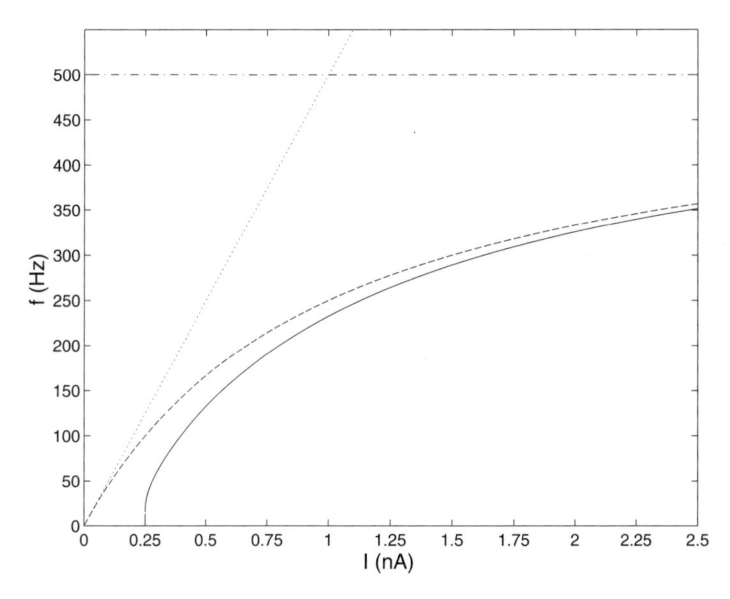

FIGURE 9.1 Relation between synaptic current I and firing rate f in three different integrate-and-fire models. In a basic integrate-and-fire model without refractory period, f is given by $I/CM\ V_{th}$ (dotted line). Adding a refractory period t_{ref} results in a f-I curve that saturates at $f_{max} = 1/t_{ref}$ (broken line). In a leaky integrate-and-fire model (solid line), I has to exceed a threshold current I_{th} in order to trigger spikes. The following parameters were used: CM = 0.1 nF, V_{th} = 20 mV, t_{ref} = 5 ms and RM = 80 MΩ (resulting in f_{max} = 500 Hz and I_{th} = 0.25 nA).

In this *leaky integrate-and-fire* model voltage will decay back to rest with a time constant $RM \cdot CM$ if no input is present. As demonstrated in Figure 9.1 leaky integrate-and-fire models show a current threshold: below this value ($I_{th} = V_{th}/RM$) no spiking occurs. Above the threshold the mean firing frequency is now given by:

$$f = \frac{1}{t_{ref} - \left(RM\,CM\,ln\left(1 - \frac{V_{th}}{I\,RM}\right)\right)} \tag{9.4}$$

Additional refinements to integrate-and-fire models have been proposed, for example an adapting conductance which mimics the effect of potassium currents on spike adaptation.[7] An excellent recent review of these models and their properties can be found in Reference 8.

9.2.2 Conductance Based Models

Though one can extend integrate-and-fire models to mimic complex firing patterns, it is not obvious how to relate the necessary parameters to the voltage-clamp data which are often available. Conductance based models are more closely based on the biophysical properties of neurons. In a conductance based model, the change in membrane voltage $V(t)$ is determined by the sum of n ionic currents and an externally applied current $I(t)$:

$$CM\frac{dV(t)}{dt} = -\sum_{i=1}^{n} g_i(t)\left(V(t)-E_i\right)+I(t) \tag{9.5}$$

Each Ohmic current is computed as the product of a conductance g_i which may be variable (voltage or ligand gated) or not (the leak) and a driving force $(V(t) - E_i)$.

An electrical circuit representation of such a model can be found in Figure 9.2. Membrane conductances can be divided into three groups: the leakage conductance g_{leak}, voltage-dependent conductances and ligand-activated ones which are usually synaptic. Voltage-dependent conductances are often represented using the Hodgkin–Huxley formalism (see Section 5.2) as $g_{vd}\ (V,t) = \bar{g}_{vd}\ m^p h^q$ where $\bar{g}_{vd}$ is the maximum conductance, given by the density of the channels in the membrane multiplied by the single channel conductance, and m and h are the activation and inactivation variables. In the case of ligand-gated channels, g_{syn} is usually computed using an alpha or double exponential equation (see Section 10.2). The reversal potential E_i is often assumed to be constant and given by the Nernst equation (Chapter 5, Equation 5.2). Sometimes it is necessary to simulate the effect of changes in ionic concentration on the current, which often implies using Goldman–Hodgkin–Katz equations (see Section 9.4.5). Notice that only one $V(t)$ is computed thus it is assumed that this model is isopotential (no voltage gradients exist).

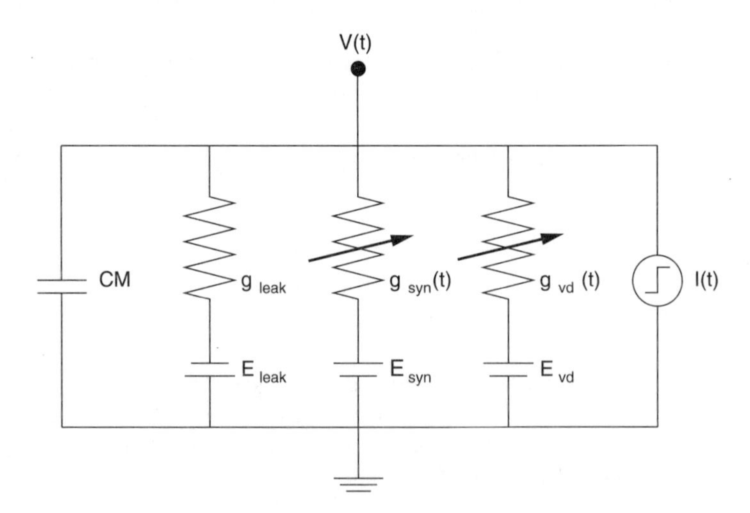

FIGURE 9.2 Circuit diagram of an isopotential compartment in a conductance based model. The total current across the membrane is given by the sum of five different types of current: a capacitive current, a leakage current, synaptic and voltage dependent channel currents, and an injection current.

The classic example is still the Hodgkin–Huxley model (see Reference 9 and Section 5.2.1) where $n = 3$ (fast sodium current, delayed rectifier and leak). Another simple model often used are the Morris–Lecar equations (see Chapter 10, Box 10.3). The properties of the Hodgkin–Huxley model have been extensively studied (reviewed in References 10 and 11). Unfortunately it has also been abused in the modeling literature. One should remember that this is a model of an axon, not of a neuron. Therefore it is not realistic to plug the Hodgkin–Huxley equations into a neuron model and call this a "realistic first approximation." Many vexing problems exist with the original Hodgkin–Huxley equations which limit their usefulness for biophysically realistic neuron models. For example, they have a minimum firing rate under conditions of constant current input;[12] the model cannot fire at less than 50 Hz using standard parameters. While one can "linearize" the Hodgkin–Huxley model (i.e., make the initial part of the $f(I)$ curve linear with an origin at zero) by adding noise,[12] it is more realistic to use equations derived for the particular neuron that is being modeled. The first such set of equations was derived for a gastropod neuron. To model this neuron's firing pattern three sets of equations (fast sodium current, delayed rectifier and A current) were needed,[13] giving rise to a linear $f(I)$ curve for small current amplitudes. Finally, the Hodgkin–Huxley model works only at low temperatures. If one corrects the rate constants for body temperature (compared to the 6°C temperature at which the measurements were made; see Chapter 5, Equation 5.41) no action potential generation is possible.

The art of making good conductance based point neuron models is described in detail in the next chapter.

9.3 MODELS WITH A REDUCED MORPHOLOGY

Point neuron models assume that all currents flow across a single patch of membrane. This seriously limits the number of phenomena they can reproduce. In particular point models cannot simulate dendritic spike initiation or dendritic synaptic integration which both depend on the existence of voltage gradients within the neuron. It may not be necessary, however, to simulate the complete morphology of the neuron. Modeling a reduced morphology greatly diminishes the number of model parameters which need to be defined and increases computation speed. But choosing the best approach requires trade offs.

9.3.1 LIMITATIONS OF REDUCED MORPHOLOGIES FOR ACTIVE MODELS

A set of simple rules can be used to reduce some passive dendritic trees to an equivalent cable.[1,2] This is generally not possible for active dendrites. The basic problem is that the local impedance in the "reduced" dendrite is different from that in the branches of the original one. Take for example the once popular *ball and stick* model where the soma is approximated by a membrane patch (equivalent to a spherical compartment of the correct size; using Equation 9.5) which is attached to a single equivalent cylinder representing the dendrite (Figure 9.3A). The somatic and dendritic compartment are each isopotential and connected to each other via the dendritic axial conductance (see below, Equation 9.8). In such a model the load that the dendrite imposes upon the soma is represented accurately giving a correct approximation of the somatic input impedance. But this is not the case in the dendrite.

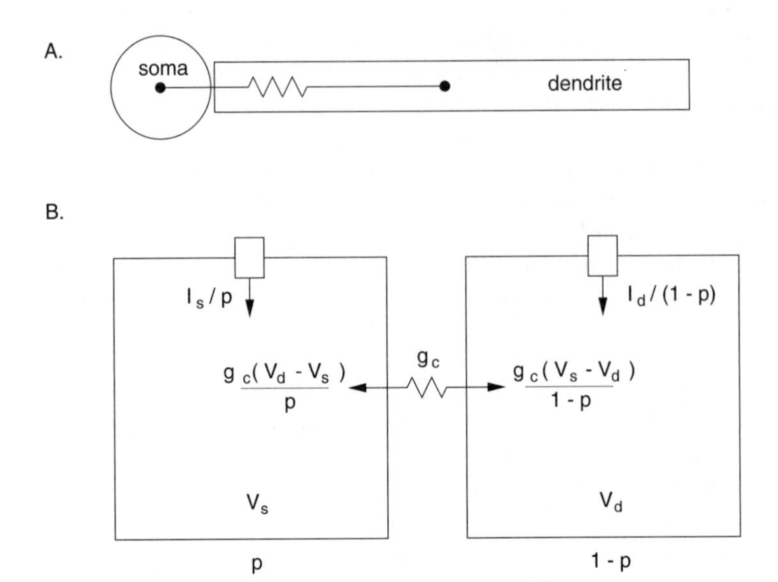

FIGURE 9.3 Schematic representation of ball and stick model (A) and Pinsky and Rinzel model (B). The ball and stick model is a morphological approximation. The Pinsky and Rinzel model is a phenomenological rather than a morphological model.

The cylindrical dendritic compartment now has quite a large membrane surface and thus a low impedance, while small branches of the complete dendrite have a very high impedance.

Consequently, when ionic channels are added to such a ball and stick model their density (expressed as $\bar{g}_i$) can be close to the real density in the soma, but in the equivalent dendrite much lower densities have to be used to obtain realistic voltage transients. This is not simply a scaling problem. If the interaction between many channels is required to replicate, for example, dendritic spike generation all the $\bar{g}_i$ values will need to be tuned together specifically for this reduced model (see below). This also includes synaptic conductances ($\bar{g}_{syn}$) which, if the model is to be used in a network simulation, have additional restrictions because of network scaling requirements (Section 11.3.4). While the final model may replicate the firing behavior of the neuron, it will be restricted in its use. For example, it cannot be expected to generate realistic voltage clamp responses because the $\bar{g}_i$ parameters do not have the true values. Nor can these models replicate the full range of rise times for somatic postsynaptic potentials as all synaptic input is received at the same electrotonic distance from the soma (see Reference 14 for an analytical method to compute this in passive dendritic trees).

One solution to these problems would be to use an intermediate class of models, not representing the full morphological complexity but with still more than two compartments. While a number of recipes on how to do this have been proposed,[15,16] all suffer from the local impedance problems described above. An advantage of using more than one compartment to represent the dendrite is that this allows for a better spatial separation of synaptic inputs and thus enhances the realism of a model that receives layered inputs (e.g., on the apical dendrite in a pyramidal neuron, see Section 11.4 and References 17 and 18 for examples).

9.3.2 Phenomenological Reduced Models

The previous section may have seemed an advisory not to use reduced models. In fact, we believe that these can be quite useful provided one realizes their intrinsic limitations. Therefore it is better to think of them as phenomenological models instead of as models with a reduced morphology.

A nice example can be found in the work of Pinsky and Rinzel[19] who could approximate the full firing behavior of an active membrane 19-compartment cable model of a CA3 pyramidal neuron[18] with just two active compartments and a coupling conductance (Figure 9.3B). This was achieved through a separation of voltage-gated channels over two regions: a somatic region with currents supporting a classic action potential and a dendritic one with calcium and calcium-activated currents able to generate a dendritic spike. The interaction between these two compartments allows the model to generate burst firing which is not possible in the isolated compartments because coupling current has to flow back and forth. A somatic action potential triggers the dendritic spike, current flowing back from the dendrite then triggers the second action potential, etc. This scheme will not work in a single compartment expressing all the voltage-gated channels, because a somatic after-hyperpolarization is required to remove the inactivation of the sodium channel at the same time that the dendritic spike is depolarizing the dendrite (see Reference 19 for a more detailed analysis).

The behavior of the model is governed by two equations describing the change of somatic and dendritic voltage V_s and V_d:

$$CM_s \frac{dV_s(t)}{dt} = -\sum_{i=1}^{n} g_{i,s}(t)\left(V_s(t) - E_i\right) + \frac{g_c\left(V_d(t) - V_s(t)\right) + I_s(t)}{p} \tag{9.6}$$

$$CM_d \frac{dV_d(t)}{dt} = -\sum_{i=1}^{k} g_{i,d}(t)\left(V_d(t) - E_i\right) + \frac{g_c\left(V_s(t) - V_d(t)\right) - I_{syn}(t) + I_d(t)}{1-p} \tag{9.7}$$

The two compartments are connected to each other by a coupling conductance g_c. All currents are scaled according to the relative contribution of somatic and dendritic compartment (p and $1 - p$; the scaling for the voltage dependent currents is factored in g_i).

In itself this model may not seem very different from the ball and stick model of Figure 9.2A, but conceptually it is because it does not try mimic morphology. The second compartment exists only to generate more complex firing behavior than possible in a single compartment model. The spatial components of the model are embedded in the parameters p and g_c which have no direct relation with the actual size or shape of the neuron. In fact, Pinsky and Rinzel point out that a morphologically realistic value for g_c is too weak to produce bursting behavior. A typical value for p is 0.5 which indicates that the somatic compartment represents more than the soma alone. Besides the compactness of the model it also has the advantage that its behavior can be related to simple parameters. Depending on the value of g_c different bursting and firing patterns can be observed.[19]

9.4 COMPARTMENTAL MODELS WITH ACTIVE MEMBRANE

Ideally all models used to study the integration of synaptic input and its effect on neuronal firing should represent both the complete dendritic morphology and all voltage and ligand gated channels. This is best done with compartmental models (Section 8.2, References 1, 2, and 20) because of their great flexibility in specifying morphologies (Chapter 6) and channel equations (Chapter 5). These models can then be used to investigate the mechanisms by which synaptic currents and voltage-gated channels interact, and how morphology influences these processes. Many examples can be found where these models were used to quantitatively confirm a conceptual hypothesis.[4,21,22] But one can also use them to perform experiments "*in computo*,"[23] where the model is used as an easily accessible preparation whose complex properties are investigated. An example of the benefits of such an approach is given in Box 9.1. Whatever the use of the model, a big advantage is that all the variables can be directly related to experimentally measurable quantities which allows for easy interaction between simulation and experiment (see also Section 10.4).

Unfortunately the wider use of large active membrane compartmental models is hampered by the difficulty in constraining their parameters (Sections 9.4.6 and 9.4.7) and by the heavy computational load they impose. The latter is becoming less of a problem because popular simulation packages (like GENESIS and NEURON, see Chapter 10, Box 10.2) are optimized for these computations and because fast computers are not expensive anymore. As a consequence simulations of the Purkinje cell model described in Box 9.1 which took over an hour to compute a decade ago now take a minute. However, the time needed to construct such a model has diminished much less as it still requires a lot of hand tuning. The automatic parameter search routines described in Section 1.5 have not yet proved their use for such large models.

9.4.1 BASIC EQUATIONS AND PARAMETERS

A compartmental model is a spatial discretization of the cable equation (Section 8.2). The dendritic structure is subdivided into many compartments so that both its morphology is faithfully replicated and that each compartment is small enough to be considered isopotential (see below). The change of voltage in each compartment is described by an ODE that is derived from Equation 9.5 by adding the current flow between connected compartments:

$$CM_j \frac{dV_j(t)}{dt} = -\sum_{i=1}^{n} g_{i,j}(t)\left(V_j(t) - E_i\right) + \sum_{k=1}^{m} \frac{V_k(t) - V_j(t)}{RI_{j,k}} + I_j(t) \tag{9.8}$$

Note the addition of the j subscript to all terms denoting that this equation refers to one compartment of many, with parameters specific to each (e.g., CM_j), and that this compartment is connected to m other compartments k with a connecting *cytoplasmic resistance* $RI_{j,k}$.

Box 9.1 Predictions Made by a Purkinje Cell Model

We will briefly describe results obtained with our Purkinje cell model to demonstrate the power of the *in computo* experimental approach. The model is a fully active compartmental model of a cerebellar Purkinje cell[53] based on morphological data provided by Rapp et al.[30] The basic model contains 1600 electrically distinct compartments[37] to which a variable number of spine compartments can be added. Ten different types of voltage-gated channels are modeled, resulting in a total number of 8021 channels in all compartments. Channel kinetics are simulated using Hodgkin–Huxley-like equations based on Purkinje cell specific voltage clamp data or, when necessary, on data from other vertebrate neurons. The soma contains fast and persistent Na^+ channels, low threshold (T-type) Ca^{2+} channels, a delayed rectifier, an A-current, non-inactivating K^+ channels and an anomalous rectifier. The dendritic membrane includes P-type and T-type Ca^{2+} channels, two different Ca^{2+}-activated K^+ channels and a non-inactivating K^+ channel. The P-type Ca^{2+} channel is a high-threshold, very slowly inactivating channel, first described in the Purkinje cell. Ca^{2+} concentrations are computed in a thin submembrane shell using fast exponential decay. See Section 9.4.8 and Reference 37 for a description on how the model was tuned to reproduce the *in vitro* firing behavior (Figure B9.1A).

In the simulations presented here we reproduce the typical *in vivo* firing pattern of Purkinje cells (Figure B9.1B) by applying combined random activation of excitatory and inhibitory contacts, mimicking the background activity of parallel fiber and stellate cell synapses respectively. Under these conditions the model shows a normal and robust irregular firing behavior over a wide range of firing frequencies which can be evoked by many different combinations of the background input frequencies.[32] The background inhibition is essential for the generation of *in vivo* Purkinje cell firing patterns in the model, and the temporal average of the total inhibitory current (i.e., the sum of all individual synaptic currents) must exceed that of the total excitatory current.[32,54] The inhibition suppresses spontaneous dendritic spiking (which is not observed *in vivo*) and reduces the fast regular somatic spiking caused by intrinsic plateau currents in the model.[23] Note that the requirement of net synaptic inhibition does not fit the integrate-and-fire model of neuronal spiking: the Purkinje cell dendrite acts most of the time as a current sink, not as a current source. Subsequent experimental studies have confirmed that background inhibition is essential to suppress dendritic spiking[55] and to obtain irregular somatic spiking.[56] A study using dynamic voltage clamp protocols (Chapter 10, Box 10.1) in slice showed that the Purkinje cell firing pattern could be changed to a typical *in vivo* one only if the simulated synaptic current had an inhibitory component larger than the excitatory one,[57] again confirming a model prediction.

In another series of studies the *in vivo* effect of synchronous activation of a small number of excitatory parallel fiber inputs was simulated. The first such study demonstrated that parallel fiber input can activate voltage-gated Ca^{2+}

(continued)

Box 9.1 (continued)

channels on the spiny dendrite[58] (Color Figure 9.2*). This activation of Ca^{2+} channels has an important functional consequence: it amplifies the somatic response to synchronous input. First, even though the Purkinje cell receives over 150,000 parallel fiber inputs, activating about 100 of them synchronously is enough to reliably evoke a spike (Color Figure 9.1*).[59] Second, the amplification is not the same for each synchronous input of the same size. It depends on its location on the dendritic tree: distant focal inputs get amplified more than proximal ones, which effectively cancels the passive cable attenuation (Section 8.2) and makes the somatic response largely independent of input location.[58] This differential amplification is caused by the differences in local morphology of proximal and distal dendrite and by the P-type Ca^{2+} channel activation threshold of about –45 mV being close to the dendritic membrane potential *in vivo*. In the distal dendrite the input impedances of the small branches favor the recruitment of additional Ca^{2+} channels, caused by a spreading of partial Ca^{2+} channel activation to neighboring branches which leads to a larger amplification. Conversely, in the proximal dendrite, the soma and smooth dendrite act as current sinks preventing the depolarization of neighboring branches, which limits the amplification to that by the Ca^{2+} channels at the location of synaptic input only. Finally, the graded amplification also depends on the excitability of the dendritic tree which varies over time due to the background inputs.[59]

Is experimental confirmation for these predictions available? When the first predictions were made[58] no experimental evidence for parallel fiber activated Ca^{2+} influx existed, but confocal imaging experiments confirmed this shortly afterwards.[60] However, these studies also seemed to contradict the increased distal amplification by spreading of Ca^{2+} channel activation. In these experiments, larger parallel fiber inputs caused more local Ca^{2+} influx without any spreading of the Ca^{2+} signal into neighboring dendrites, contrary to what the model predicts. This difference may be explained, however, by the relatively hyperpolarized state of the neuron in the experimental slice conditions. As shown in Color Figure 9.2A, the model shows identical behavior when simulated without background input (resting membrane potential of –68 mV) while in the presence of such input (mean membrane potential –54 mV) the Ca^{2+} signal does spread with larger focal excitation (Color Figure 9.2B). Therefore final validation of some model predictions will require experimental techniques which allow visualization of dendritic activity under *in vivo* conditions.

* Color Figures 9.1 and 9.2 follow page 140. *(continued)*

The basic properties of such a compartment in the case where it is passive have been described in Chapter 8 (Figures 8.1 and 8.2). Equation 9.8 extends this to the active case, represented by the first term of its right-hand side.

Box 9.1 (continued)

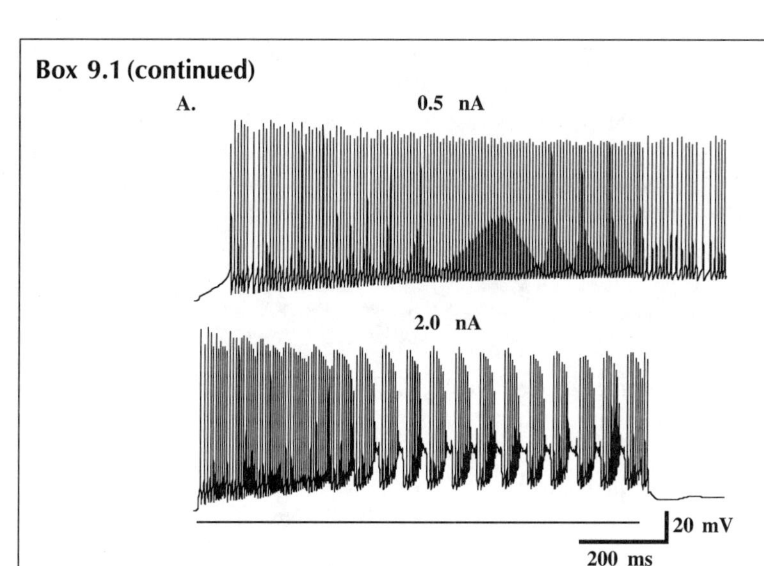

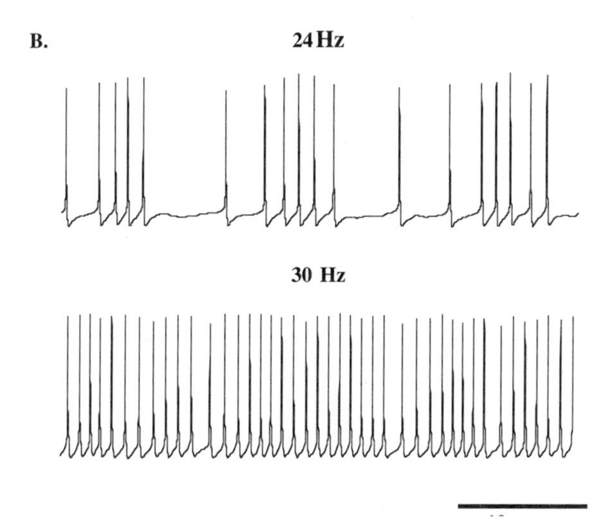

FIGURE B9.1 Simulation of Purkinje cell firing pattern under two very different conditions. (A) The model was tuned to reproduce the typical firing pattern in slice which is demonstrated here for two different somatic current injection levels (sharp electrode, duration of current injection shown by the bar). Notice the regular firing and the delayed appearance of bursting for the 2 nA current injection. See Reference 37 for details. (B) Simulated *in vivo* firing pattern due to combined random excitatory (rates indicated, only 1% of parallel fibers simulated) and inhibitory (1 Hz) background input. Notice the irregular firing and absence of bursting. See References 32 and 54 for more details.

We will briefly develop the additional equations needed to compute the parameters for Equation 9.8. Note that in practice these computations are usually handled by the simulation software which reads in the neuron morphology from a file (Chapter 6, Box 6.2). As described in Section 8.2, the passive properties of a neuron can be described by three cell specific parameters: C_m, R_m, and R_i which are defined for units of surface area and length. Assuming that the compartment is cylindrical with length l_j and diameter d_j gives the following equations:

$$\textit{surface:} \quad S_j = \pi d_j l_j \tag{9.9}$$

$$\textit{membrance capacitance:} \quad CM_j = C_m S_j \tag{9.10}$$

$$\textit{leak conductance:} \quad GM_j = \frac{S_j}{R_m} \tag{9.11}$$

$$\textit{cytoplasmic resistance:} \quad RI_{j,k} = \frac{4R_i l_j}{\pi d_j^2} \tag{9.12}$$

Two remarks need to be made about the above equations. First, we have defined the leak conductance GM_j in Equation 9.11 instead of its inverse, the *membrane resistance* RM_j. This is more convenient as the leak current now becomes one of the currents of Equation 9.8 with a fixed $g_{leak} = GM_j$.

Second, one has to be careful how to compute cytoplasmic resistance. There are two ways in which one can approach this (Figure 9.4): using asymmetric compartments (GENESIS nomenclature,[24] called 3-element segments in Reference 25) or symmetric compartments (4-element segments; this is the only type available in the NEURON simulator). The difference between these two approaches is *where* to compute V_j: at the boundary of the compartment (asymmetric) or at its center (symmetric). As V_j is assumed to be constant across the compartmental membrane this may seem rather irrelevant, but it does determine how the cytoplasmic resistance should be computed (Figure 9.4). For the asymmetric case this resistance is taken across one resistor with diameter d_j and length l_j (Equation 9.12). For symmetric compartments it is taken across two resistors in series with diameters d_j, d_k, and lengths $l_j/2$, $l_k/2$ (Equation 9.12). In the case of an unbranched cable $RI_{j,k}$ is the sum of the two values. In the case of branch points on a symmetric compartment Equation 9.12 should be replaced by:

$$RI_{j,k} = \frac{2R_i l_k}{\pi d_k^2}\left(1 + \frac{l_j}{d_j^2}\sum_{p=1}^{q}\frac{d_p^2}{l_p}\right) \tag{9.13}$$

where the summation is over all q compartments at the *same* side as compartment k. Using symmetric compartments gives a more accurate spatial discretization but results in a more complex solution matrix.

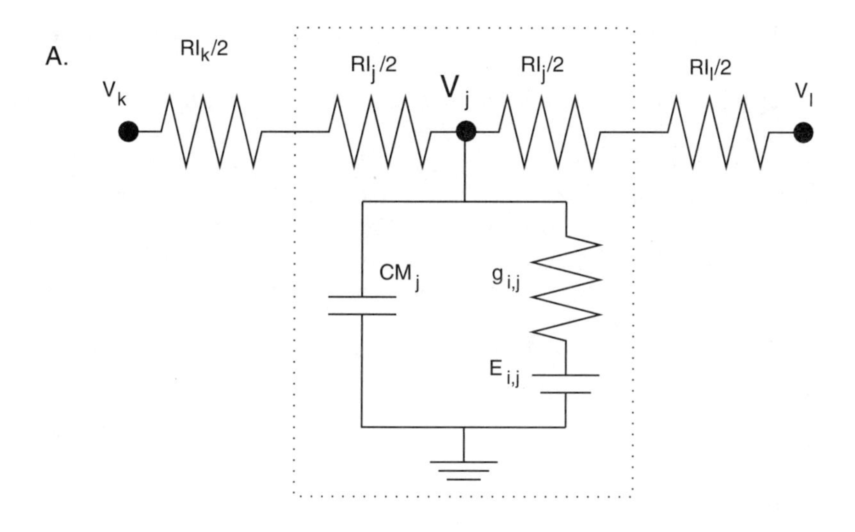

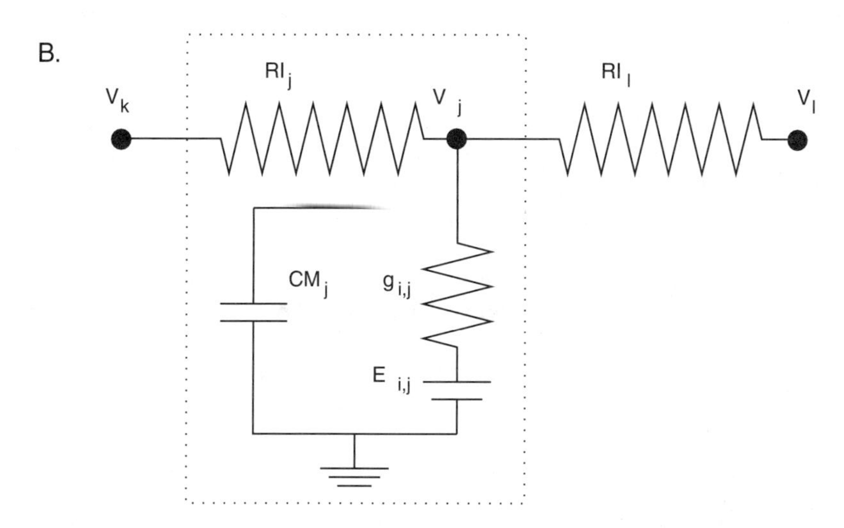

FIGURE 9.4 Circuit diagram of a symmetric (A) and an asymmetric (B) compartment in an unbranched section of a multi-compartmental conductance based model. In the symmetric compartment, the cytoplasmic resistance between two compartments $RI_{j,k}$ is given by the sum of half of the individual resistances RI_j and RI_k. In the asymmetric compartment, the compartmental voltage V_j is measured at the compartment boundary and $RI_{j,k}$ equals RI_j.

Equation 9.8 is repeated for each compartment leading to a system of coupled ODEs (Section 1.2). In general, this is a stiff system for which implicit solution routines like the Crank–Nicholson method are preferred. Ordering the compartments in a specific way — introduced by Hines,[26] and performed automatically by most simulation packages — reduces the number of off-diagonal elements and decreases the complexity of the solution matrix.[27] An extensive analysis of the Crank–Nicholson solution method, its boundary conditions and its accuracy can be found in Reference 28.

To solve Equation 9.8 the conductances $g_{i,j}$ (and, if variable, the reversal potentials $E_{i,j}$) must also be computed. Computing $g_{i,j}$ introduces additional sets of ODEs with voltage-dependent rate factors (Equation 5.7 in Chapter 5). As V_j appears in the equations describing the rate factors (see Table 5.1) this could make integration cumbersome and require iterative solutions of both sets of ODEs for each time step until V_j converges. Because the rate constants determining the voltage dependence of ionic channels are usually several order of magnitudes larger than Δt, one can assume for their calculation that V_j is constant over one integration step. This makes it possible to solve the two sets of ODEs in parallel, a solution approach implemented in GENESIS and NEURON (see References 28 and 29 for more detail).

9.4.2 Adding Spines

A large part of the membrane surface of most neurons consists of spines, over 50% in the case of Purkinje cells.[30] It is not possible to measure these faithfully using light microscopy (Chapter 6) and omitting them will lead to large errors in the fitted cable parameters (Section 8.7), typically leading to unphysiologically large values for C_m.

Spines can be included in two ways: explicitly by adding them as separate compartments or implicitly by collapsing their membrane into that of the compartment they are attached to. Modeling all spines explicitly is not practical as this can easily increase the number of compartments by an order of magnitude or more. Nevertheless it is better to explicitly model spines at the sites of excitatory synaptic contact in the model because the input impedance at a spine is much higher then at the dendritic shaft (Figure 9.5). Many spines have voltage-gated channels but these must be included in the model only if the Ca^{2+} concentration in the spine is of interest (Section 9.4.5) or if the density or properties of the channels are different from those on the dendritic shaft. Otherwise the activation of the channels on the dendritic shaft will result in almost identical results as those in a model with active spines (Figure 9.5). If the only goal is to approximate the voltage transfer function it is sufficient to use two electrical compartments for the spine: a spine head coupled to a much thinner spine neck.

The spines which are not represented as separate compartments should be collapsed into the membrane surface of the compartments. For cylindrical compartments Equation 9.9 is replaced by:

$$S'_j = \pi d_j l_j + S_{sp}\left(f_{sp} S_j - N_{sp}\right) \tag{9.14}$$

where S_{sp} is the mean surface area of the spines (which should match that of the explicitly modeled spines), f_{sp} is the spine density and N_{sp} the number of explicitly modeled spines. Equation 9.14 is implemented by specific commands in GENESIS morphology files (*add_spines, *fixed_spines). The spine density is best expressed in spines/μm² as this takes into account that small diameter branches have less spines.[31]

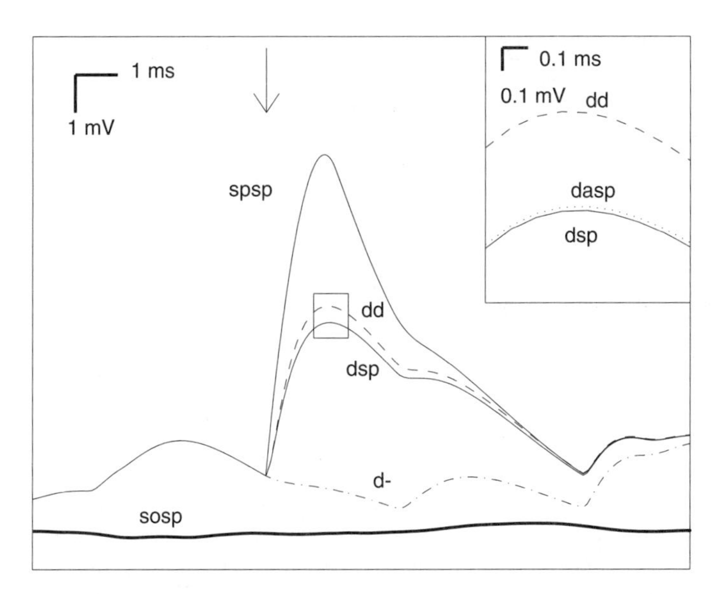

FIGURE 9.5 Voltage traces in the Purkinje cell soma, a spine and the attached dendritic shaft for activation of a synaptic input at different locations. There is a notable difference between the activation of spines and the activation of dendritic shafts in the model, but only a very small effect of Ca^{2+} channels on the spine head (inset). The dendritic branch receives many random synaptic inputs (mimicking the *in vivo* situation, see Section 9.4.3) resulting in a very noisy trace, in contrast to the somatic trace which is much less noisy because of dendritic filtering; *sosp*, somatic voltage during activation of a spine; *spsp*, voltage in the activated spine; *dsp*, dendritic voltage for activation of the attached spine; *dd*, dendritic voltage for direct activation of the dendritic shaft; *dasp*, dendritic voltage for the activation of an active spine with CaP channels; *d-*, dendritic voltage without activation of either the shaft or an attached spine. Synaptic input was on the spine or the dendritic shaft ($\bar{g}_{syn}$ = 2 nS, time indicated by arrow). The active spine had a CaP channel on its head with the same $\bar{g}$ as on the dendrite. Rest of the model as in Reference 32.

The spine collapsing procedure only affects CM_j and RM_j (Equations 9.10–9.11), not $RI_{j,k}$ (Equations 9.12–9.13), as it is assumed that spines do not participate in the axial current flow inside the compartment.

9.4.3 The Number of Synaptic Contacts

How many spines should be modeled explicitly? A better question is to ask how many synaptic contacts should be modeled. One is usually forced to simulate only a fraction of the inputs because modeling all of them (~10,000 excitatory contacts for pyramidal cells, ~200,000 for cerebellar Purkinje cells) is not practical. Even if computer memory and speed allow for it (which is now the case even for inexpensive computers) it does not make much sense to model all synapses. First, the number of compartments modeled is usually much smaller so that 10–100 synaptic contacts will have the same isopotential postsynaptic target. Second, how should the activation of so many synapses be computed? If the model is embedded in a network simulation, the other neurons will control synaptic activation, but in general, network models do not represent the full density of cells so that the number of contacts will be small (see Section 11.3.4). If a single cell model is simulated, one is usually forced to use random activation of most of the synaptic contacts. For conditions of little synaptic saturation due to temporal summation (small $\overline{g}_{syn}$ and low activation rates f_{syn}) the total synaptic drive A onto the cell can be approximated by:

$$A \approx N_{syn}\,\overline{g}_{syn}\,f_{syn} \tag{9.15}$$

This relation suggests that one can compensate changes in the number of synapses simulated (N_{syn}) by inversely scaling $\overline{g}_{syn}$ or f_{syn}. For the Purkinje cell model described in Box 9.1 it has been shown that, using this procedure, interspike interval distributions remain constant over a range of N_{syn} values.[32] Therefore, assuming low $\overline{g}_{syn}$ and low f_{syn}, Equation 9.15 can be used to change the number of synaptic contacts as desired or to estimate how the synaptic activation rates in the model compare to physiological ones. Note that Equation 9.15 holds only for asynchronous activation of synapses: the N_{syn} synapses should be activated independently from each other. A rule of thumb is that it is not useful to model more than one synaptic contact of a particular receptor type on each compartment. To simplify the model and increase computation speed most simulators allow multiple connections to a single postsynaptic target. All these connections have the same $\overline{g}_{syn}$ but they can have independent weights and be activated independently in the context of a network simulation. Computing multiple activations requires convolution of the synaptic conductance over time.[17,33]

9.4.4 What About the Axon?

Until recently most single cell models did not include a representation of the axon. Nowadays it is customary to include at least the axon hillock. There are several reasons for doing so. As slice experiments have demonstrated that action potentials tend to originate in the axon hillock or even further down in the axon[34] it is better to include the first axonal segments in the model. Usually only the unmyelinated first part is included, but, using the proper low capacitance values for myelin and high Na^+ channel densities and low R_m for the nodes of Ranvier,[35] it is possible to include longer segments.

Including the axon hillock is especially important if one wants to model back propagating action potentials,[22] though it is not trivial to make such models fit measured values for Na^+ channel densities.[36] If one does not simulate an axon hillock spikes are initiated in the soma of the model. Because of the low impedance of the soma this requires unphysiologically large densities of somatic Na^+ channels (e.g., Reference 37).

Another reason to include the axon may be that unmyelinated axons impose a large capacitative load on the neuron which must be included in the model if its C_m is to be correct. An equivalent cable approach[1,2] can be used to model the axon beyond the spike initiation zone. However, in some cases it may be of interest to model the morphology of the axon in some detail. For example invertebrate axons often have multiple spike initiation sites (Section 10.4). Very few detailed axon models exist as it is often assumed that spike conduction can be simply represented as delays (Section 11.3.2). Existing axonal models demonstrate how the morphological complexity of axons can influence their filtering properties[38] and determine the distribution of time delays.[39]

9.4.5 Modeling Ionic Concentrations

The activation or inactivation of many channels depends on ionic concentrations, for example calcium for the ubiquitous Ca^{2+}-activated K^+ channels. Moreover, in some cases the flux through channels may change the local ionic concentration significantly so that E_i is not constant (Equation 9.8). As a consequence, models of changes in ionic concentration may have to be included. General approaches are covered extensively in Chapter 3. Here we will consider the choice of models only and focus on simulating the intracellular calcium concentration.

Until recently little experimental data were available on calcium concentrations inside neurons. Therefore these were often modeled using an "exponentially decaying pool." This corresponds to the excessive buffer approximation of Section 3.2.1.2 and requires only two parameters: an effective volume v to scale the Ca^{2+} influx J (Chapter 3, Equation 3.6) and a decay time constant τ (Section 3, Equation 3.11). This purely phenomenological approach is used in most compartmental models because it is often adequate to replicate the behavior of Ca^{2+}-dependent channels (see also Box 9.1). If several kinds of Ca^{2+}-dependent processes co-exist it may not be possible to tune the model with a single type of calcium pool, in this case one can opt to model multiple pools with different decay time constants.[40]

The exponentially decaying pool can not be expected to give realistic values for the intracellular calcium concentration, or at its best only for the concentration in a thin submembrane shell where it can not be measured accurately in most experiments. In order to compare the simulated concentrations to experimental data, to study the interaction with intracellular release processes or to compute the effect of buffering, a full model of buffered calcium diffusion has to be implemented. The techniques for doing so are described in Section 3.3.1.

The dimensionality of the calcium model is determined by that of the gradients simulated. The distances over which Ca^{2+} gradients can exist are much smaller than for voltage gradients.[41] As a consequence, if the sources of calcium influx are distributed smoothly along the length of the simulated dendrite, it is usually sufficient to model a one-dimensional gradient from the membrane to the core of the compartment using concentric shells with a thickness of 0.1–0.2 μm. The assumption of a one-dimensional gradient will not be correct if calcium influx occurs at "hot spots" which create additional gradients along the length of dendrite. A practical guide to both one and two-dimensional modeling of calcium concentrations in neuronal models can be found in Reference 42.

Calcium concentrations have also been modeled in spines, assuming that a correct approximation can be obtained by one-dimensional diffusion along the length axis of the spine.[43,44] While such models have provided insights into the relation between calcium influx and LTP induction one should realize that they are very crude approximations of reality. As for transmitter diffusion (Chapter 4), stochastic models of three-dimensional calcium diffusion[45] may be needed to simulate these processes correctly, given that a spine contains an average of only one or two Ca^{2+} ions at rest.

Besides controlling Ca^{2+}-dependent processes, calcium concentrations should also be used for the computation of calcium currents. Because of the rectifying effect of the large Ca^{2+} gradient across the cell membrane calcium currents can not be considered Ohmic[35] and should therefore not be modeled as the product of a voltage-dependent g and a driving force $(V - E)$ (Equation 9.8). The best way to model calcium currents is to use the Goldman–Hogkin–Katz equation (GHK, Chapter 5, Equation 9.38 and Reference 42). The Nernst equation (Chapter 5, Equation 5.2) does not give a correct approximation for calcium currents because it still assumes a linear current, but it can be used to model changes in external potassium concentrations[46] as the transmembrane K^+ gradient is much smaller. To have the current through the Ca^{2+} channels reverse at the correct voltage (about 50 mV) the small outward K^+ current through these channels[35] needs to be modeled with a second GHK equation. The use of GHK equations to compute the voltage-dependent fractional Ca^{2+} current through NMDA channels is described in Chapter 11, Box 11.2.

9.4.6 How to Start Constructing the Model

Good active membrane compartmental models constitute a synthesis of data about a particular neuron type. This implies that the first step in constructing such a model is to become completely familiar with all available data. Traditionally this involves searching the literature, but it is hoped that neuroinformatics databases as a source of original data will aid modelers in the future. Soon it may no longer be necessary to measure data points with a ruler on a small published graph.

Ideally one would want to use only data obtained from the specific neuron type in one animal species. For active membrane models it is not possible to obtain all the data in a single recording, as can be done for passive membrane models (Chapter 8). At least one should obtain a good morphological reconstruction (Chapter 6) and basic passive cable parameters for this particular morphology (Chapter 8) and as many current clamp and voltage clamp (Chapter 5) data as possible. But after all this is done it seems to be inevitable that, even in the case of extensively studied preparations, some of the necessary data are missing. Therefore model building can already be useful before the model is finished as it may guide future experimental research (see Section 10.4). However, often it is not feasible to wait for the missing data or the data may be technically difficult to acquire. Examples of the latter are the equations for the fast Na^+ current, which is difficult to voltage clamp accurately because of its fast kinetics and large amplitude, and spine densities and sizes which can only be obtained using electron microscopy. In such instances one is forced to improvise, which usually involves using data from different animal species or from other neuron types. For example many modelers have implemented Na^+ channel equations derived from the original Hodgkin–Huxley equations (for example References 18, 37, 46, and 47). However, as pointed out before (Section 9.2.2), these equations describe a squid axon so it would seem more logical to use data from mammalian neurons.[48,49] As a rule of thumb one should try to complement missing data with data obtained in the most closely related system.

At this stage of model construction familiarity with the experimental literature is particularly important. It may be that somebody voltage clamped the channel you need but only studied activation, not inactivation. Data from different sources will need to be combined and it may not be trivial to select the particular data set which will be used for the inactivation data. If the choice is not clear it may be worthwhile to implement and compare different solutions or to consider the inactivation kinetics a free parameter. If possible one should get a first hand experience of the *in vitro* or *in vivo* firing pattern of the neuron (record yourself or assist during a recording) because published traces sometimes overemphasize particular aspects of the neuron's normal behavior.

If good experimental data are available one may be faced with the opposite problem: which data set to select for parameterization? It would seem logical to fit to the mean values and curves, but in practice this does not always work. For example, in some models channel kinetics equations based on a particular experiment gave better results than those based on the mean over all experiments, presumably because the voltage-dependences of activation and inactivation were related in a way that was lost in the averaging.[37]

A related issue is how much of the data to include in the model. Typical active membrane models incorporate five to ten different ionic channel types[18,21,37,40,46] while recent advances in molecular technology suggest that to include all expression variants and different states of regulation the number should be at least one order of magnitude higher. More quantitative studies of the effects of incorporating multiple variants of the same channel type are needed, but at present this does not seem necessary unless it is the specific focus of the modeling study (e.g., neuroendocrine modulation of firing patterns). With about ten different channels active membrane models are already very complex systems which are sometimes difficult to tune and little is to be expected from adding an extra channel equation which differs only slightly in its kinetics. The channels which are implemented in the model should be seen as representative members of a larger family.

As already mentioned, it is generally not possible to get all data from the experimental literature. The remaining data become the free parameters of the model; usually these include the densities ($\bar{g}_i$) of the channels and the time constants of the calcium pool(s). Patch clamp experiments can provide $\bar{g}_i$ values for the soma and thicker parts of dendrites. In practice, for reasons not completely understood, these values often need to be modified to make the model reproduce physiological behavior.[4,36] Converting patch clamp data into $\bar{g}_i$ also requires estimating the size of the patch which introduces a source of error. Nevertheless patch clamp measurements are useful to constrain the range of acceptable $\bar{g}_i$ values. Voltage clamp experiments can give an estimate of the maximum total conductance of a particular channel type (summated over all compartments; see also Reference 5 for an interesting experimental approach), but it is important to be aware that space clamp problems and side effects of pharmacological blocks may influence these measurements. The passive cable parameters are often found separately (Chapter 8) but as it may not be possible to constrain them completely some of them can also be free parameters (within a range) during the tuning of the active membrane model.

9.4.7 Parameterizing the Model

Compartmental modeling may seem plug-and-play. Once you have collected your morphology, cable parameters and channel kinetics you plug them into your favorite simulator and the model is ready. In practice things are not so simple. The data obtained are often not in a form suitable for modeling. They first need to be converted into equations and as these usually take standard forms this means finding the correct parameters. Many of these issues have been covered in preceding chapters. In this section we will give some practical advice on parameterization and in the next one we will consider methods to test the model's accuracy.

An important issue in parameterization is standardization. Everything should be in the same units (unless the simulator software converts it for you) and obtained under the same conditions. Many experimental data are obtained at room temperature. Will this also be the temperature for the model or should it operate at body temperature? In the latter case the channel kinetics need to be scaled (Section 5.4.2) accordingly. Were the channel kinetics obtained at physiological concentrations? If not, it may be possible to estimate the concentration effects on voltage-dependency.[35,50] Was the shrinkage identical for the stained neuron and the preparation which was used to measure spine densities? Were they both corrected for shrinkage (Section 6.4.3)? The list of possible errors or omissions is long and if one is not familiar with the experimental techniques used it may be useful to ask assistance from an experienced colleague.

At the level of the model itself it is important to decide in advance which spatial and temporal scales need to be represented accurately. The differential equations used in these models are not very suitable to simulate events which occur in the μsec range or on a sub-μm scale (see Chapter 4) so it does not make much sense to try to reach parameter accuracy at this level. Conversely, very long time constants (in the sec range) may be omitted from the model if other processes on this time scale (e.g., second messenger regulation of channels) are not simulated.

To obtain accurate results the equations need to be discretized correctly (Section 1.3.3). The rule is to use the largest possible values of Δx and Δt for which a decrease does not change the model behavior, because this increases computation efficiency. Compartment size (Δx) is determined by two factors: the morphology and a maximum electrical size (they are assumed to be isopotential). One tries to represent all the branch points in a morphology accurately, but it may not be necessary to keep all small differences in diameter considering inaccuracies in the measurements (Chapter 6, Table 6.1). If one combines multiple measurements along an unbranched segment into a single compartment care should be taken to use the same total membrane surface (Equation 9.9) and RI (Equation 9.12) as for the corresponding number of smaller compartments. A rule of thumb for passive compartments is to take a maximum electrical size of 0.1 λ (Section 8.5, Reference 20), but it is not entirely clear what should be taken for active ones as the voltage-gated channels make λ variable over time.[37] A possibility is to start discretizing at 0.05 $\lambda_{passive}$ and to check if the model behavior changes if smaller Δx are used. NEURON handles spatial discretization automatically, but GENESIS is less flexible in this respect. Similarly, models are usually run with time steps Δt in the order of ten μsec (if calcium diffusion processes are included they need to be in the one μsec range), but the effect of using smaller time steps should be checked. Note that it is necessary to check for accurate use of Δx and Δt repeatedly while the model parameters are being tuned.

Finally, not every parameter needs to be resolved at the compartmental scale. For example, in a large model it is not practical to tune all conductances $\bar{g}_{ij}$ for each individual compartment. Instead the model can be subdivided into functionally defined zones where the $\bar{g}_i$ values are equal,[37] reducing the number of different parameters by more than an order of magnitude.

9.4.8 How to Test the Model

Ideally the accuracy of the model should be checked with two separate sets of tests. The first set will be used to tune the parameters during the development of the model, the second to evaluate the faithfulness of the resulting model.

It is important to clearly define which behaviors the model is required to reproduce. If one wants to use automatic parameter search methods (Section 1.5) these requirements have to be formalized by a fitness function (Section 1.4) which may be the weighted sum of many separate fitness scores.[51] Designing a good fitness function can be very difficult and may need a lot of hand tuning. It is important not to overfit the model either. As it will be based on data from many different sources it is of little value to tune it until it reproduces one particular spike train with perfect precision. An alternative are statistical approaches like the density trajectory-based fitness function of Section 1.4. In general one should not expect active models to attain the same degree of accuracy as passive membrane models (Chapter 8). They should represent a class of neurons, not reproduce one particular experiment.

The most commonly used method is to manually tune the model and visually evaluate the simulation traces. For complex models with many free parameters this may be the only practical solution as automatic search methods have not yet been shown to work for models with thousands of compartments. An advantage of manual tuning is that the modeler develops a feeling for the model properties and its parameter sensitivities at the same time. A hybrid approach may also be useful, first finding a proper range of parameter values manually and then fine tuning them automatically.

As described in Section 10.4.3 one should perform a parameter range study at the end of the tuning process, whether it was done automatically or not. Such a study can not validate a model but by giving insights into the importance of the different parameters it may indicate potential problems. For example, the size of the leak conductance should not be a critical parameter, if it is the model that is probably lacking one or more K^+ channels. This procedure will also provide insight into the uniqueness of the model. In particular, if two disjoint regions of parameter space both give good results one should try to understand if this is realistic. In the best case this might indicate that further experiments are needed to better constrain the model, in the worst case it may point to a fundamental flaw.

In the older literature it was considered important for the model behavior to be robust against large changes in parameter values. This is not a realistic requirement for active membrane models as it may be important for a neuron to keep its operational range close to a bifurcation point (like a switch from regular firing to bursting) which can only be achieved in a tight parameter range. In biological neurons many intracellular processes exist to continually adapt channels properties and densities and these may be employed by nature to keep some parameters in a small range.[52]

The best way to validate the final model is to test if it can reproduce behaviors for which it was not tuned. For example, the Purkinje cell model (Box 9.1, References 32 and 37) was tuned to reproduce the responses to intracellular current injection *in vitro*. Subsequently it was demonstrated that the same model could also reproduce responses to synaptic activation of four different kinds of inputs obtained either *in vitro* or *in vivo*. Although the latter required the tuning of a few additional free parameters (the $\bar{g}_{syn}$) it indicated that the basic active membrane model remained faithful under different circumstances. For example, the model was able to reproduce the complex spike in response to climbing fiber excitatory input which involves widespread activation of voltage-gated channels in dendrite and soma. We think that this approach is employed too rarely. While the exact procedure will depend on the preparation, it should always be possible to divide the experimental data into a set that is used for the tuning and a set for subsequent testing. The disadvantage may seem that the model is less accurately tuned but a careful check of the model's faithfulness is an essential prerequisite for it to produce reliable insights into the functioning of the neuron that is being simulated.

REFERENCES

1. Segev, I., Rinzel, J., and Shepherd, G. M., Eds., *The Theoretical Foundation of Dendritic Function. Selected Papers of Wilfrid Rall with Commentaries*, MIT Press, Cambridge, 1995.
2. Rall, W. and Agmon-Snir, H., Cable theory for dendritic neurons, *Methods in Neuronal Modeling: From Ions to Networks*, Koch, C. and Segev, I., Eds., 2nd ed., MIT Press, Cambridge, 1998.
3. Stuart, G., Spruston, N., and Häusser, M., Eds., *Dendrites*, Oxford University Press, Oxford, 1999.
4. Hoffman, D. A., Magee, J. C., Colbert, C. M., and Johnston, D., K^+ channel regulation of signal propagation in dendrites of hippocampal pyramidal neurons, *Nature*, 387, 869, 1997.
5. Destexhe, A., Contreras, D., Steriade, M., Sejnowski, T. J., and Huguenard, J. R., *In vivo, in vitro* and computational analysis of dendritic calcium currents in thalamic reticular neurons, *J. Neurosci.*, 16, 169, 1996.
6. Zador, A., Impact of synaptic unreliability on the information transmitted by spiking neurons, *J. Neurophysiol.*, 79, 1219, 1998.

7. Wehmeier, U., Dong, D., Koch, C., and Van Essen, D., Modeling the mammalian visual system, in *Methods in Neuronal Modeling: From Synapses to Networks*, Koch, C. and Segev, I., Eds., MIT Press, Cambridge, 1989.
8. Koch, C., *Biophysics of Computation: Information Processing in Single Neurons*, Oxford University Press, New York, 1999.
9. Hodgkin, A. L. and Huxley, A. F., A quantitative description of membrane current and its application to conduction and excitation in nerve, *J. Physiol.*, 117, 500, 1952.
10. Cronin, J., *Mathematical Aspects of Hodgkin–Huxley Neural Theory*, Cambridge University Press, London, 1987.
11. Jack, J. J. B., Noble, D., and Tsien, R. W., *Electric Current Flow in Excitable Cells*, Clarendon Press, Oxford, 1975.
12. Stein, R. B., Some models of neuronal variability, *Biophys. J.*, 7, 37, 1967.
13. Connor, J. A. and Stevens, C. F., Prediction of repetitive firing behaviour from voltage clamp data on an isolated neurone somata, *J. Physiol.*, 213, 31, 1971.
14. Agmon-Snir, H. and Segev, I., Signal delay and input synchronization in passive dendritic structures, *J. Neurophysiol.*, 70, 2066, 1993.
15. Bush, P. C. and Sejnowski, T. J., Reduced compartmental models of neocortical pyramidal cells, *J. Neurosci. Meth.*, 46, 159, 1993.
16. Stratford, K., Mason, A., Larkman, A., Major, G., and Jack, J. J. B., The modelling of pyramidal neurones in the visual neurones in the visual cortex, *The Computing Neuron*, Dubin, R., Miall, C., and Mitchison, G., Eds., Addison-Wesley, Wokingham, UK, 1989.
17. Wilson, M. A. and Bower, J. M., The simulation of large-scale neuronal networks, Methods in Neuronal Modeling: From Synapses to Networks, Koch, C. and Segev, I., Eds., MIT Press, Cambridge, 1989.
18. Traub, R. D., Wong, R. K. S., Miles, R., and Michelson, H., A model of a CA3 hippocampal pyramidal neuron incorporating voltage clamp data on intrinsic conductances, *J. Neurophysiol.*, 66, 635, 1991.
19. Pinsky, P. F. and Rinzel, J., Intrinsic and network rhythmogenesis in a reduced Traub model for CA3 neurons, *J. Comput. Neurosci.*, 1, 39, 1994.
20. Segev, I. and Burke, R. E., Compartmental models of complex neurons, in *Methods in Neuronal Modeling: From Ions to Networks*, Koch, C. and Segev, I., Eds., 2nd ed., MIT Press, Cambridge, 1998.
21. Mainen, Z. F. and Sejnowski, T. J., Influence of dendritic structure on firing pattern in model neocortical neurons, *Nature*, 382, 363, 1996.
22. Mainen, Z. F., Joerges, J., Huguenard, J. R., and Sejnowski, T. J., A model of spike initiation in neocortical pyramidal neurons, *Neuron*, 15, 1427, 1995.
23. De Schutter, E., Using realistic models to study synaptic integration in cerebellar Purkinje cells, *Rev. Neurosci.*, 10, 233, 1999.
24. Bower, J. M. and Beeman, D., *The Book of GENESIS: Exploring Realistic Neural Models with the GEneral NEural SImulation System*, TELOS, New York, 1998.
25. Segev, I., Fleshman, J. W., Miller, J. P., and Bunow, B., Modeling the electrical behavior of anatomically complex neurons using a network analysis program: passive membrane, *Biol. Cybern.*, 53, 27, 1985.
26. Hines, M., Efficient computation of branched nerve equations, *Int. J. Biomed. Comput.*, 15, 69, 1984.
27. Eichler West, R. M. and Wilcox, G. L., A renumbering method to decrease matrix banding in equations describing branched neuron-like structures, *J. Neurosci. Meth.*, 68, 15, 1996.

28. Mascagni, M. V. and Sherman, A. S., Numerical methods for neuronal modeling, in *Methods in Neuronal Modeling: From Ions to Networks*, Koch, C. and Segev, I., Eds., 2nd ed., MIT Press, Cambridge, 1998.
29. Moore, J. W. and Ramon, F., On numerical integration of the Hodgkin and Huxley equations for a membrane action potential, *J. Theor. Biol.*, 45, 249, 1974.
30. Rapp, M., Yarom, Y., and Segev, I., The impact of parallel fiber background activity on the cable properties of cerebellar Purkinje cells, *Neural Comput.*, 4, 518, 1992.
31. Larkman, A. U., Dendritic morphology of pyramidal neurones of the visual cortex of the rat: III. Spine distributions, *J. Comp. Neurol.*, 306, 332, 1991.
32. De Schutter, E. and Bower, J. M., An active membrane model of the cerebellar Purkinje cell: II. Simulation of synaptic responses, *J. Neurophysiol.*, 71, 401, 1994.
33. Lytton, W. W., Optimizing synaptic conductance calculation for network simulations, *Neural Comput.*, 8, 501, 1996.
34. Stuart, G., Spruston, N., Sakmann, B., and Häusser, M., Action potential initiation and backpropagation in neurons of the mammalian CNS, *Trends Neurosci.*, 20, 125, 1997.
35. Hille, B., *Ionic Channels of Excitable Membranes*, Sinauer Associates, Sunderland, 1992.
36. Rapp, M., Yarom, Y., and Segev, I., Modeling back propagating action potential in weakly excitable dendrites of neocortical pyramidal cells, *Proc. Natl. Acad. Sci. U.S.A.*, 93, 11985, 1996.
37. De Schutter, E. and Bower, J. M., An active membrane model of the cerebellar Purkinje cell. I. Simulation of current clamps in slice, *J. Neurophysiol.*, 71, 375, 1994.
38. Manor, Y., Koch, C., and Segev, I., Effect of geometrical irregularities on propagation delay in axonal trees, *Biophys. J.*, 60, 1424, 1991.
39. Tettoni, L., Lehmann, P., Houzel, J. C., and Innocenti, G. M., Maxsim, software for the analysis of multiple axonal arbors and their simulated activation, *J. Neurosci. Meth.*, 67, 1, 1996.
40. Tegnér, J. and Grillner, S., Interactive effects of the GABABergic modulation of calcium channels and calcium-dependent potassium channels in lamprey, *J. Neurophysiol.*, 81, 1318, 1999.
41. Kasai, H. and Petersen, O. H., Spatial dynamics of second messengers: IP_3 and cAMP as longe-range and associative messengers, *Trends Neurosci.*, 17, 95, 1994.
42. De Schutter, E. and Smolen, P., Calcium dynamics in large neuronal models, *Methods in Neuronal Modeling: From Ions to Networks*, Koch, C. and Segev, I., Eds., 2nd ed., MIT Press, Cambridge, 1998.
43. Zador, A., Koch, C., and Brown, T. H., Biophysical model of a Hebbian synapse, *Proc. Natl. Acad. Sci. U.S.A.*, 87, 6718, 1990.
44. Holmes, W. R. and Levy, W. B., Insights into associative long-term potentiation from computational models of NMDA receptor-mediated calcium influx and intracellular calcium concentration changes, *J. Neurophysiol.*, 63, 1148, 1990.
45. Bormann, G., Brosens, F., and De Schutter, E., Modeling molecular diffusion, *Computational Modeling of Genetic and Biochemical Networks*, Bower, J. M. and Bolouri, H., Eds., MIT Press, Cambridge, 2000.
46. Yamada, W. M., Koch, C., and Adams, P. R., Multiple channels and calcium dynamics, *Methods in Neuronal Modeling: From Ions to Networks*, Koch, C. and Segev, I., Eds., 2nd ed., MIT Press, Cambridge, 1998.
47. Destexhe, A. and Pare, D., Impact of network activity on the integrative properties of neocortical pyramidal neurons *in vivo*, *J. Neurophysiol.*, 81, 1531, 1999.

48. Hamill, O. P., Huguenard, J. R., and Prince, D. A., Patch-clamp studies of voltage-gated currents in identified neurons of the rat cerebral cortex, *Cereb. Cortex*, 1, 48, 1991.
49. Alzheimer, C., Schwindt, P. C., and Crill, W. E., Modal gating of Na^+ channels as a mechanism of persistent Na^+ current in pyramidal neurons from rat and cat sensorimotor cortex, *J. Neurosci.*, 13, 660, 1993.
50. Johnston, D. and Wu, S. M.-S., *Foundations of Cellular Neurophysiology*, MIT Press, Cambridge, 1995.
51. Eichler West, R. M., De Schutter, E., and Wilcox, G. L., Using evolutionary algorithms to search for control parameters in a nonlinear partial differential equation, *Evolutionary Algorithms*, Davis, L. D., De Jong, K., Vose, M. D., and Whitley, L. D., Eds., Vol. III, IMA Volumes in Mathematics and its Applications, Springer-Verlag, New York, 1998.
52. Turrigiano, G., Abbott, L. F., and Marder, E., Activity-dependent changes in the intrinsic properties of cultured neurons, *Science*, 264, 974, 1994.
53. Llinás, R. R. and Walton, K. D., Cerebellum, in *The Synaptic Organization of the Brain*, Shepherd, G. M., Eds., 4th ed., Oxford University Press, NY, 1998.
54. Jaeger, D., De Schutter, E., and Bower, J. M., The role of synaptic and voltage-gated currents in the control of Purkinje cell spiking: a modeling study, *J. Neurosci.*, 17, 91, 1997.
55. Jaeger, D. and Bower, J. M., Prolonged responses in rat cerebellar Purkinje cells following activation of the granule cell layer: an intracellular *in vitro* and *in vivo* investigation, *Exp. Brain Res.*, 100, 200, 1994.
56. Häusser, M. and Clark, B. A., Tonic synaptic inhibition modulates neuronal output pattern and spatiotemporal integration, *Neuron*, 19, 665, 1997.
57. Jaeger, D. and Bower, J. M., Synaptic control of spiking in cerebellar Purkinje cells: Dynamic current clamp based on model conductances, *J. Neurosci.*, 19, 6090, 1999.
58. De Schutter, E. and Bower, J. M., Simulated responses of cerebellar Purkinje cells are independent of the dendritic location of granule cell synaptic inputs, *Proc. Natl. Acad. Sci. U.S.A.*, 91, 4736, 1994.
59. De Schutter, E., Dendritic voltage and calcium-gated channels amplify the variability of postsynaptic responses in a Purkinje cell model, *J. Neurophysiol.*, 80, 504, 1998.
60. Eilers, J., Augustine, G. J., and Konnerth, A., Sub-threshold synaptic Ca^{2+} signaling in fine dendrites and spines of cerebellar Purkinje neurons, *Nature*, 373, 155, 1995.

10 Realistic Modeling of Small Neuronal Circuits

Ronald L. Calabrese, Andrew A.V. Hill, and Stephen D. van Hooser

CONTENTS

10.1 INTRODUCTION

Small networks of neurons, particularly those from invertebrate preparations where it is possible to work with identified neurons, are attractive objects for modeling using conductance-based or "realistic" neuron models. Often there is the possibility of using voltage clamp techniques to characterize the membrane currents of the component neurons and their synaptic interactions, so that sensible conductance-based model neurons can be constructed. Moreover, for small networks it is often possible to quantify precisely the network behavior that one wishes to simulate in

0-8493-2068-2/01/$0.00+$.50

the network model so that a benchmark for model performance can be established. The aim of this chapter is to provide some tools useful in modeling synaptic interactions in small network models and to lay out a strategy for using these tools in constructing network models. Particular attention is given to the processes of parameterization and model testing. There is above all an emphasis on the interactive nature of realistic modeling with the model being developed from experimental observations and in turn generating predictions that are experimentally testable. The chapter also presents an example from our own work that we hope illustrates this strategy. The chapter ends with a discussion of how abstract models can inspire, supplement, and illuminate the more realistic models developed with the tools and strategy presented here.

10.2 PHENOMENOLOGICAL MODELING OF SYNAPTIC INTERACTIONS

The detailed methods for modeling synaptic interactions developed in Chapters 4 and 5 are not practical in most network models. The level of experimental detail necessary for such models is often not available and the computational overhead is steep. Several approaches have been used to simplify models of synaptic interactions for use in network simulation. Since we are considering only conductance based models here the relevant approach boils down to relating some measure of presynaptic activity to fractional transmitter release, f_{syn}, activating the postsynaptic conductance, ($\bar{g}_{syn}$). For the parameter, $\bar{g}_{syn}$, to have physiological meaning as the maximal postsynaptic conductance (f_{syn}) should be constrained to vary between 0 and 1, but this need not be the case in practice. Then postsynaptic current is calculated in the normal way described in Chapter 5,

$$I_{Syn} = f_{Syn}\bar{g}_{Syn}\left(V_{post} - E_{Syn}\right), \tag{10.1}$$

where E_{Syn} is the reversal potential of the postsynaptic current and V_{Post} is the postsynaptic membrane potential. The trick of course is to pick the right measure of presynaptic activity and to correctly specify the dynamics of f_{Syn}.

10.2.1 SPIKE-MEDIATED SYNAPTIC TRANSMISSION

In networks where spiking activity dominates synaptic interactions, the appropriate measure of presynaptic activity is the occurrence of a presynaptic spike. Spike occurrence and thus spike timing (t_{spike}) is easily specified by setting a voltage threshold and an appropriate synaptic delay (t_{delay}) can then be implemented. (Spike detection and synaptic delays are standard parts of software packages such as GENESIS.) For spike mediated synaptic transmission, Equation 10.1 becomes

$$I_{Syn}(t,V) = \bar{g}_{Syn}\left(V_{post} - E_{Syn}\right)\sum_{spikes} f_{Syn}\left(t - \left(t_{spike} + t_{delay}\right)\right). \tag{10.2}$$

The detection of a presynaptic spike then triggers an appropriate time course for $f_{Syn}(t)$. The dynamics of $f_{Syn}(t)$ must then be specified of course. The simplest method is to use the so called alpha function,[1]

$$f_{Syn}(t) = \frac{t}{t_p}\, e^{\left(1-t/t_p\right)}. \tag{10.3}$$

This function rapidly rises to its peak at t_p and falls more slowly after that. It can be used to roughly fit the time course of most spike mediated synaptic potentials, if some estimate of the time to peak (t_p) of the postsynaptic current is available.

If voltage clamp data is available on the time course of individual spike mediated postsynaptic currents (PSCs), then it is possible to fit it with exponential functions to derive rising and falling time constants. One can then use a dual exponential function to specify the dynamics of the postsynaptic conductance. A common form used e.g. by the GENESIS simulator[2] is

$$f_{Syn}(t) = \frac{A}{\tau_1 - \tau_2}\left(e^{-t/\tau_1} - e^{-t/\tau_2}\right), \text{ for } \tau_1 > \tau_2. \tag{10.4}$$

Here τ_1 corresponds to the falling time constant of the postsynaptic current and τ_2 corresponds to the rising time constant and this equation can be fitted directly to PSCs and the parameter A adjusted so that f_{Syn} approaches unity at the peak of the PSC. A simpler and more intuitive form is used in our own simulations:[3]

$$f_{Syn}(t) = \left(1 - e^{-t/\tau_{rise}}\right)e^{-1/\tau_{fall}}, \text{ for } \tau_{fall} > \tau_{rise}. \tag{10.5}$$

Here τ_{rise} is explicitly the rising and τ_{fall} the falling time constant extracted by fitting single exponential functions to the rise and fall of PSCs.

In using all these methods it must be emphasized that the relevant parameters are to be derived from voltage clamp data, i.e., individual PSCs. The rise and fall of postsynaptic potentials will be governed by the prevailing membrane time constant (see Section 8.2), which for the fall of individual subthreshold PSPs is the resting membrane time constant.

10.2.2 Graded Synaptic Transmission

In many invertebrate neuronal networks and in the vertebrate retina, the slow wave of polarization of a neuron is effective in mediating presynaptic Ca^{2+} entry and synaptic release. In fact some neurons may not spike at all or if they do, spikes may be relatively ineffective in causing synaptic release. In such networks, presynaptic spike activity is obviously an inadequate measure for gating release in a model network. One commonly used strategy is to relate the postsynaptic activation variable to presynaptic membrane potential in a graded fashion. An example of this method can be found in a recent model of the crab stomatogastric system,[4] which we adapt here to be consistent with the notation developed above.

$$I_{Syn} = \bar{g}_{Syn} f_{Syn}\left(V_{post} - E_{syn}\right) \tag{10.6a}$$

$$\tau_f\left(V_{pre}\right)\frac{df}{dt} = f_\infty\left(V_{pre}\right) - f_{Syn} \tag{10.6b}$$

$$f_\infty\left(V_{pre}\right) = 1/\left(1 + \exp\left[\kappa\left(V_{pre} - V_\kappa\right)\right]\right) \tag{10.6c}$$

$$\tau_f\left(V_{pre}\right) = \tau_{rise} + \tau_{fall}/\left(1 + \exp\left[\lambda\left(V_{pre} - V_\lambda\right)\right]\right) \tag{10.6d}$$

where κ determines the steepness, and V_κ the midpoint of the sigmoidal relation between presynaptic potential (V_{pre}) and the fractional transmitter release, f_{Syn}, activating the maximal postsynaptic conductance, ($\bar{g}_{Syn}$). τ_{rise} rise determines the rise time of the synaptic current when the V_{pre} is high, and τ_{fall} determines its decay time when V_{pre} is low, and λ and V_λ determine the steepness and midpoint of this sigmoidal relation respectively. This form is indeed quite flexible and can accommodate both graded (predominant in the Int1 to LG synapse) and spike-mediated (predominant in the LG to Int1 synapse) synaptic transmission with suitable adjustment of the parameters. For example, to implement rapid spike-mediated transmission κ and V_κ are adjusted so that the relation between f_{Syn} and V_{pre} is steep (large κ) and has a depolarized midpoint (positive V_κ). Thus spikes, and spikes only, will reach potentials that result in f_{Syn} being greater than zero and a spike will push f_{Syn} nearly to unity. Now spikes will produce discrete PSPs which rise with τ_{rise} and fall with τ_{fall}, given adjustments to λ and V_λ that parallel those to κ and V_κ. If κ and V_κ are adjusted so that the relation between f_{Syn} and V_{pre} is shallow (small κ) and has a midpoint relatively near the presynaptic resting potential (V_κ relatively near $v_{pre-Rest}$), transmission will be smoothly graded with V_{pre}. Making τ_{rise} suitably long will prevent much contribution to transmission from spikes, while making τ_{rise} and τ_{fall} vanishingly small will make transmission an instantaneous function of V_{pre} and allow both spikes and subthreshold changes in membrane potential to contribute to transmission.

The great simplicity and flexibility of Equation 10.6 for modeling synaptic transmission make them ideal for incorporation into dynamic current clamping programs[5,6] (Box 10.1).

In our own work on reciprocally inhibitory heart interneurons from the leech, we faced a neuronal circuit in which both spike-mediated and graded synaptic transmission were substantial.[3,7–9] A simple model in which presynaptic membrane potential was the factor that governed graded synaptic transmission like Equations 10.6, would not have allowed us to express the complex dynamics of this transmission. More importantly, such a model would not have allowed us to effectively separate spike-mediated and graded transmission so that their relative

Box 10.1 Dynamic-Clamp and Voltage-Clamp with Realistic Waveforms

Dynamic-clamp is a hybrid technique in which an artificial conductance is added to a neuron by computing in real time the current that would flow through a given conductance and then injecting that current into the neuron in real time.[5] The current is calculated using the standard current equation $I = g(V_m - E_{rev})$ and the dynamics of the conductance change can be implemented in any form according to the type of current (e.g., voltage-gated or synaptic) simulated. For example, to create an artificial synaptic interaction using Equations 10.6, two neurons *A* and *B* that have no synaptic interaction are recorded with sharp microelectrodes or with whole cell patch electrodes. An artificial synapse is then created from cell *A* to cell *B* via a computer by assigning cell *A*'s membrane potential to V_{pre} and cell *B*'s membrane potential to V_{post} and assigning appropriate parameters to Equations 10.6. These potentials (digitized using an *A* to *D* converter) are then used to calculate I_{Syn} on the fly. I_{Syn} is injected (using a *D* to *A* converter) into cell *B*. These types of artificial synapses have been successfully used to study the oscillatory properties of reciprocally inhibitory two cell networks[22] and the interaction between pattern generating networks that operate at different periods.[16]

Voltage-clamp with realistic waveforms has also been used to study pattern-generating networks.[8] This technique is not really different from ordinary voltage-clamp methods, but the emphasis is on the choice of voltage command waveforms that mimic physiological activity. For example, in our analysis of oscillation by reciprocally inhibitor pairs of heart interneurons, we imposed a waveform that closely approximated the slow wave of membrane potential of these neurons during normal activity. This waveform allowed us to assess the amount of low-threshold Ca^{2+} currents that flow into a presynaptic neuron and the resulting graded synaptic transmission during normal activity (Figure 10.2A1). Traversing the same voltage range with a slightly altered waveform greatly augmented the Ca^{2+} currents and graded transmission (Figure 10.2A2).

contributions under different conditions could be assessed. We had data relating low-threshold Ca^{2+} currents measured in the presynaptic neuron to graded synaptic transmission measured as postsynaptic currents, but no such data relating high-threshold Ca^{2+} currents to spike-mediated release. On the other hand, we did have data on the time course of individual spike-mediated IPSCs. We thus chose to model spike-mediated transmission using Equations 10.2 and 10.5. To model graded transmission we chose to relate presynaptic Ca^{2+} entry via low-threshold channels to synaptic release. Thus the graded synaptic current is given by

$$I_{SynG} = \bar{g}_{SynG} f_{SynG}\left(V_{post} - E_{Syn}\right) \quad (10.7a)$$

$$f_{SynG}(t) = \frac{P^3}{C + P^3} \quad (10.7b)$$

$$\frac{dP}{dt} = I_{Ca} - BP \quad (10.7c)$$

$$I_{Ca} = \max(0, -I_{CaF} - I_{CaS} - A) \quad (10.7d)$$

$$\frac{dA}{dt} = \frac{A_\infty(V_{pre}) - A}{0.2} \quad (10.7e)$$

$$A_\infty(V_{pre}) = \frac{10^{-10}}{1 + e^{-100(V_{pre} + 0.02)}} \quad (10.7f)$$

(Numerical constants derived from data on leech heart interneurons. All units are SI, i.e., MKS).

C is a parameter adjusted so that $f_{Syn}(t)$ comes appropriately close to unity during peak low-threshold Ca^{2+} currents. P is the internal "concentration" of Ca^{2+} (in Coulombs) in an unspecified volume that directly contributes to synaptic release. Presynaptic low-threshold Ca^{2+} currents (I_{Ca}) govern the build-up of P. Two currents contribute to I_{Ca}, I_{CaF}, and I_{CaS}, both of which are specified by Hodgkin–Huxley equations (see Section 5.2.1). The voltage-dependent variable A is a variable threshold reflecting Ca^{2+} entry via I_{Ca} that does not directly contribute to P, and B is a buffering rate constant that governs the removal of P.

Figure 10.1 shows how an earlier version of this model of graded transmission[7] fitted the voltage clamp records upon which it was based. As can be seen the fit is quite accurate. This earlier version however failed under network model conditions, yielding transients during slow changes in membrane potential that are never observed in physiological recordings. The fault lay in the voltage dependent variable A because it was given instantaneous dynamics. This fault was ameliorated in Reference 3 and the model was further refined in this current version (Hill et al., unpublished) by additional data[8] as discussed in Section 10.4.3 below. The older version is illustrated here to emphasize that model output may be very accurate in simulating the data upon which it is based but fail to yield meaningful results under other conditions. The implications of using independent models for spike-mediated and graded synaptic transmission in simulating leech heart interneurons will be discussed. Suffice it to say at this point that the separation of these two modes of transmission led to biological insights that would not have been possible had they been fused into a single model of the type embodied in Equations 10.6.

10.2.3 Synaptic Plasticity

Synaptic plasticity is easily accommodated in any of the models of synaptic transmission discussed above by introducing a plasticity variable M_{Syn}, which is usually an

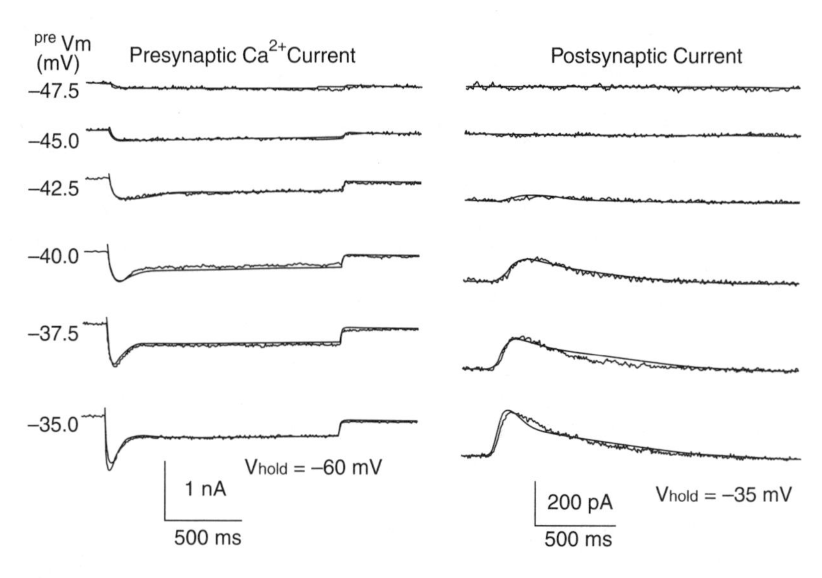

FIGURE 10.1

Graded inhibitory synaptic transmission in leech heart interneurons: fitting the relation between the presynaptic low-threshold Ca^{2+} currents and postsynaptic conductance. Both cells of a reciprocally inhibitory pair were voltage-clamped. Presynaptic low-threshold Ca^{2+} currents (left) and postsynaptic currents (right) evoked by voltage steps (1.5 s) in the presynaptic cell from a holding potential of –60 mV (V_{hold}) to the step potential indicated (preV_m). The postsynaptic cell was held at –35 mV (V_{hold}). The saline contained 0 Na^+ and 2 mM Ca^{2+}. Superimposed lines are simulated values of current (from Reference 7).

explicit function of a pre and/or postsynaptic variable and time. Thus for spike mediated synaptic transmission (and ignoring the synaptic delay, t_{delay}) Equation 10.1 becomes

$$I_{Syn}(t,V) = \bar{g}_{Syn}\left(V_{post} - E_{Syn}\right)\sum_{spikes} M_{Syn} f_{Syn}\left(t - t_{spike}\right). \tag{10.8}$$

The essence of any model of synaptic plasticity will of course be the choice of pre and post synaptic variable upon which M_{Syn} depends. (The product $M_{Syn}\, f_{Syn}$ is often lumped into a single function.) There is a rich literature of models of synaptic plasticity at neuromuscular junctions.[10] Moreover, several sophisticated models of synaptic plasticity suitable for use in network simulations have been developed which use physiologically important variables such as presynaptic Ca^{2+} and cAMP levels[11] as variables which influence presynaptic release, i.e. influence release functions such as $f_{Syn}\, M_{Syn}$. A more general, albeit less physiological method has been employed by Abbott and co-workers.[12] In this method, f_{Syn} describes the postsynaptic response to an individual isolated presynaptic spike, e.g., using Equation 10.4. Now a series of scaling factors which are functions that depend only on the timing of a particular spike with respect to previous spikes is used to adjust the amplitude of individual postsynaptic responses (PSCs) in response to a presynaptic spike train. In its most general form:

$$M_{Syn} = A_1^{p1} A_2^{p2} A_3^{p3} \cdots \tag{10.9}$$

Each scaling factor recovers with a single exponential time constant between spikes. These scaling functions are derived from data generated with random presynaptic spike trains by a learning algorithm that steadily improves the quality of the prediction. Scaling functions are then tested with an arbitrary presynaptic spike train different from the training one. In practice, the number of scaling functions needed will depend on the number of processes that contribute to plasticity, e.g., only one scaling function was needed to describe short-term facilitation at a crab neuromuscular junction[13] but three to describe plasticity at rat cortical synapses, where both facilitation and two forms of depression are present.[12]

Spike-mediated transmission between leech heart interneurons displays a simple form of short-term plasticity that illustrates how a phenomenological approach can be applied to modeling such plasticity. As illustrated in Figure 10.2B, during normal bursting activity, the amplitude of spike-mediated IPSCs varies throughout a burst of action potentials. Previous work[14] had shown that this plasticity was related to the membrane potential from which each presynaptic spike arose and that this effect waxed and waned with an exponential time course. Similar modulation of spike-mediated transmission by baseline presynaptic membrane potential has been observed at other synapses. We chose to model this phenomenon as simply as possible, because we had no direct data on the cellular mechanisms involved. The voltage dependence expressed in standard Hodgkin–Huxley equations for voltage-gated currents (see Section 5.2.1) seem easily adapted to this need and easily implemented in our modeling software (GENESIS). Thus M_{Syn} in Equation 10.8 is governed by

$$\tau_{Syn} \frac{dM_{Syn}}{dt} = M_{Syn\infty}\left(V_{pre}\right) - M_{Syn} \tag{10.10a}$$

$$M_{Syn\infty} = A + \frac{B}{1 + e^{-C(V_{Pre}+D)}} \tag{10.10b}$$

(All units are SI, i.e., MKS.)

This simple adaptation of a pre-existing form adequately modeled the observed voltage-dependent modulation observed at these synapses (Figure 10.2B). Obviously it should be possible to develop expressions for M_{syn} like Equations 10.10 to incorporate synaptic plasticity into graded synaptic transmission models like Equations 10.6.

The decision to incorporate synaptic plasticity into a network model ultimately rests on the question to be asked. Obviously if one is modeling a neuronal network to explicate the cellular mechanisms for learning, then it is imperative to incorporate synaptic plasticity. On the other hand, in rhythmically active networks it may not be important to incorporate plasticity, if such plasticity is only expressed at transients and only steady-state behavior is being considered.[4] In our own work, we chose to incorporate synaptic plasticity (Equations 10.10) in considering the steady-state behavior of the leech heartbeat motor pattern generator, because such plasticity occurs on a cycle by cycle basis.

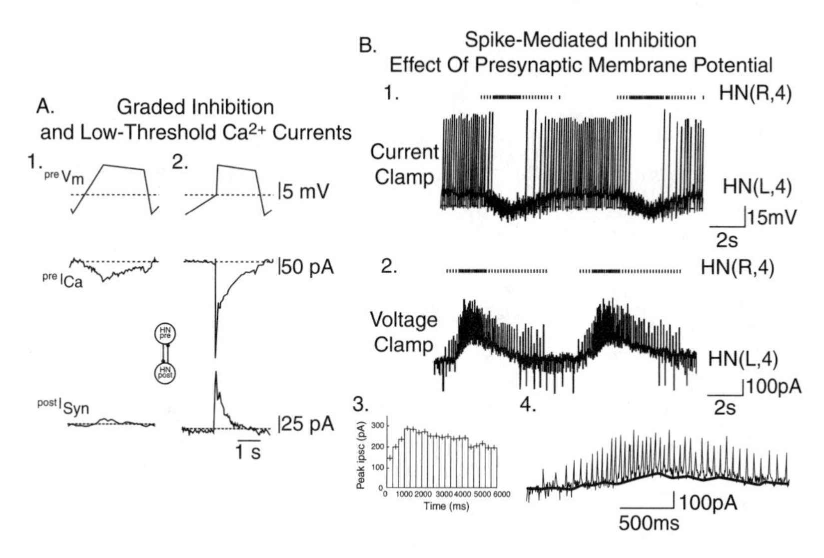

FIGURE 10.2 A. Graded inhibitory synaptic transmission in leech heart interneurons. Presynaptic Ca^{2+} currents and postsynaptic currents evoked by realistic waveforms in heart interneurons. Both cells of a reciprocally inhibitory pair were voltage-clamped and one (pre) was driven with a realistic waveform repetitively (three-cycle average illustrated). The other was held at –40 mV (post). The waveform on the left closely approximates the slow-wave of normal activity in heart interneurons, while the one on the right goes through the same range of voltage but jumps into the depolarized state. The slow upward trajectory of the left-hand waveform inactivates low-threshold Ca^{2+} currents, which are thus subdued, as is the postsynaptic response. The right-hand waveform evoked more Ca^{2+} current and a larger postsynaptic response. The saline contained 0 Na^+ to suppress spikes. Dotted lines mark 0 in the current traces and –50 mV in the voltage traces, as adapted from Reference 8. B. Spike-mediated transmission in leech heart interneurons varies in amplitude throughout a burst according to the baseline level of depolarization. (B1) Normal bursting activity in a reciprocally inhibitory pair of heart interneurons. Only the spikes of the right cell HN(R,4) are indicated as small vertical lines. Dashed line indicates –50 mV. (B2) The left cell HN(L,4) cell was held at –55 mV in voltage clamp. Spike-mediated IPSCs from the opposite cell are apparent on top of a graded IPSC (slow wave) and EPSCs (downward blips between the two barrages of IPSCs) resulting from electrical coupling with the HN(L,3) cell are apparent. Rhythmic activity persists in the HN(R,4) cell despite the lack of spike activity in the HN(R,4) cell, because this preparation contains more anterior ganglia and thus the HN(4) cells are linked in the pattern-generating network of Figure 10.4. Again, the spikes of the right cell HN(R,4) are indicated as small vertical lines. (B3) Histogram of 9 barrages of IPSCs from the HN(L,4) cell in the HN(R,4) cell. The time indicated is from the first spike in the HN(L,4) burst. (B4) Expansion of the first part of the record of B2, illustrating individual IPSCs. Thick line indicates the waveform of the underlying graded IPSC, as adapted from Reference 8.

10.2.4 Electrical Coupling

In some neuronal networks electrical coupling contributes significantly to network function and it must be included. In it simplest form electrical coupling, which is non-rectifying, can be expressed by

$$I_{elec1} = g_{elec}\left(V_1 - V_2\right), \quad I_{elec2} = -I_{elec1} \tag{10.11}$$

where V_1 is the membrane potential of the referent (receiving) cell (cell 1) and V_2 is the membrane potential of the coupled cell, cell 2. Rectification is easily handled by conditional statements, e.g., specifying that if $(V_1 - V_2) > 0$ then $I_{elec} = 0$ will make an electrical synapse rectifying from cell 2 to cell 1.

10.3 BASIC MODELING STRATEGY FOR SMALL CIRCUITS

Small networks afford the opportunity to use realistic conductance based modeling techniques (described in Chapter 5) without straining the computational power and speed of current PC's. Moreover, there are a number of software packages (including GENESIS, NEURON, and SNNAP) available to facilitate such modeling on PC's (Box 10.2). Nevertheless, there are still compromises to be made on how much detail is incorporated into any particular network model; not enough detail will certainly lead to false conclusions, whereas too much detail wastes valuable time and diverts attention from the significant parameters that affect network function. Another important consideration is the generality which one wishes to achieve. The addition of idiosyncratic details will limit applicability to other networks. The choices that the scientist makes about what to include and what to exclude will ultimately determine usefulness of the model both as an experimental tool and in terms of the insights into basic biological mechanisms that it can afford.

The realistic modeling of small networks basically boils down to solving the following current balance equation for any given single-compartment neuron model or any single compartment within a neuron model:

$$C\frac{dV}{dt} = -\sum\left(I_{Ion}\right) - \sum\left(I_{Syn}\right) + \sum\left(I_{Inject}\right) \tag{10.12}$$

where I_{Ion} represents membrane currents (e.g., voltage-gated and leak currents), I_{Syn} synaptic currents, and I_{Inject} injected currents (or the current from connected compartments). Many of the details of which we speak above are basically the compartmental structure of the component neuron models, and the details of the equations that describe I_{Ion} and I_{Syn}.

Box 10.2 Choice of Modeling Software

The choice of a software package for model development is a difficult step. Often this choice is dictated by local expertise with a particular package or in computer programming. Our model is now in its third generation, with each generation implemented using different software. Generation I was implemented in Nodus,[23] a well-conceived general simulation package which now has limited developmental support and expandability. Generation II was implemented in Neurolab, a home grown product. The current generation III is implemented in GENESIS [2] (*http://www.bbb.caltech.edu/GENESIS*) a general simulation package with a large user base and user support group (BABEL) and near infinite expandability owing to its object oriented nature. Our experience with these three packages clearly inclines us towards the use of a widely used and supported package like GENESIS, NEURON,[24] or SNNAP.[25] Each of these packages has its strengths and limitations. GENESIS has a strong tutorial base that makes learning to use the package step-by-step, easy, and fun. The tutorials can be used as the bases of research grade simulations, and network models are easily implemented. The tutorial on the leech heart interneuron network associated with this chapter was developed in GENESIS. The user interface beyond the tutorials leaves much to be desired (point-and-click-ers need not bother), and the software runs only under UNIX or LINUX. NEURON (*http://www.neuron.yale.edu*) has a well developed user interface, numerous user workshops, and runs under both Windows and UNIX/LINUX, but has limited facility with networks and is not as easily expanded to fit particular needs. SNNAP runs under UNIX/LINUX, Windows and has been recently ported to JAVA. It has a very well-developed graphical user interface, and is particularly useful for modeling network where synaptic plasticity is a focus of attention. LINUX on PC's while initially off-putting to a generation of Windows or Mac OS users is a superior environment for network modeling, owing to its stability, fully developed multitasking (including usage of dual processors), infinite regulated expandability, cheap availability, and ease of installation on all current PCs (Red Hat or VA). Fast dual processor PCs running GENESIS (or NEURON) are extremely economic, fast, and efficient vehicles for small network modeling.

10.3.1 Neuronal Morphology

The full morphological detail of the component neurons (see Chapters 6, 8, and 9) is almost never implemented in small network models; rather simplifications are made to increase computational efficiency and to focus attention on circuit interactions. In our own work modeling the leech heartbeat motor pattern generator network, we chose to ignore the morphological complexity of the heart interneurons and treat each cell as a single isopotential compartment. This simplifying assumption was taken for several reasons.

1. We lacked detailed knowledge of the distribution of voltage-gated and synaptic conductances within the neurons.
2. The available data suggested that these neurons are electrically compact and receive a limited number of inputs with widely distributed synaptic contacts (as in many invertebrate ganglion cells synaptic inputs and outputs occur cheek by jowl on the same neuritic branches).
3. The large number of voltage-gated and synaptic currents that were to be incorporated into each model neuron demanded computational efficiency. Similar considerations have often caused other workers modeling motor pattern generating networks to use model neurons with one or a few compartments.[4]

Ultimately, the level of morphological detail included must depend both on the available data and on the experimental questions asked of the model. If there is evidence for example of segregation of electrical activity within neurons that influences circuit behavior, then morphological detail adequate to accommodate such segregation must be incorporated (see below). Nevertheless, it may be beneficial to start by using neurons comprising few compartments, with the view that additional morphological complexity can be added as data becomes available and when comparison to experimental data points out model discrepancies that may have a morphological basis. This was certainly our own strategy.

10.3.2 Intrinsic Membrane Currents

The necessary experimental data on the important intrinsic membrane properties of the component neurons that form the network must be gathered and appropriately described mathematically. Much of these data are likely to be voltage clamp records for which there are established procedures for obtaining accurate mathematical descriptions (see Chapter 5). In some cases, voltage clamp data may not be available for all cells or all currents in particular cells, and current equations must be bootstrapped from current clamp recordings. It is then efficient to start from an existing mathematical description of a related current or the same current from a related neuron. For example, it is often the case in invertebrate neuronal networks that no data on the fast Na^+ current (I_{Na}) that mediates spikes is available, because of space clamp limitations of voltage clamps applied to the cell body. In our own work, our response to this problem was to adapt the fast Na^+ current from the squid (i.e., the original Hodgkin–Huxley equations for I_{Na}) to fit the types of spikes observed in leech heart interneurons.

In several preparations, there are particular neurons for which there are accurate mathematical descriptions of voltage-gated currents that can serve as bootstrapping resources for modeling voltage-gated currents in neurons for which voltage clamp data is unobtainable or fragmentary. For example, for the well-studied stomatogastric neuronal networks of crustaceans, the crab LP neuron serves this role.[15] Often these current descriptions are incorporated into libraries in simulation packages such as GENESIS and are easily modified to bootstrap related currents.[2]

10.3.3 Synaptic Connections

In Section 10.2, we presented methods for modeling synaptic interactions in network models and discussed considerations that would dictate the choice of one method over another. Ultimately the data available on a particular synaptic connection will dictate the detail that can be incorporated into its description. Often this data is limited to current clamp measurements which do not justify going beyond the simplest descriptions (e.g., Equations 10.2 and 10.3 or 10.6).

10.3.4 Parameterization

Once an appropriate level of detail has been incorporated into a model then parameterization is of paramount importance. These parameters include the values such as those that determine voltage-dependence of time constants and activation variables for voltage-gated currents, and the maximal conductances and reversal potentials for all currents. Parameterization can be thought of as tuning to experimental data, which in principle seems simple, but in practice is difficult. Much of the data on intrinsic membrane currents and synaptic interactions that is incorporated into the model will be generated in biophysical experiments under artificial experimental conditions, but the network model will ultimately be compared to current clamp records during normal physiological activity. The procedures for establishing whether the model conforms well to the observed behavior of the biological network are of necessity ad hoc and dictated by the experimental question which the model is to address.

First, as many model parameters for intrinsic membrane properties and synaptic interactions as possible should be fully specified by the data. This process will inevitably leave some parameters unknown or unconstrained. These free parameters must then be tuned to produce realistic cellular and network activity and produce a canonical or benchmark model. It is then important to determine the sensitivity of the model's output to the values chosen for these free parameters and if feasible to map the entire free parameter space. This procedure will allow the modeler to determine whether model behavior is robust, i.e., not overly sensitive to the choice of particular values chosen for free parameters. It is possible that the values of the chosen parameters are very close to a bifurcation in system behavior. By bifurcation is meant a point in parameter space where system behavior spits along different fundamental mechanisms of network operation. Alternatively, the model might be extraordinarily sensitive in some way to a particular parameter so that careful measurement of that parameter must be undertaken to constrain better the parameter. Sensitivity analysis can also aid in the identification of parameters critical to proper network function. In performing sensitivity analyses, it is necessary to first determine an output variable of the model that is important to proper network function in the biological system and then determine its sensitivity to free parameter changes. In a rhythmically active network, such a measure would be cycle period which can be expected to vary smoothly over a physiologically relevant range.

To map the parameter space, free parameters are individually varied over a range of $\pm 100\%$ from the initial chosen value and the effect on the output variable determined. In a rhythmically active network, for example, there may be regions of such a parameter space where rhythmicity becomes erratic or breaks down altogether (the point at which this occurs is a bifurcation), whereas in other regions the period remains constant or varies smoothly. Moreover, one must realize that by varying a single parameter at a time one is only taking sections of a multidimensional parameter space controlling that output variable. (The dimension of the space is equal to the number of free parameters, N, plus 1.) Within the whole multidimensional space, changes in one parameter may compensate for changes in a second, have little or no effect on changes in the second, or exacerbate the action of changes in the second. To ensure that a canonical model can serve as a useful benchmark for study of the biological system which it models, the values of the free parameters should be well within the range where the models and the biological system are similar. It is then useful to determine the sensitivity of the canonical model output variable to a small (around ±5–10%) change in each parameter from the canonical value as a percent change in output variable per percent change in the parameter. For example, in a model of the interaction of the pyloric and gastric networks in the crustacean stomatogastric ganglion[4] maximal conductances of all synaptic currents (and certain voltage-gated currents) were free parameters and the output variable used was gastric period. Parameter changes of +10% from canonical values were imposed and sensitivity was assayed as:

$$S_{output} = \frac{\Delta output \,/\, output}{\Delta parameter \,/\, parameter} \tag{10.13}$$

This analysis showed that model period was not very sensitive to most parameters (the absolute value of $S_{period} < 1.25$) but identified a key synapse that was important in determining model period. Large parameter sweeps then could be used to assay the role of this synapse in controlling model period. The order in which one performs these assays — mapping parameter space and sensitivity analysis — is arbitrary and the process in often iterative. Regardless, some similar assessment of parameterization seems a necessary precondition to producing a useful model of any small network.

10.3.5 Model Testing

An overriding principle that has guided our own use of realistic network modeling is that it should interact directly with biological experiment. Thus an important aspect of model development is model testing through experiment. Model testing is of necessity idiosyncratic to the model. Sensitivity analysis can help identify key parameters that influence network function and generate experimentally testable predictions. In small networks, it is often possible to alter key parameters such as $\bar{g}$ of voltage-gated and/or synaptic currents using dynamic current clamping[5] (Box 10.1) and thus test directly model predictions generated during parameter sweeps.[16] Alternatively, voltage clamping with realistic voltage waveforms (i.e., waveforms that mimic activity measured with intracellular or whole-cell voltage recordings during normal network function) can be used to test directly current flows predicted in the model (Figure 10.2A).[8] Equally important as adequate model testing, however, is the realization that such network models are a means to help understand network function and not in themselves an end. The type of modeling described here is an iterative process: experiment, model, experiment, model, experiment.

10.3.6 Model Expansion and Compromises

In modeling a small neuronal network it is useful to take a step by step approach. Most experimentalists have a strong sense of how the network, upon which they work, functions, and these hunches can help in model development. One may start by identifying key elements in the network and focusing initial modeling attempts on them. For example, in the work on the interaction of the pyloric and gastric mill networks,[16] the entire pyloric circuit was represented as a rhythmic synaptic input to key gastric model neurons. This simplification was tenable because the purpose of the model was to help explicate how pyloric input influences gastric period. Thus feedback from the gastric circuit to the pyloric circuit was not directly relevant. This simplification increased computational efficiency and did not influence model performance or interpretation. Future expansion of this model might now involve inclusion of the modulatory feedback to explore its functional consequences and will necessitate modeling key pyloric neurons. Let the model fit the question; detail for the sake of detail can impede progress, but all the details necessary to answer the questions asked of the model must be included.

10.4 LEECH HEARTBEAT TIMING NETWORK: A CASE STUDY

To further explicate the modeling process described we will illustrate the process by example. There is a tremendous literature on modeling small networks, to which the reader may refer to find examples from his/her own field. We will pass over this rich literature in this chapter and use only our own work, in the area that we know best, motor pattern-generation, to exemplify an approach which we believe can be particularly fruitful. The essence of this approach is the close interplay of modeling and experiment, so that insights derived from experiments are used to ameliorate the model and model prediction guide that choice of experiment.

10.4.1 Experimental Background

A network of seven bilateral pairs of segmental heart (HN) interneurons produces rhythmic activity (at about 0.1 Hz) that paces segmental heart motor neurons, which in turn drive the two hearts (see References 17 and 18 for a review). The synaptic connections among the interneurons and from the interneurons to the motor neurons are inhibitory. The first four pairs of heart interneurons can reset and entrain the rhythm of the entire pattern-generating network of interneurons. Thus this network is designated the *timing network* (eight-cell network) for heartbeat. The other three pairs of heart interneurons are followers of these anterior pairs. Two foci of oscillation in the timing network have been identified in the third and fourth ganglia, where the oscillation is dominated by the reciprocal interactions of the third and fourth pair of HN interneurons respectively. Reciprocally inhibitory synapses between the bilateral pairs of HN interneurons in these ganglia (Figure 10.3A), combined with an ability of these neurons to escape from inhibition and begin firing, pace the oscillation. Thus each of these two reciprocally inhibitory heart interneuron pairs can each be considered an *elemental oscillator* (two-cell oscillator) and these interneurons are called *oscillator interneurons*. The HN interneurons of the first and second ganglia act as *coordinating interneurons*, serving to couple these two elemental oscillators, thus forming the (eight-cell) timing network for the system (Figure 10.4). When the third and fourth ganglia are separated by cutting the connective, each isolated 3rd and 4th ganglia contain a *segmental oscillator* (six-cell oscillator) that consists of a pair of reciprocally inhibitory oscillator interneurons (elemental oscillator) and the active "stumps" of the coordinating neurons which provide additional inhibition. Invertebrate axons may emit neuritic branches (input and output) at several sites within the CNS and thus have multiple (electrotonically distant) sites of synaptic input, spike initiation, and synaptic output. Under normal conditions, the coupling between the segmental oscillators causes the activity of the HN(3) oscillator to lag that of the HN(4) oscillator on the same side by about 15% in phase (Figure 10.4).

Coupling between segmental oscillators by the coordinating neurons is quite strong. Repetitive current pulses that alter the cycle period of one segmental oscillator can entrain the entire beat timing network to a range of frequencies faster and slower than the system's natural state. For example, imposing a train of current pulses on an HN(3) cell can lock the HN(4) pair to an artificially high frequency. The segmental oscillators are mutually entraining. The phase lag between the segmental oscillators is a function of the entrainment conditions. For example, when an HN(3) pair entrains an HN(4) pair to a low frequency, the driving pair (HN(3)) lags behind the driven pair. Conversely when one pair entrains the other to a high frequency, the driving pair leads.

Several ionic currents have been identified in single electrode voltage clamp studies that contribute to the activity of oscillator heart interneurons.[18] These include two low-threshold Ca^{2+} currents (one rapidly inactivating (I_{CaF}) and one slowly inactivating (I_{CaS}), three outward currents (a fast transient K^+ current (I_A) and two delayed rectifier-like K^+ currents, one inactivating (I_{K1}), and one persistent (I_{K2}), a hyperpolarization-activated inward current (I_h) (mixed Na^+/K^+, $E_{rev} = -20$ mV), a low-threshold persistent Na^+ current (I_P) and a leakage current (I_L). The inhibition between oscillator interneurons consists of a graded component (I_{SynG}) that is associated with the low-threshold Ca^{2+} currents and a spike-mediated component (I_{SynS}) that appears to be mediated by a high-threshold Ca^{2+} current. Spike-mediated transmission varies in amplitude throughout a burst according to the baseline level of depolarization[8] (Figure 10.2B). Graded transmission wanes during a burst owing to the inactivation of low-threshold Ca^{2+} currents. Blockade of synaptic transmission with bicuculline leads to tonic activity in oscillator heart interneurons and Cs^+, which specifically blocks I_h, leads to tonic activity or sporadic bursting.

This wealth of detailed information demanded modeling to both organize the plethora of data, facilitate insights into the interactions between membrane properties

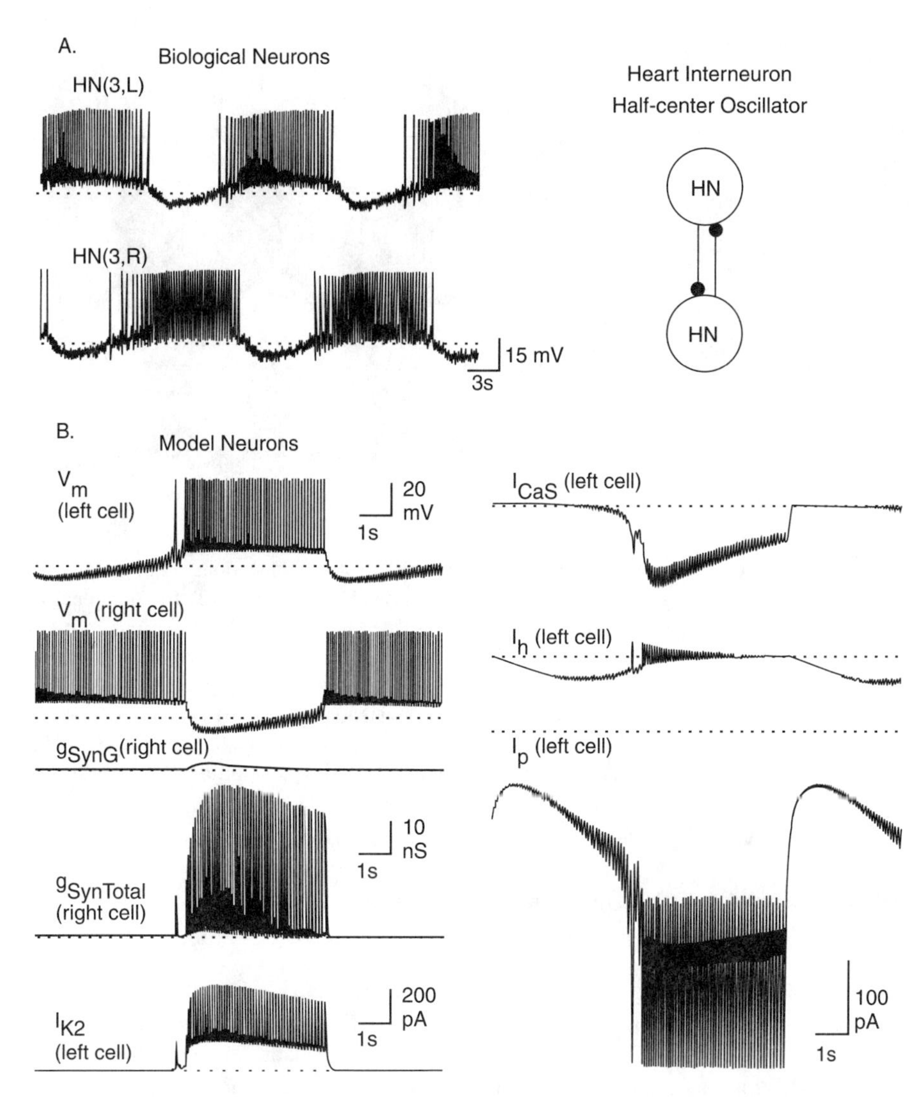

FIGURE 10.3 (A) Simultaneous intracellular recordings showing the normal rhythmic activity of two reciprocally inhibitory heart (HN) interneurons that compose an elemental (half-center) oscillator in an isolated ganglion preparation. Heart interneurons are indexed by body side (R, L) and ganglion number. (B) Synaptic conductances and some major intrinsic currents that are active during a single cycle of the third generation model of a heart interneuron elemental oscillator model (half-center). The graded synaptic conductance (g_{SynG}) is shown at the same scale as the total synaptic conductance ($g_{SynTotal}$) which is the sum of the graded and spike-mediated conductances. The slow calcium current (I_{CaS}), the hyperpolarization-activated current (I_h), and the persistent sodium current (I_P) are shown to the same scale. Note that I_P is active throughout the entire cycle period. Dashed lines indicate –50 mV in voltage traces, 0 nA in current traces, and 0 nS in conductance traces. Inset: elemental oscillator circuit. Black balls indicate inhibitory synapses.

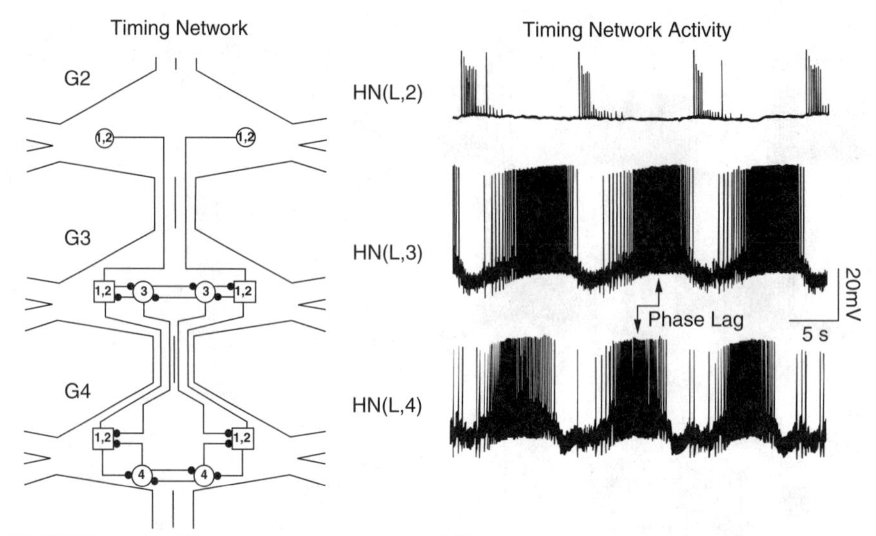

FIGURE 10.4 Circuit diagram showing inhibitory synaptic connections among the HN interneurons of the timing network. Coordinating neurons HN(1) and HN(2) are functionally equivalent and are lumped together in the diagram. The HN(1) and HN(2) neurons receive synaptic inputs, initiate action potentials, and make synaptic outputs at sites located in the third and fourth ganglia (open squares). Timing network activity recorded intracellularly from left coordinating neuron HN(2) and oscillator neurons HN(3) and HN(4).

and synaptic connections that determine network function and guide future experimental exploration. The general availability of voltage clamp data for oscillator interneurons made conductance based modeling an obvious choice.

10.4.2 Two-Cell Oscillator Model: The Elemental Oscillator

Our approach in modeling this system was to start with an elemental oscillator and then build up to the entire timing network (Figure 10.4). This decision was influenced by three considerations. First, very little voltage clamp data was available on the coordinating interneurons, whereas there was strong database for the oscillator interneurons. Second, we wished to explore the potential of the two-cell oscillator, both for its own intrinsic interest and its ability to provide general insights into how oscillations are generated by reciprocally inhibitory neurons. Third, we were convinced we would need the insights into how oscillations were generated in the two-cell oscillator to guide our interpretations of activity within the timing network (eight-cell network).

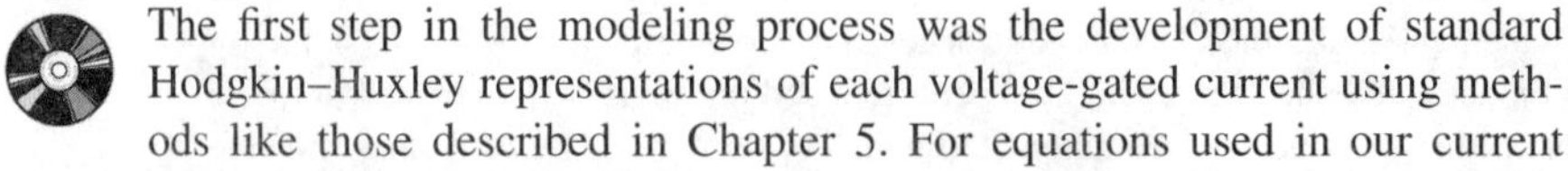
The first step in the modeling process was the development of standard Hodgkin–Huxley representations of each voltage-gated current using methods like those described in Chapter 5. For equations used in our current implementation of this model (third generation) see the CD-ROM. Synaptic transmission between the oscillator neurons involves spike-mediated and graded transmission that both show complex dynamics which depend on membrane potential

and low-threshold Ca^{2+} currents respectively. Equation 10.5 fitted to unitary IPSCs was used to describe the postsynaptic conductance associated each action potential. The fraction of the postsynaptic maximal conductance reached after each spike was made a function of the past membrane potential to reflect the fact that spike-mediated transmission varies in amplitude throughout a burst according to the baseline level of depolarization (Equations 10.10). Graded synaptic transmission was represented by a synaptic transfer function, which related postsynaptic conductance (the result of transmitter release) to presynaptic Ca^{2+} build-up and decline, via low-threshold Ca^{2+} currents and a Ca^{2+} removal mechanism respectively (Equations 10.7). Each cell was modeled as a single isopotential compartment with a leak conductance and a membrane capacitance (Section 10.3.1). Thus each cell in our model Equation 10.11 becomes:

$$C\frac{dV}{dt} = -\begin{pmatrix} I_{Na} + I_P + I_{CaF} + I_{CaS} + I_h + I_{K1} + I_{K2} \\ +I_A + I_L + I_{SynG} + I_{SynS} - I_{inject} \end{pmatrix} \tag{10.14}$$

10.4.3 Parameterization and Model Testing

The development of our second generation model[3,19] best illustrates our process of parameterization and model testing. All the characterized voltage-gated and the synaptic currents (this generation of the model did not include the dependence of spike-mediated transmission on baseline membrane potential) were incorporated in the model. Free parameters in the model were the maximal conductance ($\bar{g}_{ion}$) for each current (voltage-gated or synaptic). The $\bar{g}_{ion}$'s were adjusted to be close to the average observed experimentally. The reversal potential, E_{ion}, for each current was determined experimentally and they were considered fixed. The final parameter values, used in the canonical model, were chosen so that the model activity closely approximated that of biological neurons, specifically during normal oscillations, in the passive response to hyperpolarizing current pulses, and in the reaction to current perturbations. The model cells were also required to fire tonically when all inhibition between them was blocked.[18]

At this point in the modeling process, we explored the dependence of model behavior to the choice of free parameters. We chose cycle period of the oscillation as the output variable for this analysis. In the biological system, period is variable depending on the preparation and its modulatory state but normally varies only over a constrained range (7–14 sec). First, free parameters were individually varied over a range of ±100% from the canonical value and cycle period was determined. We found that some parameter settings gave rise to oscillations that were fundamentally different from those observed both in the biological system and in the canonical model. These oscillations had periods that were relatively insensitive to further variation in the parameter that gave rise to them, whereas the period during the more biologically relevant oscillations varied smoothly with the varied parameter. Moreover, in the first type of oscillation synaptic transmission was dominated by its graded component, whereas in the more biological similar oscillation it was dominated by its spike-mediated component. We had revealed a bifurcation in model behavior and delineated a region of parameter space where model behavior diverges fundamentally from that of the biological system. This observation focused our attention on such parameters as $\bar{g}_{SynG}$ and $\bar{g}_{CaS}$ for future experimental study. The former determines the strength of graded transmission directly and the latter determines it indirectly by being the primary avenue for presynaptic Ca^{2+} entry and thus graded transmitter release. To ensure that the canonical model would serve as a useful benchmark for study of the biological systems, we made sure the values of the free parameters were well within the range where the models and the biological system were similar. We then determined the sensitivity of the canonical model period to a $\pm 5\%$ change in each parameter from the canonical value as a percent change in period per percent change in the parameter (see Equation 10.13). Only in the case of $\bar{g}_{K1}$ and $\bar{g}_{SynS}$ was this sensitivity greater than 0.5. This rather low value means that a 1% change in a parameter causes only a 0.5 % change in period. The observed sensitivity to $\bar{g}_{K1}$ and $\bar{g}_{SynS}$ was expected because g_{K1} controls spike rate and thus indirectly the amount of inhibition and g_{SynS} directly controls the amount of inhibition. Thus the canonical model was parameterized in a way that small parameter differences did not alter model behavior in fundamental ways.

Analysis of current flows during this activity (see Figure 10.3B for generation III model activity and currents) indicated that graded transmission to the antagonist cell occurs only at the beginning of the burst period. The low-threshold Ca^{2+} currents that mediate this inhibition inactivate significantly during the later part of the inhibited period due to the slow rise of the membrane potential trajectory to the burst phase. This observation constituted a significant model prediction. Thus graded inhibition helps turn off the antagonist neuron, but sustained inhibition of the antagonist neuron is spike-mediated. The inward currents in the model neurons act to overcome this inhibition and force a transition to burst phase of oscillation. I_P is active throughout the activity cycle providing a persistent excitatory drive to the system. I_h is slowly activated by the hyperpolarization caused by synaptic inhibition, adding a delayed inward current that drives further activation of I_P and eventually the low-threshold Ca^{2+} currents (I_{CaS} and I_{CaF}). These regenerative currents support burst formation. I_P, because it does not inactivate, provides steady depolarization to sustain spiking, while the low-threshold Ca^{2+} currents help force the transition to the burst phase but inactivate as the burst proceeds, thus spike frequency slowly wanes during the burst. Outward currents also play important roles, especially the I_Ks. I_{K2}, which activates and deactivates relatively slowly and does not inactivate, regulates the amplitude of the depolarization that underlies the burst, while I_{K1}, which activates and deactivates relatively quickly and inactivates, controls spike frequency.

We gathered several insights from this effort. The broad parameter sweeps illuminated which parameters were instrumental in establishing the period of the system. In particular, the strength of synaptic inhibition ($\bar{g}_{SynS}$) and the maximal conductance of I_h ($\bar{g}_h$) appear to set oscillation period because they cause smooth monotonic changes in period over a broad range without producing bifurcations in model behavior. During normal activity graded synaptic transmission is repressed and spike-mediate transmission is dominant. The various inward currents provide excitatory drive at different phases in the oscillation; I_{CaS} during the burst phase, I_h

during the inhibited phase, and I_P during both phases. Moreover, two obvious flaws were apparent in the model. Spiking activity was at significantly higher frequency than in the biological neurons and the transitions to the burst phase were too abrupt with no overlap in the firing of the two model neurons. Spike rate had to be adjusted by reworking the fudged equation for I_{Na}. The abruptness of the transition to the burst phase and the lack of overlapping spiking in the mutually inhibitory model neurons indicated that plasticity (sensitivity to baseline membrane potential) in spike-mediated transmission might play a significant role. These realizations led to a new phase of model testing and experimental measurement.

We applied voltage clamp waveforms that directly mimicked the slow wave of their membrane potential oscillation to individual oscillator neurons under conditions where individual voltage-gated and synaptic currents were isolated (Figure 10.2A; Box 10.1). These tests showed that the first model prediction was correct; graded transmission was repressed through inactivation of low-threshold Ca^{2+} currents during the inhibited period by the slow rise of the membrane potential trajectory to the burst phase. They also gave us more accurate estimates of $\bar{g}_{SynG}$ and $\bar{g}_{CaS}$. We reassessed the amount of the other inward currents active during the realistic waveforms and came to realize that I_P was even more active during the inhibited phase than suggested by the model. We also directly measured the plasticity in IPSC amplitude during normal activity in oscillator neurons and quantitatively related these changes to baseline membrane potential (Figure 10.2B). The confirmation of model prediction experimentally increased our confidence in the model and the measurements led directly to the generation III canonical model illustrated in Figure 10.3. The new better estimates of $\bar{g}_{SynG}$ and $\bar{g}_{CaS}$ restrict greatly the volume of parameter space in which the non-biological oscillations occur. The modulation of spike-mediated transmission by presynaptic potential decreases the abruptness of the transition between the inhibited and burst phases and increases the amount of overlap between spiking activity in the two model neurons.

10.4.4 Building the Timing Network

We constructed a model of the beat timing network (the four anterior pairs of heart interneurons) based on our third generation two-cell elemental oscillator model that corresponds to the reciprocally inhibitory pairs of interneurons in the 3rd and 4th ganglia (Figure 10.3). The coordinating interneurons of the 1st and 2nd ganglia that couple the segmental oscillators were modeled, like the oscillator interneurons, as single compartments. Less physiological data is available for the coordinating interneurons than for the oscillator interneurons. Therefore, these neurons were modeled simply so that they exhibit spiking behavior similar to the biological neurons. They contain only I_{Na}, I_{K1}, I_{K2}, leak, and receive and send only spike-mediated synaptic inhibition. The coordinating neurons are capable of initiating spikes at sites in both the 3rd and 4th ganglia (Figure 10.4). In the present network model we have not included this detail and have considered the coordinating fibers to be single compartments that receive inhibition from ipsilateral HN(3) and HN(4) neurons. In this first generation timing network model, we specifically chose not to include every biological detail in order to better understand the basic operation of the system. In the future we can include these details if they appear to be important to the function of the system.

This model was developed to assist and augment our experimental analysis of the mechanism by which the intersegmental phase lag seen in Figure 10.4 is generated. To explore the generation of this phase difference, we first created two six-cell oscillators that correspond to single elemental oscillators in the 3rd and 4th ganglia and their coordinating fibers (processes of HN(1) and HN(2) cells) within one ganglion (Figure 10.5). These six-cell models each comprise a segmental oscillator, because each contains all the neuronal elements involved in oscillation in a single isolated 3rd or 4th ganglion. The six-cell oscillators behave very similarly to the two cell elemental oscillators except the additional spike-mediated inhibition from the coordinating neurons slow the oscillator down (~25%). Second, two such segmental oscillators were combined to form a single eight-cell oscillator that is equivalent to the beat timing network (Figure 10.5). An advantage of this approach is that it is possible to determine the inherent cycle periods of the segmental oscillators in isolation before coupling them to determine the ensemble behavior of the eight-cell network.

Preliminary results suggest that phase lag may be generated by differences in the inherent cycle periods of the segmental oscillators (Figure 10.5). The inherent cycle periods of the segmental oscillators were varied by altering the maximal conductance ($\bar{g}_h$) of I_h, the hyperpolarization-activated current in model HN(3) and HN(4) oscillator neurons. A decrease in I_h from the canonical level led to an increase in the cycle period of a segmental oscillator. (We can also generate differences in the period of the segmental oscillators by manipulating other maximal conductances such as I_P.) When two segmental oscillators with different inherent periods were combined the cycle period of the eight-cell network was very close to that of the faster oscillator's period. The faster oscillator consistently led the slower one and as the difference in inherent oscillator cycle periods was increased the phase lag between the oscillators also increased. Preliminary results also suggest that phase lag may also be generated by oscillators with asymmetric inhibitory synaptic connections. For example, a reduction in the maximal conductance for the synapses from the coordinating neurons to the HN(4) oscillator neurons created a system in which the HN(4) segmental oscillator led the HN(3) segmental oscillator.

Important experimental predictions of these analyses are that there will be inherent period differences between the HN(3) and the HN(4) segmental oscillators and that the coupled systems will have the period of the faster segmental oscillator. These predictions can be tested in biological experiments in which the segmental oscillator are reversibly uncoupled and recoupled by reversible blockade of action potential conduction between the ganglia.

10.5 ABSTRACT MODELS

A drawback of the approach that we have outline above is that often the networks that are modeled seem very idiosyncratic to the organism or circuit phenomenon studied. Thus realistic models appear to lack relevance to other workers. This drawback can be circumvented by the creative modeler through model reduction or

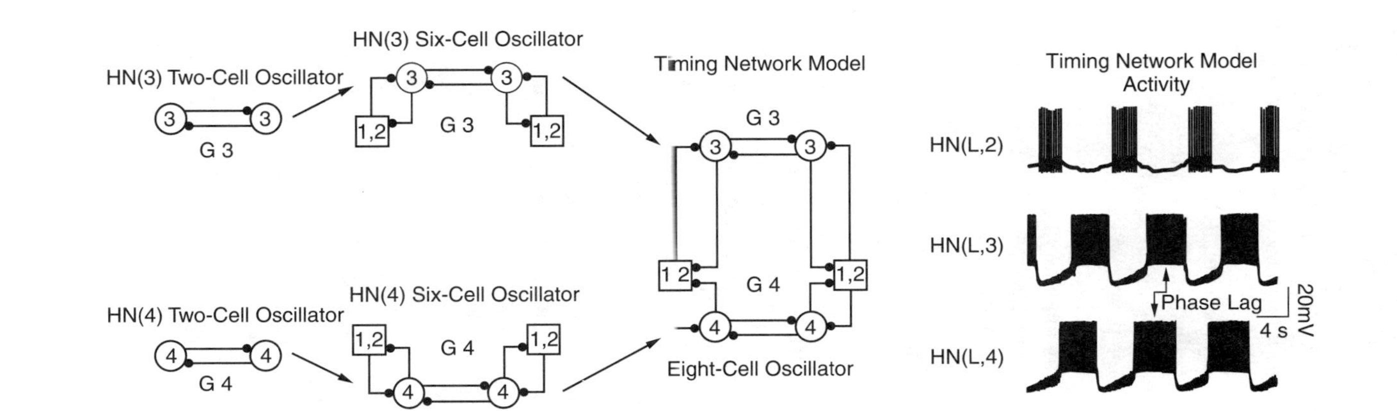

FIGURE 10.5 Strategy for construction of the timing network model and its activity. The eight-cell model of the timing network was constructed by combining two six-cell segmental oscillators that in turn are composed of single elemental oscillators and their synaptic connections with the coordinating fibers. Compare the model network and its activity to Figure 10.4. The phase lag between activity in the model HN(3) and HN(4) oscillators is about 14%, and the timing network's activity has a cycle period of about 8.8 s. The period of the uncoupled HN(4) segmental oscillator is 8.8 s and that of the uncoupled HN(3) oscillator is 9.8 s. The activity of the model HN(1) cell (not shown) is very similar to that of the model HN(2) cell. Black balls indicate inhibitory synapses.

by attacking the same network in parallel with more abstract models. As an example, we will explore the development of an abstract model of a reciprocally inhibitory elemental oscillator that paralleled the development of our own model of the leech heartbeat elemental oscillator. We also refer the reader to References 4, 16, and 20 for an experimentally based and tested realistic model that was used as the basis of an abstract model to explicate basic principles of oscillation in reciprocally inhibitory networks.

The configuration of two reciprocally (or mutually) inhibitory neurons (or groups of neurons) that produce alternating oscillation, observed in the elemental heartbeat oscillator from the leech, is a basic circuit building block which is thought to form the basis of oscillation in many motor pattern-generating networks. This circuit configuration has come to be known as a half-center oscillator. A theoretical framework for understanding how reciprocally inhibitory neurons oscillate (i.e., how half-center oscillators work) was developed in Reference 21. Their model neurons are minimal. Each contains a synaptic conductance which is a sigmoidal function of presynaptic membrane potential with a set threshold and instantaneous kinetics, a constant leak conductance, and a voltage-gated postinhibitory rebound current, I_{pir}. I_{pir} was originally envisioned to be a T-like Ca^{2+} current (low-threshold, inactivating) but its expression in the model can accommodate an h current (hyperpolarization activated inward current) also. Two different modes of oscillation appear in the model, release and escape. For the release mode to occur the synaptic threshold must be above the steady state membrane potential of the uninhibited neurons. In the release mode the inactivation of I_{pir} erodes the depolarized or active phase of a neuron so that it falls below threshold for synaptic transmission. Consequently its partner is released from inhibition and rebounds into the active depolarized state. For the escape mode to occur the synaptic threshold must be below the steady-state voltage of the neurons when uninhibited. This condition can be accomplished simply by increasing $\bar{g}_{pir}$. In the escape mode, once inactivation of I_{pir} is removed by the hyperpolarization associated with inhibition, it activates and overcomes the maintained synaptic current so that the neuron escapes into the active phase and thus inhibits its partner. This analysis is extended in Reference 21 (see Box 10.3).

Real neurons, such as leech heart interneurons, display more complicated intrinsic membrane properties and plastic synaptic interactions, but nevertheless these simplified models serve as a useful vantage point from which to view richer biological systems and realistic models. Thus it appears that the leech heart interneuron half-center oscillators (both the biological oscillators and the model oscillators) operate predominantly in the escape mode although forces that promote release are at work. Spike-mediated transmission gradually wanes during a burst because of the slowly declining envelope of depolarization during the burst phase which slows spike frequency and down modulates IPSP amplitude (Figures 10.2B and 10.3B). Nevertheless, whenever I_h is sufficiently activated to overcome the waning synaptic current, a transition from the inhibited (active) state to the burst (inactive) state occurs, which is consistent with an escape mechanism.

Box 10.3 Morris–Lecar Equations and Phase Plane Analysis

Skinner et al.[22] have extended the analysis by Wang and Rinzel[21] of an half-center oscillator using model neurons based on the Morris–Lecar equations.[26] These equations represent a simplified mechanism for producing regenerative potentials of various durations, single or repetitive, spontaneous or evoked (current injection). These model neurons have proved very useful in the dynamical systems analysis of small neural networks. Each Morris–Lecar neuron contains a low-threshold, non-inactivating Ca^{2+} (inward) current with instantaneous dynamics, a slow K^+ (outward) current and a leak current.

For each model cell:

$$C\frac{dV}{dt} = I_{ext} - \left(g_L\left(V - V_L\right) + g_{Ca}M_\infty\left(V - V_{Ca}\right) + g_K N\left(V - V_K\right)\right) \tag{1}$$

$$\frac{dN}{dt} = \lambda_N\left(N_\infty - N\right) \tag{2}$$

and

$$M_\infty(V) = \frac{1}{2}\left(1 + \tanh\left(\frac{V - V_1}{V_2}\right)\right) \tag{3}$$

$$N_\infty(V) = \frac{1}{2}\left(1 + \tanh\left(\frac{V - V_3}{V_4}\right)\right) \tag{4}$$

$$\lambda_N(V) = \phi_N \cosh\left(\frac{V - V_3}{V_4}\right) \tag{5}$$

The $V_\#$'s are parameters that determine the voltage-dependence of the currents while g_A's are the maximum conductances of the ionic currents. N and V are the state variables. V is the membrane potential, and N is the activation variable of the slow K^+ current. M_∞ is the instantaneous activation variable of the Ca^{2+} current, i.e., it is assumed that Ca^{2+} activation instantly reaches M_∞. M_∞ is not a state variable because it is not time dependent. Both M_∞ and N_∞ show a sigmoidal dependence on voltage. The relative simplicity of this system makes it a favorable subject for dynamical systems analysis. Because it has just two state variables it is possible to apply a powerful geometrical approach called *phase-plane analysis*.[27,28]

(continued)

Box 10.3 (continued)

A complete description of phase-plane analysis is beyond the scope of this chapter. In this analysis, the state variables of the systems are plotted in a two dimensional space called the phase-plane. Thus phase-plane analysis is limited to systems consisting of no more than two state variables. Some techniques permit the reduction of systems of three (or more) state variables if two of the state variables share a similar time dependence while the third changes much more rapidly or slowly. For an excellent example see Reference 20. Thus only highly simplified models of neuronal networks can be analyzed as such, and examples are limited. In the case of the Morris–Lecar model neurons, the phase plane is the *V–N* plane. The differential equations that describe the state variables are set equal to 0 and fixed curves called *nullclines* are plotted in the phase plane. For the Morris–Lecar model neurons, the *N* nullcline defined by Equation 2 is

$$dN/dt = 0,\ N = N_\infty = \frac{1}{2}\left(1 + \tanh\left(\frac{V - V_3}{V_4}\right)\right) \tag{6}$$

Thus the *N* nullcline is the steady state activation curve of the slow K^+ current. The *V* nullcline defined by Equation 10.1 is

$$dV/dt = 0,$$
$$N = \left(I_{ext} - \left(g_L(V - V_L) + g_{Ca}M_\infty(V - V_{Ca})\right)\right)/g_K(V - V_K) \tag{7}$$

Thus the *V* nullcline is essentially the steady state *I-V* relation of the currents other than the K+ current in the model neuron, normalized by the maximal K^+ current at any given voltage. Biologically relevant parameterizations of Morris–Lecar neurons have steady-state *I-V* relations that are distinctly *N*-shaped (cubic-like) reflecting a strong contribution of the inward Ca^{2+} current. The intersections of these nullclines define equilibrium points (stable or unstable) that govern the trajectories through time of the state variable in the phase plane. Many oscillatory neural models demonstrate isolated stable orbital trajectories in the phase plane called stable *periodic orbits* or *limit cycles*. A description of the techniques and application of phase plane analysis can be found in Reference 27. Phase-plane analysis is supported by the software package XPP (*http://www.pitt.edu/~phase*).

In Skinner et al.'s[22] model, two identical Morris–Lecar neurons are connected with each other through an inhibitory synaptic conductance, which is described by a steep sigmoidal function of presynaptic membrane potential with a set threshold and instantaneous kinetics.

For this system Equation 1 is modified to

$$C\frac{dV}{dt} = I_{ext} - \begin{pmatrix} g_L(V - V_L) + g_{Ca}M_\infty(V - V_{Ca}) \\ +g_K N(V - V_K) + g_{syn}S_\infty(V - V_{syn}) \end{pmatrix} \tag{8}$$

(continued)

Box 10.3 (continued)

and

$$S_\infty(V') = \frac{1}{2}\left(1 + \tanh\left(\frac{V' - V_{thresh}}{V_{slope}}\right)\right) \tag{9}$$

V is the membrane potential of the "transmitting or presynaptic" cell, V_{thresh} and V_{slope} are parameters that determine the voltage-dependence of the synaptic current, g_{syn} is the maximum conductance of the synaptic current, and S is the activation variable of the inhibitory synapse. Such model neurons can oscillate between a depolarized plateau and a sustained inhibitory trough. These oscillations can be thought of as reflecting the underlying slow wave of membrane potential when fast changes are ignored (e.g., spikes and spike mediated synaptic potentials). Thus they correspond to systems where there is a plateau of depolarization during a spike burst and dominant graded synaptic transmission. Each neuron is two-dimensional, i.e., has two state variables V and N. To use phase-plane analysis, each neuron is treated separately and one takes advantage of the instantaneous kinetics and steep voltage-dependence of the synaptic interaction. In the limit when the synaptic activation is infinitely steep and instantaneous, the model neurons have two mutually exclusive states (inhibited and not inhibited) with instantaneous transitions between them (Figure B10.1). Two mutually exclusive V nullclines can be drawn in the $V - N$ plane, one with the synaptic current activated defined by Equation 8

$$dV/dt = 0,$$

$$N = \left(I_{ext} - \left(g_L(V - V_L) + g_{Ca}M_\infty(V - V_{Ca})g_{syn} + S_\infty(V - V_{syn})\right)\right) / g_K(V - V_K) \tag{10}$$

and one with the synaptic current deactivated corresponding to Equation 7. The N nullcline remains unchanged and is given by Equation 6. Now the system moves through time on trajectories that jump from one V nullcline to another as the synapse onto the model neuron analyzed activates and deactivates. In practice, the requirement for infinite steepness can be relaxed somewhat without fundamentally altering model behavior.

This phase-plane analysis allows a further subdivision of the oscillation modes depending on whether the escape or release[21] is intrinsic or synaptic.[22] If the release is due to a cessation of synaptic transmission (crossing synaptic threshold), it is synaptic release, but if it is due to termination (deactivation of the inward current, activation of the delayed rectifier, or both) of the depolarized plateau, it is intrinsic release. If the escape is due to the commencement of synaptic transmission (crossing synaptic threshold), it is synaptic escape, but if it is due to expression of the depolarized phase (crossing plateau threshold), it is intrinsic escape. Varying the synaptic threshold causes transitions between the modes.

(continued)

Box 10.3 (continued)

Figure B10.1A illustrates the phase-plane for a parameterization of the system that leads to oscillation by the intrinsic release mechanism. There are two V nullclines, the upper corresponding to the free uninhibited model neuron (Equation 7) and the lower corresponding to the inhibited model neuron (Equation 10). There is one *N* nullcline, which is the same for both the free and inhibited model neurons. The movement of the system through time follows the trajectory marked by arrowheads. Figure B10.1B shows the alternating voltage oscillations of the two model neurons. During the oscillation movement through the phase plane is such that the free cell reaches the end of its depolarized plateau at k^+ and jumps down in membrane potential (*V*) to *k* releasing the inhibited cell. At that moment the opposite inhibited cell would be jumping to its depolarized plateau corresponding to a jump from *p* to p^+ in its phase plane, therefore, the cycle continues.

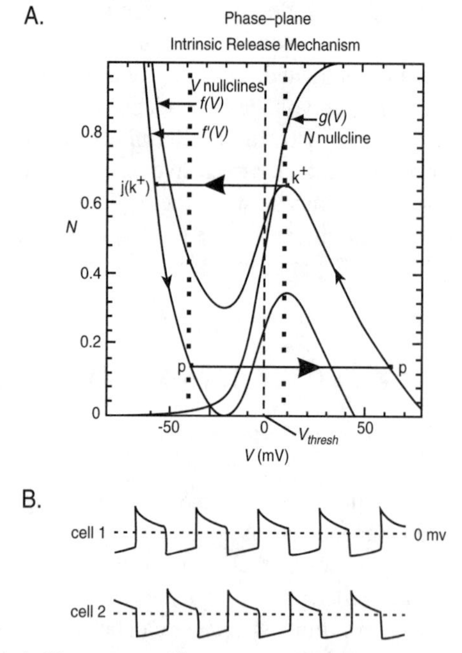

FIGURE B10.1 A half-center oscillator comprising two mutually inhibitory Morris–Lecar model neurons. The parameterization of the system illustrated leads to oscillation by a mechanism of intrinsic release. (A) Above is the phase-plane portrait, *N* versus *V*, for one model neuron. The dashed vertical line marks the synaptic threshold (V_{thresh} Equation 9) and the dotted vertical lines mark the transition voltages at which the model cell jumps from one *V* nullcline to the other. Small arrowheads mark movement of the system along the *V* nullclines and large arrowheads mark jumps from one *V* nullcline to the other. (B) Below is voltage (*V*) versus time for the two model cells (horizontal dashed line marks the synaptic threshold V_{thresh}). Model parameters were $g_{syn} = 6$ μS/cm², $I_{ext} =$ 0.4 μA/cm², $\phi_N = 2 \times 10^{-6}$ msec^{-1}, $V_{slope} = 0.001$ mV, [g_K 20, g_{Ca} 15, g_L 5 — all μS/cm²], [V_{Ca} –100, V_K –80, V_L –50, V_{syn} –80, V_1 0, V_2 15, V_3 0, V_4 15 – all mV], C 1 μF/cm² (from Skinner et al., *J. Compu. Neurosci.*, 1, 69, 1994).

10.6 CONCLUSIONS

In our approach to modeling small networks, there is a close interplay of modeling and experiment, so that insights derived from experiments are used to ameliorate the model and model prediction guide that choice of experiment. We have tried to provide the tools and the rationale for following this approach in this chapter. Part of this approach is also to use more abstract models, to focus thought, and concentrate insight. The use of abstract modeling in conjunction with more realistic models has an added value. By concentrating the essence of network function into more abstract models and by using these models to interpret results from realistic models, portability is achieved that is the hallmark of good theoretical science.

REFERENCES

1. Rall, W., *The Theoretical Foundation of Dendritic Function. Selected Papers of Wilfred Rall and Commentaries*, Segev, I., Rinzel, J., and Shepherd, G. M., Eds., MIT Press, Cambridge, 1995.
2. Bower, J. M. and Beeman, D., *The Book of GENESIS*, Springer-Verlag, New York, 1998.
3. Nadim, F., Olsen, Ø. H., De Schutter, E., and Calabrese, R. L., Modeling the leech heartbeat elemental oscillator: I. Interactions of intrinsic and synaptic currents, *J. Comput Neurosci.*, 2, 215, 1995.
4. Nadim, F., Manor, Y., Nusbaum, M. P., and Marder, E., Frequency regulation of a slow rhythm by a fast rhythm, *J. Neurosci.*, 18, 5053, 1998.
5. Sharp, A. A., O'Neil, M. B., Abbott, L. F., and Marder, E., Dynamic clamp: artificial conductances in biological neurons, *Trends Neurosci.*, 16, 389, 1993.
6. Jaeger, D. and Bower, J. M., Synaptic control of spiking in cerebellar Purkinje cells: dynamic current clamp based on model conductances, *J. Neurosci.*, 19, 6090, 1999.
7. De Schutter, E., Angstadt, J. D., and Calabrese, R. L., A model of graded synaptic transmission for use in dynamic network simulations, J. Neurophys., 69, 1225, 1993.
8. Olsen, Ø. H. and Calabrese, R. L., Activation of intrinsic and synaptic currents in leech heart interneurons by realistic waveforms, *J. Neurosci.*, 4958, 1996.
9. Lu, J., Dalton, J. F., IV, Stokes, D. R., and Calabrese, R. L., Functional role of Ca^{2+} currents in graded and spike-mediated synaptic transmission between leech heart interneurons, *J. Neurophysiol.*, 77, 1179, 1997.
10. Delaney K. R. and Tank, D.W., A quantitative measurement of the dependence of short-term synaptic enhancement on presynaptic residual calcium, *J. Neurosci.*, 14, 5885, 1994.
11. Gingrich K. J. and Byrne J. H., Single-cell neuronal model for associative learning, *J. Neurophysiol.*, 57, 1705, 1987.
12. Abbott, L. F., Varela, J. A., Sen, K., and Nelson, S. B., Synaptic depression and cortical gain control, *Science*, 275, 220, 1997.
13. Sen, K., Jorge-Rivera J. C., Marder, E., and Abbott, L. F., Decoding synapses, *J. Neurosci.*, 16, 6307, 1996.
14. Nicholls, J. G. and Wallace, B. G., Modulation of transmission at an inhibitory synapse in the central nervous system of the leech, *J. Physiol.*, 281, 157, 1978.

15. Golowasch, J., Buchholtz, F., Epstein, I. R., and Marder, E., Contribution of individual ionic currents to activity of a model stomatogastric ganglion neuron, *J. Neurophysiol.*, 341, 1992.
16. Bartos, M., Manor, Y., Nadim, F., Marder, E., and Nusbaum, M. P., Coordination of fast and slow rhythmic neuronal circuits, *J. Neurosci.*, 19, 6650, 1999.
17. Calabrese, R. L. and Peterson, E., Neural control of heartbeat in the leech, Hirudo Medicinalis, *Symp. Soc. Exp. Biol.*, 37, 195, 1983.
18. Calabrese, R. L., Oscillation in motor pattern generating networks, *Curr. Opin. Neurobiol.*, 5, 816, 1995.
19. Olsen, Ø. H., Nadim, F., and Calabrese, R. L., Modeling the leech heartbeat elemental oscillator: II. Exploring the parameter space, *J. Comput. Neurosci.*, 2, 237, 1995.
20. Manor, Y., Nadim, F., Epstein, S., Ritt, J., Marder, E., and Kopell, N., Network oscillations generated by balancing graded asymmetric reciprocal inhibition in passive neurons, *J. Neurosci.*, 19, 2765, 1999.
21. Wang, X.-J. and Rinzel, J., Alternating and synchronous rhythms in reciprocally inhibitory model neurons, *Neural Comput.*, 4, 84, 1992.
22. Skinner, F. K., Kopell, N., and Marder, E., Mechanisms for oscillation and frequency control in reciprocally inhibitory model neural networks, *J. Compu. Neurosci.*, 1, 69, 1994.
23. De Schutter, E., Computer software for development and simulation of compartmental models of neurons, *Comput. Biol. Med.*, 19, 71, 1989.
24. Hines, M. L. and Carnevale, N. T., The NEURON simulation environment, *Neural Comput.*, 9, 1179, 1997.
25. Ziv, I., Baxter, D. A., and Byrne, J. H., Simulator for neural networks and action potentials: description and application, *J. Neurophysiol.*, 71, 294, 1994.
26. Morris, C. and Lecar, H., Voltage oscillations in the barnacle giant muscle fiber, *Biophys. J.*, 35, 193, 1981.
27. Rinzel, J. and Ermentrout, B., Analysis of neural excitability and oscillations, , in *Methods in Neuronal Modeling: From Ions to Networks*, Koch, C. and Segev, I., Eds., 2nd ed., MIT Press, Cambridge, 1998, 251.
28. Ermentrout, B., Type I membranes, phase resetting curves, and synchrony, *Neural Comput.*, 8, 979, 1996.

11 Modeling of Large Networks

Michael E. Hasselmo and Ajay Kapur

CONTENTS

0-8493-2068-2/01/$0.00+$.50

11.1 INTRODUCTION

Linking molecular and cellular phenomena to behavior requires the development of large scale network models, to understand how cellular phenomena influence the functional dynamics of neural networks. These types of network models will be essential for understanding the mechanism for changes in behavior caused by experimental manipulations such as the administration of drugs or knocking out of genes.

Models of large networks of neurons have not achieved the same level of experimental constraint obtained in models of single cell processes or small networks. This results from the multitude of additional variables at the network level, including the strength and pattern of both excitatory and inhibitory synaptic connectivity, as well as the relative properties and numbers of different types of neurons. At the same time that the parameter space expands, the experimental constraints on parameter space decrease due to sparse information about network dynamics. However, the renewed focus on local field potentials and the rapid development of simultaneous recording from multiple units[1,2] will provide experimental constraints for more accurate development of network models.

This chapter will provide an overview of available techniques for combining realistic simulations of individual neurons into large networks, with an emphasis on linking properties of more abstract network models to those of networks of compartmental simulations. The chapter will start with a review of some existing types of models, with a particular emphasis on scaling problems relating small networks to large biological systems. The latter portions of the chapter focus on specific examples of two types of large scale models — one type focused on the dynamics of encoding and retrieval of sequences of spiking activity, and the other type focused on the lumped dynamics of large populations. Both types are drawn from recent work on the hippocampal formation.

11.2 PRACTICAL ISSUES

11.2.1 Software

For development of network simulations utilizing realistic compartmental representations of single neurons, the GENESIS simulation package;[3] (*http://www.bbb.caltech.edu/GENESIS*) provides a script language with extensive functions focused on implementing network simulations with complex compartmental simulations of individual neurons. As an alternative, the NEURON simulation package also provides a script language which in recent years has started to focus more on providing functions for development of network simulations;[4] (*http://www.neuron.yale.edu*).

A number of software packages are available for development of abstract network simulations. These include implementations of associative memory function and competitive self-organization as well as use of the backpropagation algorithm in the software developed with the original PDP books.[5] This software has been updated recently under the name of PDP++;[6] (*http://www.cnbc.cmu.edu/PDP++/PDP++.html*). Other packages with documentation include a recent book from the Oxford Cognitive Neuroscience course[7] and the neural networks package in MatLab.

11.2.2 Choosing Parameters

Choice of parameters for simulations of large networks often requires information obtained from anatomical studies, including numbers of different neuronal subtypes, and connection probabilities between different types of neurons. Unfortunately, many anatomical studies do not focus on providing quantitative data, and this type of information is only available in a subset of publications. For the hippocampal formation of the rat, the numbers of neurons within specific subregions and the percent connectivities for different pathways have been summarized.[8,9] Parameters for axonal conduction velocities are available from *in vivo* and *in vitro* physiological experiments. Similarly, for piriform cortex, the conduction velocities and connectivity probabilities have been determined from a range of physiological and anatomical studies.[10–12]

Unfortunately, parameters for the connectivity of large networks are often not described in sufficient detail. The role of interneurons provides a particular problem, as it is not entirely clear how to subdivide these into different classes. A number of different biochemical and morphological characteristics have been described for interneurons, but their correlation with connectivity and electrophysiological properties is not fully described.[13]

11.2.3 Evaluation of Models

Different sets of experimental data have been utilized to constrain the structure of large scale models. More abstract neural network models usually focus on data at a behavioral level,[5,14] or address the dynamics of information processing using illustrative examples that are not directly related to any specific set of experimental data.[15,16]

More realistic models must address multiple different levels of constraint. In particular, many large scale network models start with single cell simulations which directly replicate intracellular recording data from individual neurons.[17–19] These detailed single cell recordings are then combined in a network simulation to address data involving network interactions. Many models have focused on the lumped dynamics of a full population of neurons, allowing simulation of local field potential recordings at a network level. These include models of oscillatory dynamics observed using field potential recording in brain slice preparations and whole animals.[12,20–23] Other models have addressed phenomena observed during single unit recording, such as the phasic firing properties of neurons relative to the theta rhythm.[24–26] The dynamics of networks can also be analyzed in terms of experimental data on interspike intervals, auto-correlations and cross-correlations.[27,28] These type of data can demonstrate dynamical interactions within a subpopulation of neurons, even if the population is not sufficiently large for synaptic interactions to influence the local field potential. This provides another source of experimental constraints on simulations which utilize spiking neuron models, as these ISI and correlograms can be generated by the simulations and fitted with available experimental data. These techniques have been utilized to explore the mechanisms underlying generation of gamma frequency oscillations in neuronal circuits.[23]

Few models have addressed both the network physiological level as well as behavioral function. Usually, models addressing specific functional properties of a network focus on these properties without simultaneously addressing physiological data at a systems level — often because physiological data at a systems level is not available. In many cases, models of behavior are based on general anatomical features of a specific system, but use very abstract representations of individual neurons. For example, many simulations of hippocampal memory function follow the overall structure of the hippocampus without replicating single cell physiology in any detail.[29–31] More realistic neurons have been utilized in simulations of how excitatory intrinsic connections can mediate associative memory function, but data are not available to match these associative properties to physiological recordings.[19,32,33] Both behavioral and physiological data has been obtained on rat spatial navigation, and recent modeling work addresses this data on both levels.[24–26]

11.3 NETWORK VARIABLES

11.3.1 Patterns of Connectivity

The connectivity in large scale network simulations of neuronal structures usually includes several basic components summarized in the following sections, using standard terminology.[34]

11.3.1.1 Principal Cells

The principal cells in a network simulation are usually excitatory glutamatergic cells, which activate other cells through glutamatergic synaptic currents. In simulations of cortical structures, these principal cells are usually pyramidal cells.[12,17]

11.3.1.2 Afferent Input to Principal Cells

In network simulations, it is necessary to include some means by which principal neurons are excited by afferent input. The pattern of afferent connectivity can vary from single input lines for each principal neuron to very broadly distributed patterns of afferent connectivity. Simulations of the primary visual cortex usually focus on modifications in the pattern of this afferent connectivity, as these modifications provide the properties of feature detection and topography modeled in those simulations,[35] though recent models have shown that intrinsic connections make an important contribution to orientation tuning.[36,37] In contrast, models of three layered cortical structures such as hippocampus and piriform cortex do not focus on specific patterns of afferent input. These models utilize broadly distributed afferent input[12] or specific predetermined patterns of input[32,33] while emphasizing the role of intrinsic connections in mediating associative storage of the input activity.

11.3.1.3 Intrinsic Connections

The term "intrinsic" refers to connections within a region. Excitatory intrinsic connections provide a dominant synaptic influence within many regions of the brain. For example, within neocortical structures, intrinsic synapses make up a much larger percentage of excitatory synapses on an individual pyramidal cell than do excitatory afferent inputs from the thalamus.[38] The pattern of connectivity of these intrinsic inputs provides a dominant influence on the activation dynamics and functional properties of large scale networks, and many different patterns of connectivity have been utilized. In networks focused on modification of afferent input, the intrinsic connections are relegated to a minor role,[35] but these intrinsic connections play an important role in simulations of physiological network dynamics[17] and associative memory function,[33,39] as well as certain aspects of stimulus selectivity.[36,37] Many abstract simulations of network function utilize fully connected networks, in which every unit is connected to every other unit.[15,16] However, realistic simulations must address the problem of incomplete connectivity between the individual elements in a network — the sparseness of intrinsic connectivity within cortical networks. Many simulations have a probability $p = 1$ of connectivity between individual simulated neurons. Biological networks have much sparser connectivity, with values of p no larger than 0.05. This is an important parameter in our example of scaling problems discussed below.

11.3.1.4 Interneurons

Interneurons are neurons with connectivity limited to a specific neuronal region. This primarily includes inhibitory neurons, though some inhibitory neurons project to other regions[40] and some excitatory neurons remain limited to one region (e.g., spiny stellate cells in visual cortex). Most large scale network models contain only one or two classes of inhibitory interneurons with no differences in intrinsic properties of these neurons. The primary difference in these models concerns the primary source of excitatory drive on these interneurons.

11.3.1.5 Feedforward Inhibition

The term *feedforward inhibition* refers to inhibition activated predominantly by the afferent input to a region. In network models, excitatory afferent input not only contacts principal cells, but also provides distributed afferent input to inhibitory interneurons. This feedfoward inhibition serves to scale the level of inhibition to the strength of afferent input. If the network receives strong excitatory input, inhibition increases proportionately to limit the range of changes in activation, or to provide a strong inhibitory rebound after the initial excitatory input.[12]

11.3.1.6 Feedback Inhibition

Feedback inhibition refers to inhibition activated predominantly by output from the principal cells of the network. In simulations of cortical networks, pyramidal cells activate interneurons which then contact the same population of pyramidal cells.

This feedback inhibition can play a role in maintaining activity in the range of an equilibrium state,[41] or it can mediate oscillations with a time course corresponding to the feedback inhibition — for $GABA_A$ synaptic currents this corresponds to oscillations in the gamma frequency range.[12,23,42,43]

11.3.2 Axonal Delays and Synaptic Transmission

Combining realistic simulations of individual neurons in a network requires dealing explicitly with the time delays of axonal transmission and synaptic transmission. In more abstract models, changes in the output function of individual units are usually transmitted instantaneously to other units, simplifying the analysis of network dynamics. Recent research has provided some analytical tools for describing networks with time delays on individual connections,[44] but these techniques have not been utilized for more realistic simulations. Large scale network simulations of realistic neurons usually contain some representation of the spatial topography of a network and the transmission time between spike generation at a presynaptic neuron and the synaptic conductance change at a post-synaptic neuron.

In most simulations, synaptic connectivity is represented by a parameter scaling the maximal conductance ($\bar{g}_{syn}$) of the ligand-gated channel activated after generation of a presynaptic spike (see Box 11.1 and Section 10.2.1 for description of synaptic potentials). This can be initialized to a fixed value for all connections in a network or can be an exponentially decaying function of the distance between pre- and postsynaptic elements. The rate of decay of this exponential function can be used to determine the spread of connections from any given point in the network. A separate parameter may be utilized to provide different weight values for individual synapses using this type of ligand-gated channel. In our simulations we initialized all weights to a fixed low value. Weights of excitatory synapses onto pyramidal cells could then potentiated by a Hebbian learning mechanism as described below. Many simulations utilize random variation of the transmission delays and synaptic weights around a mean value. This random component can be generated from either a uniform, a gaussian or an exponential probability distribution. In the simulations of sparse networks described here, the existence of connections was determined on a random basis.

11.3.3 Synaptic Plasticity

Whereas some network simulations have been designed primarily to focus on replicating the lumped activation dynamics of distributed networks,[12,20] many network simulations focus on some aspect of the adaptive properties of a network. One major class of models focuses on simulations of associative memory function based on modification of intrinsic connections.[19,32,33] Another major class concerns formation of feature detectors and topography based on modification of excitatory afferent input connections.[35] Both these types of models implement changes in the *weight* of individual synapses — altering the $\bar{g}_{syn}$ of the synaptic current induced by an individual presynaptic spike.

Box 11.1 Modeling Hebbian Learning

In the GENESIS simulation package, learning is implemented by use of the "hebbsynchan" object. This object allows the weight of a connection to be altered in a manner similar to NMDA-dependent long-term potentiation. Long-term potentiation (LTP) is simulated by increasing the weight whenever both the pre- and postsynaptic activities are above a given threshold (separate LTP thresholds exist for pre- and postsynaptic activities). Long-term depression (LTD) is simulated by decreasing the weights whenever one activity is above the threshold and the other is below a separate threshold for induction of long-term depression (in our simulations, these thresholds were set so that only LTP, but not LTD could occur). The presynaptic activity is represented by a dual exponential waveform that is generated by a spike in the presynaptic neuron. The rise time constant of this waveform was set to be 20 msec and the decay time constant was set to 50 msec. This slow decay is consistent with the slow decay of current through the NMDA-receptor channel.[46] This allows the presynaptic activity to remain above threshold for a relatively long time (50–100ms depending on decay rate and threshold) after generation of a presynaptic spike. The postsynaptic activity is the average of postsynaptic membrane potential over a brief period (3 msecs in our simulations). Thus, if postsynaptic membrane potential crosses the postsynaptic threshold within the time window that the presynaptic activity is above threshold, the weight is increased in proportion to the pre and postsynaptic activities, and a weight change is made. Figure B11.1 shows the magnitude of potentiation as a function of the interval between pre- and postsynaptic spiking for the hebbsynchan parameters used in the simulations. Only AMPA synapses onto excitatory cells were allowed to undergo learning in our simulations.

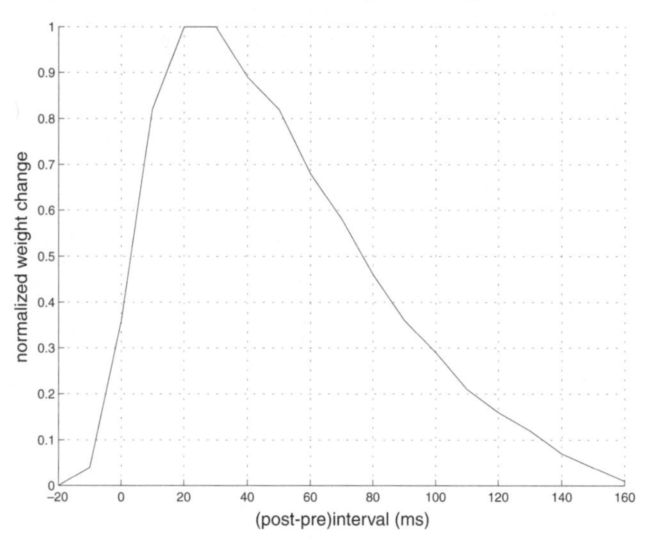

FIGURE B11.1 Magnitude of synaptic potentiation as a function of the interval between pre- and post-synaptic firing. Maximum potentiation occurred when the postsynaptic spike fired 20–30 ms after the presynaptic spike.

In more abstract simulations, useful adaptive properties arise from modification of synapses based on the Hebbian learning rules, in which the change in strength of a synapse is proportional to the product of both pre-synaptic and post-synaptic activity. Associations between stored features can be formed if the neurons activated by each individual feature are active at the same time, resulting in strengthening of the excitatory synapses between these two neurons. This property of synaptic modification can be obtained if modification depends upon calcium influx through simulated NMDA receptor channels, which are activated by a combination of presynaptic transmitter release and postsynaptic depolarization.[45,46] The functional properties of abstract networks are also greatly enhanced by the decay of connections when presynaptic activity is present without postsynaptic activity, or vice versa.[47,48] This property of synapses has also been demonstrated in the form of long-term depression.[49] More abstract learning rules based on minimizing an error between actual and desired output have been used extensively in connectionist models,[5] but have not been implemented in a manner consistent with known biological principles in realistic networks.

More biophysically realistic representations of synaptic modification commonly are based on the biophysical properties of the NMDA receptor. The computational modeling of the NMDA receptor involvement in synaptic plasticity is summarized in Box 11.2.

Box 11.2 Modeling the NMDA Conductance

The time course of synaptic conductances are commonly modeled as an alpha function or as a dual exponential equation, as described in Section 10.2.1 Since the NMDA conductance is known to depend upon postsynaptic membrane potential too, an additional control factor $k(V)$ needs to be added. So, using the dual exponential equation, we can write the NMDA conductance equation as

$$g_{NMDA}(t,V) = \bar{g} / g_{peak} * \left(\tau_1\tau_2 / (\tau_1 - \tau_2)\right)\left(e^{-t/\tau_1} - e^{-t/\tau_2}\right) * k(V)$$

where $\bar{g}$ is a scaling factor, and g_{peak} is the height of the unscaled equation at

$$t_{peak} = \left(\tau_1\tau_2 / (\tau_1 - \tau_2)\right)\ln(\tau_1 / \tau_2)$$

The voltage dependence of the NMDA conductance results from a voltage dependent block of the channel by Mg^{2+}, which blocks the channel pore at hyperpolarized potentials. The kinetic scheme for this binding reaction (see Section 2.2.1) is

$$AR^* + Mg^{2+} \underset{k^-}{\overset{k^+}{\rightleftharpoons}} ARMg^{2+}$$

(continued)

Box 11.2 (continued)

where AR^* is the agonist (glutamate)-receptor complex that is not occupied by Mg^{2+} and is in the open (conducting) state and $ARMg^{2+}$ is the agonist-receptor complex blocked by Mg^{2+} and no longer conducting. The Mg^{2+} block is thought to occur on a time scale much faster than the formation and dissociation of the agonist-receptor (AR^*) complex. Thus, the above reaction is at equilibrium and the fraction of channels in the conducting state (f_{open}) is given by

$$f_{open} = \frac{AR^*}{\left(AR^* + ARMg^{2+}\right)} = \frac{1}{\left(1 + \frac{\left[Mg^{2+}\right]}{k^- / k^+}\right)}$$

Ascher and Nowak[72] showed that the voltage dependence of the rate constants in the above kinetic scheme is given by $k-/k+ = 8.8\ 10^{-3} * \exp(V/12.5)$, where V is in mV. Thus,

$$k(V) = f_{open} = \frac{1}{\left(1 + \eta * \left[Mg^{2+}\right] * e^{-\gamma * V_m}\right)}$$

where $\eta = 0.1136$ /mM and $\gamma = 0.08$ /mV. Thus, $k(V)$ is a sigmoidal function of voltage. The time constants for the NMDA conductance waveform are usually taken to be $\tau_1 = 60$–80 ms and $\tau_2 = 0.67$ ms. Other studies have used a value of 0.33 /mM for η and 0.06 /mV for γ based on physiological measurements.[73] References 23, 46, and 74 provide examples of the use of NMDA channels in compartmental models of neurons. A complete model of the NMDA synapse requires three Goldman–Hodgkin–Katz (or constant field, see Section 5.4.2) equations to compute each of the three separate ion fluxes through the channel (Na^+, K^+ and Ca^{2+}),[46,75] with a permeability given by g_{syn}. A simpler alternative is to compute the voltage-dependent fractional Ca^{2+} current (relative to the total I_{syn}) off-line[76] or to assume it is constant.[73]

11.3.4 Scaling Issues

Current computing resources available to most researchers do not allow for simulations of mammalian brain regions to contain the same number of units as are present in the real network. Detailed simulation of such large networks would require too much time and too much memory to be practical at this stage. Thus, anyone performing large scale network simulations must address the problem of scaling down the size of the network to a smaller number of modeled units. This is a difficult issue, because many of the functional properties of networks depend upon sufficient convergence of inputs to allow individual units to spike. If a network is reduced in size without altering the connection probability or connection strength, individual units will never receive sufficient synaptic input to go over threshold.

11.3.4.1 Scaling Connection Probability

One means of compensating for the reduced synaptic drive on individual units in a smaller network is to increase the percentage connectivity. As noted above, most abstract simulations utilize full connectivity (probability of excitatory connections between individual units $p = 1$). This provides effective function in simulations of associative memory function.[19,39] Network function can still be obtained with moderate reductions in connection probability, but associative function is completely lost at levels corresponding to the actual connectivity within structures such as hippocampal region CA3. Though this region has the highest excitatory intrinsic connectivity of any cortical region, the percentage connectivity is only 4%, whereas the function of a network of 100 simulated neurons drops off rapidly below 20% (see below).

11.3.4.2 Scaling of Connection Strength

Another means of compensating for reduced synaptic drive on individual neurons is to increase the strength of individual synaptic connections. In most simulations, this requires a serious deviation from the available physiological data on strength of individual synapses. Most simulations utilize some combination of both increased connection probability and increased connection strength in order to obtain sufficient activity. If the peak conductance of synaptic currents is maintained constant, the number of simulated neurons inducing this current can in specific networks be varied over several orders of magnitude without changing the steady-state network dynamics.[23] However, when the number of synapses inducing the current drops below a certain value, the dynamical properties can change dramatically.

11.4 FUNCTIONAL DYNAMICS

11.4.1 Factors Influencing Function in Large Scale Realistic Networks. Example: Sequence Storage

In this example we discuss how sparse neural networks with biologically realistic representations of neuronal elements, connectivity patterns and learning rules can store and recall temporal sequences of spatial patterns of activity, a task which might be integral to the role of the hippocampal formation in episodic memory.[26,50] This model allows quantification of the contribution of specific physiological and anatomical parameters to the mechanisms of memory. A biologically detailed simulation of region CA3 of the hippocampus was constructed using GENESIS.[3]

11.4.1.1 Single Cell Model

In modeling large networks, simplified compartmental models of single cells make simulations more manageable in terms of computing resources and time. These simplifications include using a small number of compartments, reducing the number of branches in the dendritic tree of a neuron, omitting certain ligand- and voltage-gated conductances, and placing synaptic (ligand gated) conductances on a limited number of compartments instead of distributing them diffusely. The gain in computation speed comes at the expense of biological accuracy of the network and, therefore, the amount of simplification has to depend upon the objectives of the simulation. We include several ligand- and voltage-gated conductances found on actual CA3 pyramidal neurons in our single cell model, although we did not include realistic single cell morphology in the model. The model firing behavior is shown in Figure 11.1. Some simpler 3 compartment models containing only sodium and delayed rectifier potassium conductances were also tested for their performance in the sequence storage/recall task. For more discussion on issues related to modeling single cells see Chapter 9.

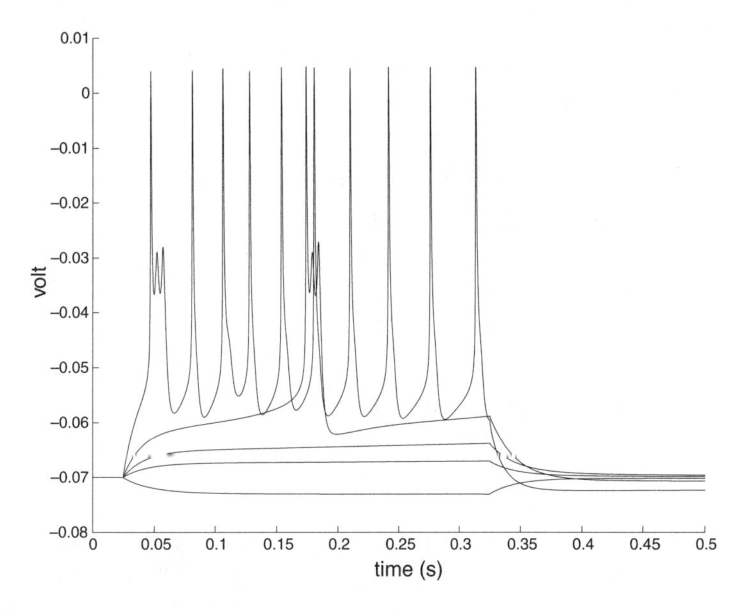

FIGURE 11.1

Response of simulated pyramidal cell to current injections of –50, 50, 100, 200 and 500 *pA*. This is the more complex cell model.

The network had populations of excitatory (pyramidal) and inhibitory (GABAergic) neurons. These were similar to the compartmental models used previously in simulations of the hippocampal CA3 region.[17,26] Pyramidal cells had five compartments and interneurons three. The input resistance and membrane time constant of these cells as measured by a hyperpolarizing current step (50 pA) were 60 MOhms and 25 ms for excitatory cells and 140 MOhms and 35 ms for inhibitory cells.

Both pyramidal cells and interneurons received excitatory synapses from other pyramidal cells in the network. On pyramidal cells these excitatory synapses activated AMPA and NMDA receptors in the proximal apical compartment, whereas on interneurons only AMPA receptors were used. Furthermore, AMPA synapses between pyramidal neurons could undergo Hebbian learning, whereas no learning was employed at AMPA synapses onto interneurons. Inhibition was provided by $GABA_A$ and $GABA_B$ type IPSPs. $GABA_B$ inputs arrived on the proximal apical compartment, and $GABA_A$ inputs on the soma compartment of both pyramidal cells and interneurons. Time constants and maximum conductances of synaptic channels are listed in Table 11.1. These parameters have the same general magnitude as parameters utilized for simulations of cortical synaptic currents in other studies,[12,51] but *in vitro* studies demonstrate differences in these time constants for different structures. These time constants are often obtained at non-physiological temperatures, requiring alteration of the time course based on estimations of the Q_{10} to obtain the faster time courses which would be present at physiological temperatures[23] (see Section 5.4.2)

TABLE 11.1
Parameters for Synaptic Channels Used in the Network Models in this Chapter

Type	$\bar{g}_{syn}$ (μS)	τ_1 (ms)	τ_2(ms)	E_{syn} (mV)
AMPA	4	1–2	2	0
NMDA	20	0.67	80	0
$GABA_A$	100	1	3–5	–70
$GABA_B$	1	30	90	–90

Note: See Chapter 10, Equations 10.1 and 10.4 for symbols used.

11.4.1.2 Network Model

The hippocampal CA3 model consisted of a sparsely connected network of excitatory and inhibitory neurons. For computing axonal delays, we allowed conduction delay to vary with distance, and assumed an axonal conduction velocity of 0.5 m/s for both excitatory and inhibitory cells. We used 20 microns as the distance between neighbouring neurons in our simulations, and computed distance based upon the location of units within a one dimensional array of units.

In all simulations, a sequence consisted of five orthogonal spatial patterns, with each pattern encoded by the firing of five neighboring excitatory (pyramidal) neurons. Thus, 25 pyramidal neurons in all were used to encode the sequence, and these will be referred to as sequence neurons, since they were directly activated by external (afferent) input to the network. The remaining cells in the network that did not receive afferent input will be referred to as background neurons. A pattern was presented to the network by causing the appropriate sequence neurons to fire by means of a large and brief (2 ms) depolarizing pulse, representing afferent input. Successive patterns in a sequence were presented at intervals of 100 ms, simulating the arrival of inputs at the theta frequency. Each sequence was presented only once to the network and, thus, very high learning rates were employed to enable single trial learning. For hippocampal circuits, physiological and behavioral data suggest that episodic information can be encoded during a single event. Thus, other models of region CA3 have also focused on single trial learning,[50] whereas models of neocortical learning commonly use multiple interleaved presentations of input stimuli.[35] A single behavioral trial may involve holding of the sequence in some buffer (perhaps in entorhinal cortex) which would allow learning to proceed with several presentations of the sequence over a longer period. At the start of a trial, weights of all excitatory synapses were initialized to a low, uniform value. The sequence was then presented and strengths of AMPA synapses were allowed to increase constituting the learning phase of the task which lasted up to 100 msec beyond the last pattern in the sequence. Activity in the network was then terminated, synaptic weights were no longer allowed to vary and the first pattern of the sequence was presented to test the network's ability to recall the stored sequence.

Output performance of the network was assessed by looking at the activity of sequence neurons that coded the spatial patterns in the sequence. There were five sequence cells per pattern. The performance measure (PM) for a pattern q was given by: $PM_q = (N_{qc} / 5) - (N_{nc} / (N_i - 5))$, where N_{qc} = number of input layer cells belonging to the pattern that fired, N_{nc} = number of sequence cells not belonging to the pattern that fired, N_i = total number of sequence cells.[25] For the sequence, PM_s was the average of the performance measures for each pattern in the sequence, excluding the first pattern that was used to cue recall, i.e.,

$$PM_s = 1/\left(N_q - 1\right)\sum_{i=2}^{N_q}\left[PM_q\right],$$

where N_q = Number of patterns in the sequence. If all sequence cells fired at some point, *PM* for that pattern would be zero. If only the sequence cells belonging to the pattern fired, *PM* would be a maximum of 1, and if only sequence cells not belonging to the pattern fired, *PM* would be a minimum of –1.

11.4.1.3 Demonstration of Scaling Issues

We began with a simple network model to show that a fully connected network could learn to store a sequence. Each cell had only three compartments with the middle (soma) compartment containing Na^+ and K^+ conductances to generate action potentials. The network consisted of 25 excitatory and 0 inhibitory cells. Initial weights of excitatory synapses were very low (0.0005 for all connections), precluding spread of activity in the network subsequent to activation of sequence cells by afferent inputs.

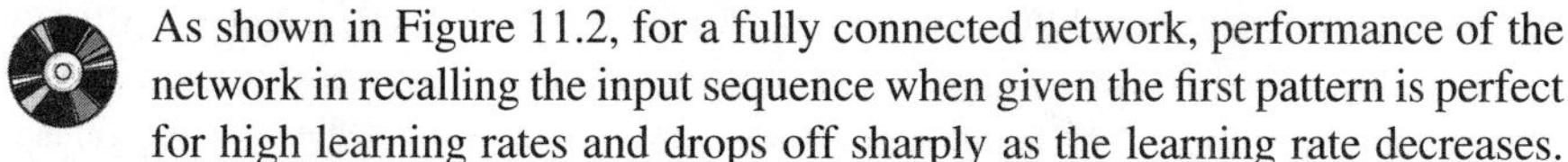

As shown in Figure 11.2, for a fully connected network, performance of the network in recalling the input sequence when given the first pattern is perfect for high learning rates and drops off sharply as the learning rate decreases. Figure 11.2A shows a raster plot of network activity for a fully connected network of 25 excitatory neurons (simple model) with learning rate = 5000, showing perfect recall (with a performance measure value of 1.0). Note that the full sequence of patterns is retrieved by the network after input of the retrieval cue, however the rate of retrieval is more rapid due to the faster conductances of the modified AMPA

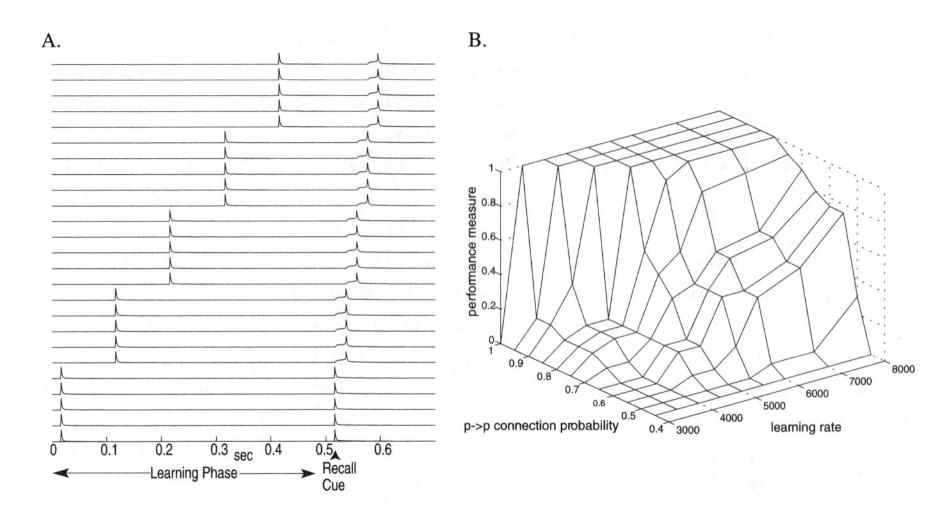

FIGURE 11.2 (A) Raster plot of network activity for the simple model of 25 fully connected excitatory neurons showing perfect recall. Each line corresponds to the somatic membrane potential for an individual neuron in the network. Pyramidal cells 0–24 were activated in five successive patterns. Weights were allowed to potentiate during the learning phase. The entire sequence could be perfectly recalled by the network after input of the retrieval cue (the first pattern in the sequence), giving a performance measure value of 1.0. (B) Network performance of the simple model in recalling a stored sequence, as a function of the sparsity of the network and the learning rate of the AMPA synapse. Performance was poor for sparse networks (low connection probability), even with high learning rates.

channels. This is consistent with available data suggesting that the place cells activated during a slow transition through different locations over several theta cycles may subsequently fire in a more rapid sequential manner.[25,27,50]

Figure 11.2B shows that the performance of the network falls dramatically as the network is made sparser (as connection probability, p, is reduced). Even for high learning rates, decreasing the connection probability to values below 0.5 causes a dramatic decrease in the performance measure. These simulations and related analytical work demonstrate that sparse networks require the participation of background neurons (those not directly receiving afferent input) to complete the functional pathways connecting one spatial pattern to another in a sequence. Given the low connectivity percentage of real biological networks, an understanding of the mechanisms for generation of stable background activity patterns is important.

To enable the participation of background neurons, network size was increased to 500 simple modeled pyramidal cells out of which only 25 received direct afferent input to code the five spatial patterns (5 cells/pattern). No synaptic inhibition was included in the model. The initial weights (weights at the start of the learning phase) of synapses between pyramidal cells were varied to activate background neurons to various extents. Connection probability of pyramidal neurons was also varied to try to find an optimum combination of these parameters. As shown in Figure 11.3, increasing the initial weight improved sequence recall, but performance was still poor for sparse networks ($p < = 0.4$). Furthermore, input cells (pyramidal neurons

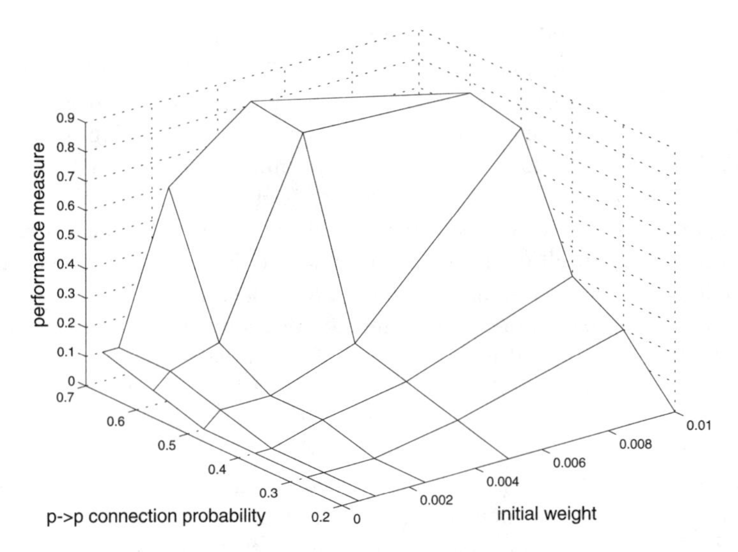

FIGURE 11.3 Network performance in a larger model for recalling a stored sequence as a function of network sparsity and initial (pre-learning) weights of excitatory synapses. Performance was poor for sparse networks ($p < 0.3$), even with high initial weights. For more dense networks ($p = 0.6$), performance improved with increasing weights. However, this resulted in sustained firing of cells and eventually a population burst during the recall phase of the trial. The network was composed of 500 excitatory cells (simple model) and no inhibitory cells. Learning rate was 4000.

that received direct afferent input to code the sequence), continued firing in a sustained manner even after completion of sequence recall. Thus, introduction of background activity appeared to necessitate the introduction of inhibitory mechanisms to control activity levels in the network. Enhancement of excitatory connections to facilitate the participation of background neurons also increases the propensity for runaway excitation in the network.

Another approach used previously,[26] could be to simulate the differential modulation of membrane potential by acetylcholine (ACh) and GABA. ACh depolarizes pyramidal neurons, thereby bringing them closer to threshold and enabling unpotentiated, weak excitatory synapses to cause them to fire. We simulated ACh-mediated depolarization in the network by reducing the amount of leakage conductance, so that the resting potential was about –60 mV during the learning phase. No ACh modulation was used during the recall phase and resting potential was –70 mV. To control the amount of activity in the network, synaptic and intrinsic inhibitory mechanisms were introduced. Synaptic inhibition was provided by two populations of feedback interneurons, one activating $GABA_A$ and the other $GABA_B$ conductances on pyramidal neurons. Both conductances had the same $\overline{g}_{syn}$ value. The probability of forming a connection between a pyramidal cell and an interneuron ($p -> i$) was set to 0.2 and the reciprocal probability ($i -> p$) to 0.3. Intrinsic inhibitory mechanisms included an AHP current and another calcium-dependent potassium current.[17]

As expected, an intermediate level of inhibition was needed to produce the optimum level of background activity that resulted in the best recall performance of the network. This inhibition could be provided by just the AHP alone, the synaptic inhibition alone, or a combination of the two (Figure 11.4A). Too little inhibition resulted in population bursts in the network which degraded performance, whereas too much inhibition prevented background cells from participating in the coding scheme. Therefore, a crucial issue is the adjustment of inhibition (synaptic and intrinsic) in the network such that the level of background activity is optimal for network performance. The ideal situation would be one where the level of inhibition dynamically adjusts itself to maintain a given activity level in the network for small perturbations in excitatory strength, connection probability ($p -> p$), or number of excitatory cells in the network. Auto-regulation of neuronal activity parameters has been suggested and utilized in some models.[52] As seen in Figure 11.4A, the best performance measure was ~0.8, and not yet perfect with the above scheme. This resulted because background firing was not uniformly distributed across the sequence and was weak during the last two patterns.

The problem of obtaining evenly distributed background activity was solved by combining the cholinergic depolarization with a phasic oscillation in membrane potential at theta frequency. This replicates the effect of combined cholinergic modulation and GABAergic input from the medial septum driving theta frequency oscillations in the hippocampus.[53] The incorporation of these theta frequency oscillations allowed the background activity to be evenly distributed across the individual elements of the sequence during learning, and therefore to contribute to effective retrieval of the full sequence, even in a very sparse network ($p = 0.1$) as shown in Figure 11.4B. The progressive increase in depolarization of neurons across the theta cycle ensured that only a specific subset of background neurons were activated, after which feedback inhibition would prevent activation of additional neurons, and intrinsic adaptation would prevent reactivation of the same neurons on the following cycle. This network contained 200 pyramidal cells, and 40 of each type of inhibitory cell. Strengths of recurrent connections were too weak, prior to learning, to cause spread of activity in the network, until depolarization was provided to the background neurons, bringing them over threshold to participate in the encoding of the sequence.

This phasic change in depolarization could also be effective in gating the output from the hippocampus. In conditions where multiple sequences are being retrieved in region CA3, the phasic increase in depolarization would allow neurons receiving a combination of layer II entorhinal input and intrinsic connectivity to fire first, before any spurious background activity. If this mechanism is coupled with sequence readout in the entorhinal cortex layer III which starts with a desired goal location, then it would allow sequence retrieval in region CA1 to be guided by a destination. An alternate approach to the use of background neurons would be the use of a more broadly distributed subthreshold afferent input, which only causes neurons to come over threshold when intrinsic connectivity matches afferent input, thereby ensuring that the only neurons activated by a sequential input are those which are interconnected. Oscillatory changes in the magnitude of long-term potentiation which are out of phase with the excitatory feedback transmission but in phase with excitatory afferent input allow an effective separation of retrieval dynamics and new encoding.

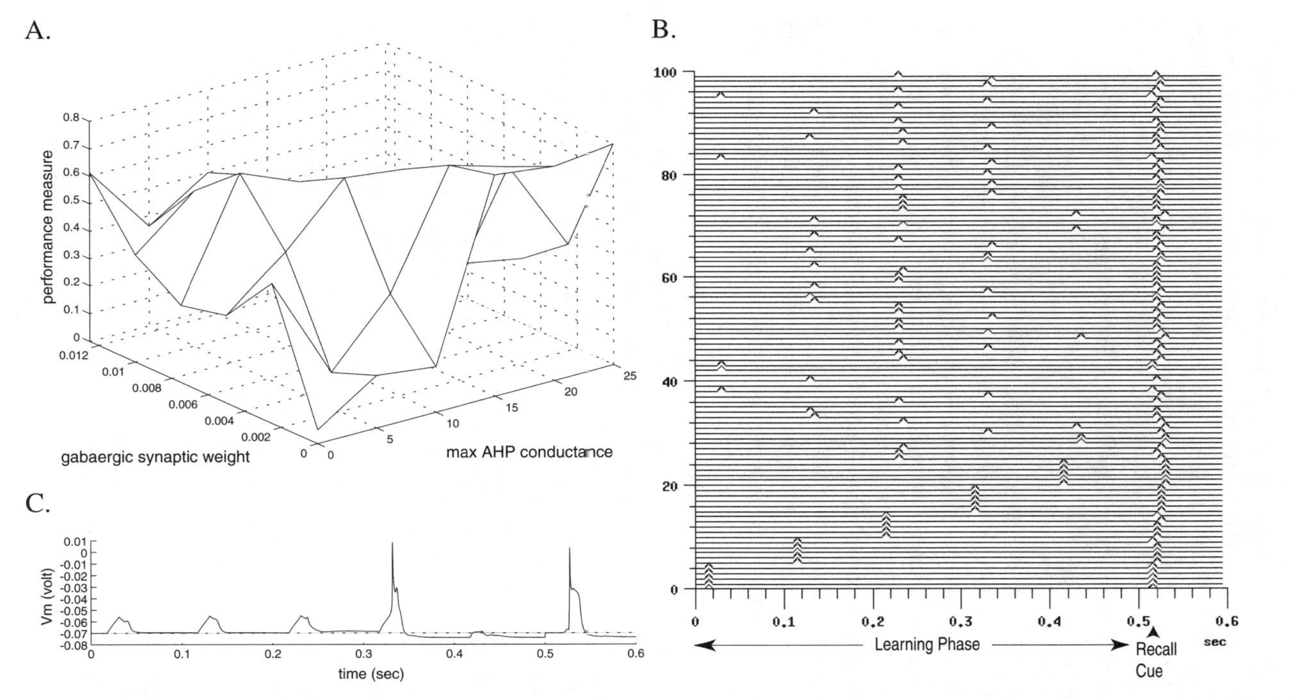

FIGURE 11.4 (A) Network performance in recalling a stored sequence as a function of synaptic inhibition ($GABA_A$ and $GABA_B$) and intrinsic inhibition (spike after-hyperpolarization). Performance was better when network activity was controlled with either one form of inhibition in large amounts, or a combination of the two in smaller amounts. The network had 100 excitatory cells (complete cell model), and 10 each of $GABA_A$ and $GABA_B$ type inhibitory cells. Background cells received ACh modulation (see text). Connection probabilities are given in the text. The learning rate was 500. (B) Perfect recall by a sparse network ($p = 0.1$) when oscillatory depolarizing input is provided to background neurons. Raster plot shows spike activity in a subset of 100 pyramidal neurons during learning (0–0.5 sec) and recall of a sequence of five patterns. Evenly distributed recruitment of background neurons enabled accurate retrieval of the sequence even in a sparse network. (C) The voltage in one of the background neurons of the sparse oscillatory network.

11.4.2 Linking Abstract Models to Realistic Models. Example: Attractor Dynamics And Oscillations

In contrast to the example of sequence storage described above, numerous simulations of large networks focus on how the average activity of a large population of neurons changes over time. These simulations often focus on attractor dynamics and oscillations which can also be described in simpler, more abstract simulations.

11.4.2.1 Attractor Dynamics in Continuous Firing Rate Models of Memory

Numerous abstract network simulations of episodic memory function have been developed.[30,31,54] These models simulate aspects of human performance in memory tasks using networks inspired by the anatomy of the hippocampal formation. The memory function of these simulations arises from attractor dynamics due to strong recurrent excitation in simulations of hippocampal region CA3 which drives the network into specific learned patterns of activity. The term *attractor* refers to the fact that input patterns which differ with regard to individual features (different initial conditions) can elicit the same final pattern of neuronal activity (the same attractor). Attractor dynamics provide a mechanism whereby the memory retrieved can be independent of variation in the retrieval cue.

Early attractor models limited the maximal output level of neurons with a sigmoid function which approached an asymptotic maximum of output no matter how high the input.[15,16] This was critized as unrealistic, since recordings from cortical structures show that neurons do not fire much above 100 Hz,[55] whereas during current injection, a neuron can be driven to rates over 300 Hz.[56] Thus, if attractor dynamics exist, they must limit firing rate without relying on the maximum firing rate of individual neurons. Newer attractor models addressed this issue in various ways, including balancing excitatory and inhibitory influences on neurons.[41,57,58] These models used a continuous firing rate representations, in which an analog variable approximately represents the number of neurons firing during one time step, and instantaneous synaptic effects, in which the output of one unit immediately affects the firing of another unit (in proportion to a synaptic weight variable). These models avoid the problem of temporal variability caused by the use of spiking output.

In describing the activity of biological networks, it is useful to model the activity of populations of excitatory units and inhibitory units separately, rather than allowing individual units to send out both excitatory and inhibitory connections,[16] which does not occur in real neural circuits. Dynamics of the mathematical representation used here were first studied by Wilson and Cowan,[58–60] and this type of representation has been used to study the dynamics of cortical networks including piriform cortex[43] hippocampus[41,61] and somatosensory cortex.[62]

Changes in the average membrane potential of the population of excitatory and inhibitory neurons are described by the following equations:

$$da/dt = A + W\left[a - \theta_a\right]_+ - \eta a - H\left[h - \theta_h\right]_+$$

$$dh/dt = A' + W'\left[a - \theta_a\right]_+ - \eta' h - H'\left[h - \theta_h\right]_+$$

or

$$\begin{bmatrix} \dfrac{da}{dt} \\ \dfrac{dh}{dt} \end{bmatrix} = \begin{bmatrix} W - \eta & -H \\ W' & H' - \eta' \end{bmatrix} \begin{bmatrix} a \\ h \end{bmatrix} + \begin{bmatrix} A - W\theta_a + H\theta_h \\ A' - W'\theta_a + H'\theta_h \end{bmatrix} \tag{11.1}$$

These equations show the change in average membrane potential (a) of the excitatory population and average membrane potential (h) of the inhibitory population (where zero is resting potential). These averages correspond to the membrane potential determined by synaptic input. η represents the passive decay of membrane potential, and is the inverse of the average membrane time constant. The summed firing rate of the excitatory population is computed by a threshold linear function $[a - \theta]_+$ of average membrane potential. These firing rates are zero when membrane potential is below θ and increases linearly ($a - \theta$) for values above θ. The same function is used for inhibitory firing rate. A represents the afferent input to a population of neurons. W represents the average strength of excitatory synapses arising from cortical pyramidal cells and synapsing on other excitatory neurons, and W' represents excitatory synapses on inhibitory interneurons. H represents the average strength of inhibitory synapses on pyramidal cells, and H' represents inhibitory synapses on inhibitory interneurons. The simplified system is summarized in Figure 11.5A. This simplified system can be analyzed using standard stability descriptions for coupled differential equations.[63] In particular, the stability of certain states depends upon values of the trace $T = (W - \eta) + (-H' - \eta')$ and the determinant $K = (W - \eta)(-H' - \eta') + W'H$.

The equilibrium states of networks of this type have been evaluated previously.[41,58] When the trace < 0 and the determinant > 0, the network has stable equilibrium states. These can be obtained by algebraically solving for a or h, after setting $da/dt = dh/dt = 0$. Real biological networks probably only enter equilibrium states for brief periods, but the network may be continuously moving toward particular stable equilibrium states. If we assume $I = A - \eta\theta$ and $A' = H' = 0$, then the average excitatory membrane potential during this equilibrium state is:

$$a_{eq} = \theta_a + \frac{I + H\theta_h}{\eta - W + \dfrac{HW'}{\eta'}} \tag{11.2}$$

An example of the response of such a network to input is shown in Figure 11.5B. The network approaches an initial equilibrium state during the presence of afferent input A, and after removal of this afferent input, it evolves to a different, lower self-sustained equilibrium state. The attractor dynamics of this system have been used in a number of simulations of modulatory influences in cortical structures.[41,64] This simplified representation also allows analysis of oscillatory dynamics. In particular, if the trace = 0 and the determinant > 0, then stable limit cycle oscillations will be obtained due to the feedback interaction between excitatory and inhibitory neurons. The fast time constant of inhibition results in oscillation frequencies in the gamma range.

The intrinsic adaptation of neurons can be represented in a manner similar to feedback inhibition, with γ representing activation of voltage-dependent calcium currents above a threshold θ, Ω representing the decay of intracellular calcium concentration, and μ representing calcium activation of calcium-dependent potassium currents. These parameters can be adjusted to model adaptation properties of real pyramidal cells.[41] The equations take this form:

$$da/dt = A + W[a - \theta_a]_+ - \eta a - \mu c$$
$$dh/dt = \gamma[a - \theta_a]_+ - \Omega c$$

or

$$\begin{bmatrix} \frac{da}{dt} \\ \frac{dc}{dt} \end{bmatrix} = \begin{bmatrix} W - \eta & -\mu \\ \gamma & -\Omega \end{bmatrix} \begin{bmatrix} a \\ c \end{bmatrix} + \begin{bmatrix} A - W\theta_a \\ -\gamma\theta_\gamma \end{bmatrix} \tag{11.3}$$

Similar dynamics appear for these equations. When the trace of these equations (W-η-Ω) equals 0, sustained limit cycle oscillations appear. The slower time constant of calcium decay results in slower oscillations which could correspond to the frequencies of theta rhythm oscillations. This could be an effective model of theta oscillations which appear in slice preparations of hippocampus,[65] though *in vivo* oscillations appear to be forced by septal input.[53] Interestingly, even for higher values of W, these slow frequency oscillations appear due to crossing of the threshold for activation of calcium channels.

If both intrinsic adaptation and feedback inhibition are incorporated in the same model, then numerous types of interactions can be obtained. Transitions between these states might commonly occur in cortical circuits due to modulatory effects of acetylcholine, which has been shown to change the strength of the excitatory feedback connections W.[41] As shown in Figure 11.5C, changing just the magnitude of W causes the network to show a number of different dynamical states. For small $W = 0.10$, the network shows strongly damped theta oscillations. For $W = 0.11$, the equations representing intrinsic adaptation have trace = 0, and sustained limit cycles at theta frequency appear in the model. For $W = 0.13$, threshold crossing results in theta cycles, with nested gamma cycle oscillations due to feedback inhibition at the peak of each theta cycle. Thus, gamma depends upon sufficient excitatory activity in the network. Gamma oscillations appear with a theta envelope in structures such as piriform cortex[66] and hippocampus,[67] possibly because the relatively homogeneous excitatory connections in these structures cause synchronous phases of excitation, thereby restricting the periods of gamma. For $W = 0.15$, the stability of the equilibrium between excitation and inhibition dominates, and theta rhythm oscillations no longer appear. For $W = 0.19$, the trace of the interaction between excitatory and inhibitory neurons is near zero, so the network should show gamma frequency limit cycles in the absence of adaptation. However, adaptation pushes activity below threshold and causes longer stretches between cycles. For $W = 0.3$, the interaction of feedback excitation and inhibition becomes unstable, and the network shows explosive bursts of activity terminated by feedback inhibition bringing the network below threshold. Adaptation influences the interval between these explosive bursts. These explosive increases could correspond to the sharp waves observed after periods of theta frequency oscillations in rats,[68] which appear during lower levels of acetylcholine. The transition from theta to sharp waves to quiescence could reflect the decrease in acetylcholine levels in hippocampus, resulting in a progressive increase in W combined with a loss of pyramidal cell depolarization.

11.4.2.2 Attractor Dynamics and Oscillations in Spiking Network Models

Is it possible to obtain attractor dynamics in networks of realistic spiking neurons? With spiking output the relative timing of spiking of individual neurons plays an important role in the temporal variability of input to other units. This makes it much more difficult to enter and maintain an attractor state. If neurons are all deterministic and have the same parameters, they will tend to spike in synchrony and will show oscillatory states rather than fixed point attractor dynamics. The ideal case is that of completely asynchronous firing, in which different neurons within a population evenly cover the full range of possible spike times, ensuring that the excitatory feedback within the population stays uniform across time. In networks of deterministic neurons, this can be approximated most effectively by distributing parameters around specific values so that neurons are less likely to fire in synchrony. For example, if neurons have mutual excitatory connections, but one neuron receives slightly stronger input than another, then the neuron receiving slightly stronger input will fire slightly in advance of the other.

Another means of dealing with the problem of spike timing is to utilize network oscillatory dynamics as a sort of clock to control the relative spike timing of individual neurons. In particular, fast feedback inhibition mediated by simulated $GABA_A$ receptors will cause networks to fire in the gamma frequency range. This can then limit the maximal firing rate of a network to the gamma frequency range. However, networks that utilize this technique have utilized structured patterns of point to point inhibitory effects, analogous to those used in Hopfield networks, by coupling each excitatory neuron with a single inhibitory neuron.[19] This is an unrealistic feature. Other networks have obtained attractor dynamics by limiting the maximal excitatory synaptic transmission between neurons by having synapses become weaker at higher frequences.[33] Further analysis of attractor dynamics in

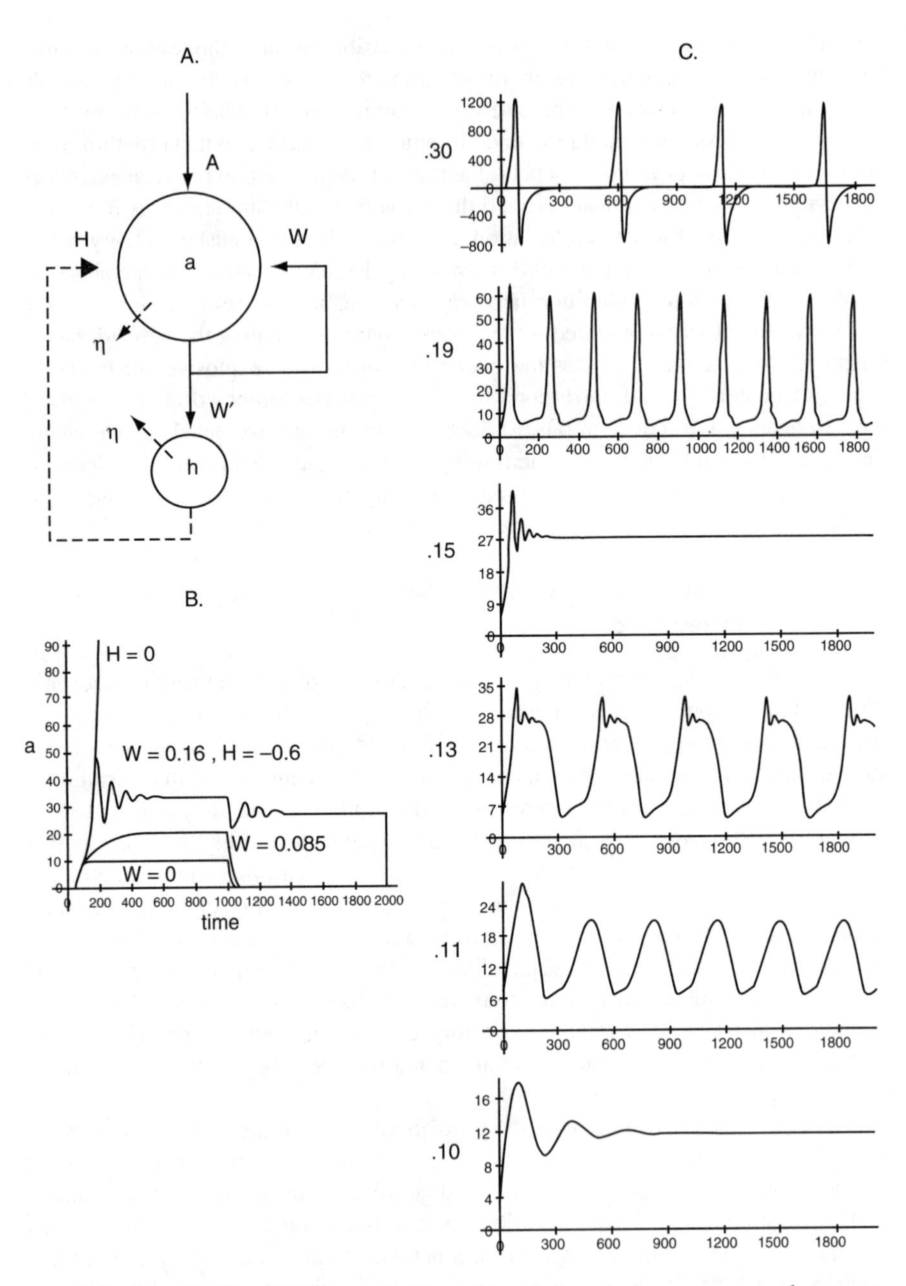

FIGURE 11.5 (A) Simplified network for mathematical analysis. *a* represents the average membrane potential of a sub-population of excitatory neurons, and *h* represents the average membrane potential of a population of inhibitory neurons. See text for other symbols used. (B) Attractor dynamics in a continuous firing rate network with substractive inhibition. Plot showing the activity (*a*) of a single neuron across time for different values of the synaptic parameters. With no excitatory feedback ($W = 0$), the network shows only a passive increase in activity when afferent input is present. With no feedback inhibition ($H = 0$), the network shows explosive growth of activity. With a proper balance of excitatory and inhibitory feedback

(continued)

spiking networks is essential to understanding the relevance of abstract attractor networks to the function of real neuronal systems.

Oscillatory dynamics similar to those observed in the abstract network also appear in networks of spiking neurons. In particular, gamma frequency oscillations have been demonstrated due to interactions of excitatory and inhibitory units in such networks.[12,23] However, many of the spiking network models focus on inhibitory interaction between interneurons.[21] In the abstract network, the trace cannot be set to zero with inhibitory connections alone,[58] but gamma frequency oscillations can be obtained due to threshold crossing. Theta frequency oscillations have also been demonstrated in networks of spiking neurons, simulating the effects of carbachol modulation in the slice preparation.[20] These theta frequency oscillations appear to result from the same interaction of feedback excitation and intrinsic adaptation resulting in oscillations in the abstract network described above.

11.5 FUTURE DIRECTIONS

Further development of large-scale network simulations is essential to understanding the link between experimental data at a cellular and molecular level and the significance of this data for behavioral function. Large-scale simulations will progress towards addressing both behavioral and systems physiological data. This will be most feasible in areas where both types of data have been gathered simultaneously from awake behaving animals. This includes models of hippocampal formation single unit activity and behavioral function during performance of spatial navigation tasks,[27] modeling of entorhinal and perirhinal single unit activity during performance of delayed non-match to sample tasks,[69] and modelling of piriform cortex single unit activity during performance of olfactory discrimination task.[70] Models must become increasingly sophisticated in addressing data at these multiple levels.

In particular, realistic models must take advantage of single cell properties beyond what is already simulated in an integrate and fire model. The bursting and adaptation properties of single neurons are important, and should contribute more substantively to the functional dynamics of network models. Initial work has demonstrated that modulation of adaptation could influence the speed of learning[39] and theta frequency oscillations[20] and that modulation of bursting properties could contribute to setting network dynamics for associative retrieval.[19] Behavioral data on the effects of various drugs demonstrate that modulatory influences on single cell properties have a strong influence on the function of cortical networks.[71] Many of these modulatory effects cannot be represented by simple integrate and fire neurons.

FIGURE 11.5 (continued)
($W = 0.016$, $H = -0.06$) the activity grows to a certain level that can be sustained in the absence of afferent input. (C) Dynamical properties of continuous firing rate network with different values for W representing cholinergic modulation of feedback excitatory transmission. These could correspond to different behavioral states associated with different levels of cholinergic modulation. For high levels of acetylcholine, synaptic transmission is weak ($W = 0.11$) and oscillations resembling theta rhythm appear. For lower levels of acetylcholine, synaptic transmission is strong ($W = 0.3$) and events resembling sharp waves appear. See text for further discussion.

Addressing these types of modulatory influences will be an important area for further simulations of large scale networks of realistic neurons.

REFERENCES

1. Wilson, M. A. and McNaughton, B. L., Dynamics of the hippocampal ensemble code for space, *Science*, 261, 1055, 1993.
2. Nicolelis, M. A., Ghazanfar, A. A., Faggin, B. M., Votaw, S., and Oliveira, L. M., Reconstructing the engram: simultaneous, multisie, many single neuron recordings, *Neuron*, 18, 529, 1997.
3. Bower, J. M. and Beeman, D., *The Book of GENESIS: Exploring Realistic Neural Models with the GEneral NEural SImulation System*, 2nd ed., TELOS/Springer-Verlag, New York, 1997.
4. Hines, M. L. and Carnevale, N. T., The NEURON simulation environment, *Neural Comput.*, 9, 1179, 1997.
5. McClelland, J. L. and Rumelhart, D. E., *Explorations in parallel distributed processing*, MIT Press, Cambridge, 1989.
6. O'Reilly, R. C., Munakata, Y., and McClelland, J. L., *Cognitive Neuroscience: A Computational Exploration*, MIT Press, Cambridge, 2000.
7. McLeod, P., Plunkett, K., and Rolls, E. T., *Introduction to Connectionist Modeling Of Cognitive Processes*, Oxford University Press, New York, 1998.
8. Amaral, D. G., Ishizuka, N., and Claiborne, B., Neurons, numbers and the hippocampal network, *Prog. Brain Res.*, 83, 1, 1990.
9. Johnston, D. and Amaral, D. G., Hippocampus, in *Synaptic Organization of the Brain*, Shepherd, G. M., Ed., Oxford University Press, New York, 1997.
10. Haberly, L. B., Neuronal circuitry in olfactory cortex: anatomy and functional implications, *Chem. Senses*, 10, 219, 1985.
11. Haberly, L. B. and Bower, J. M., Olfactory cortex: model circuit for study of associative memory? *Trends Neurosci.*, 2, 258, 1989.
12. Wilson, M. A. and Bower, J. M., Cortical oscillations and temporal interactions in a computer simulation of piriform cortex, *J. Neurophysiol.*, 67, 981, 1992.
13. Freund, T. F. and Buzsaki, G., Interneurons of the hippocampus, *Hippocampus*, 6, 347, 1996.
14. Fiala, J. C., Grossberg, S., and Bullock, D., Metabotropic glutamate receptor activation in cerebellar Purkinje cells as substrate for adaptive timing of the classically conditioned eye-blink response, *J. Neurosci.*, 16, 3760, 1996.
15. Hopfield, J. J., Neurons with graded responses have collective computational properties like those of two-state neurons, *Proc. Natl. Acad. Sci. U.S.A.*, 81, 3088, 1984.
16. Amit, D. J., *Modeling Brain Function: The World of Attractor Neural Networks*, Cambridge University Press, Cambridge, 1989.
17. Traub, R. D., Wong, R. K., Miles, R., and Michelson, H., A model of a CA3 hippocampal pyramidal neuron incorporating voltage-clamp data on intrinsic conductances, *J. Neurophysiol.*, 66, 635, 1991.
18. Barkai, E. and Hasselmo, M. E., Modulation of the input/output function of rat piriform cortex pyramidal cells, *J. Neurophysiol.*, 72, 644, 1994.
19. Menschik, E. D. and Finkel, L. H., Neuromodulatory control of hippocampal function: towards a model of Alzheimer's disease, *Artif. Intell. Med.*, 13, 99, 1998.

20. Traub, R. D., Miles, R., and Buzsaki, G., Computer simulation of carbachol-driven rhythmic population oscillations in the CA3 region of the *in vitro* rat hippocampus, *J. Physiol*, 451, 653, 1992.
21. Traub, R. D., Whittington, M. A., Colling, S. B., Buzsaki, G., and Jefferys, J. G., Analysis of gamma rhythms in the rat hippocampus *in vitro* and *in vivo*, *J. Physiol.* (London), 493, 471, 1996.
22. Liljenstrom, H. and Hasselmo, M. E., Cholinergic modulation of cortical oscillatory dynamics, *J. Neurophysiol.*, 74, 288, 1995.
23. Maex, R. and De Schutter, E., Synchronization of golgi and granule cell firing in a detailed network model of the cerebellar granule cell layer, *J. Neurophysiol.*, 80, 2521, 1998.
24. Tsodyks, M. V., Skaggs, W. E., Sejnowski, T. J., and McNaughton, B. L., Population dynamics and theta rhythm phase precession of hippocampal place cell firing: a spiking neuron model, *Hippocampus*, 6, 271, 1996.
25. Jensen, O. and Lisman, J. E., Hippocampal CA3 region predicts memory sequences: accounting for the phase precession of place cells, *Learn. Mem.*, 3, 279, 1996.
26. Wallenstein, G. V. and Hasselmo, M. E., GABAergic modulation of hippocampal activity: sequence learning, place field development, and the phase precession effect, *J. Neurophysiol.*, 78, 393, 1997.
27. Skaggs, W. E., McNaughton, B. L., Wilson, M. A., and Barnes, C. A., Theta phase precession in hippocampal neuronal populations and the compression of temporal sequences, *Hippocampus* 6, 149, 1996.
28. Gray, C. M., Konig, P., Engel, A. K., and Singer, W., Oscillatory responses in cat visual cortex exhibit inter-columnar synchronization which reflects global stimulus properties, *Nature*, 338, 334, 1989.
29. Treves, A. and Rolls, E. T., Computational analysis of the role of the hippocampus in memory, *Hippocampus*, 4, 374, 1994.
30. Hasselmo, M. E. and Wyble, B. P., Simulation of the effects of scopolamine on free recall and recognition in a network model of the hippocampus, *Behav. Brain Res.*, 89, 1, 1997.
31. O'Reilly, R. C., Norman, K. A., and McClelland, J. L., A hippocampal model of recognition memory, in *Advances in Neural Information Processing Systems 10*, Jordan, M. I., Kearns, M. J., and Solla, S. A., Eds., MIT Press, Cambridge, 1998.
32. Hasselmo, M. E. and Barkai, E., Cholinergic modulation of activity-dependent synaptic plasticity in rat piriform cortex, *J. Neurosci.*, 15, 6592, 1995.
33. Fransen, E. and Lansner, A., Low spiking rates in a population of mutually exciting pyramidal cells, *Network*, 6, 271, 1995.
34. Shepherd, G. M., *Synaptic Organization of the Brain*, Oxford University Press, New York, 1997.
35. Erwin, E., Obermayer, K., and Schulten, K., Models of orientation and ocular dominance columns in the visual cortex: a critical comparison, *Neural Comput.*, 7, 425, 1995.
36. Troyer, T. W., Krukowski, A. E., Priebe, N. J., and Miller, K. D., Contrast-invariant orientation tuning in cat visual cortex: thalamocortical input tuning and correlation-based intracortical connectivity, *J. Neurosci.*, 18, 5908, 1998.
37. Adorjan, P., Levitt, J. B., Lund, J. S., and Obermayer, K., A model for the intracortical origin of orientation preference and tuning in macaque striate cortex, *Vis. Neurosci.*, 16, 303, 1999.
38. Douglas, R. J., Koch, C., Mahowald, M., Martin, K. A., and Suarez, H. H., Recurrent excitation in neocortical circuits, *Science*, 269, 981, 1995.

39. Barkai, E., Bergman, R. E., Horwitz, G., and Hasselmo, M. E., Modulation of associative memory function in a biophysical simulation of rat piriform cortex, *J. Neurophysiol.*, 72, 659, 1994.
40. Freund, T. F. and Buzsaki, G., Interneurons of the hippocampus, *Hippocampus*, 6, 347, 1996.
41. Hasselmo, M. E., Schnell, E., and Barkai, E., Dynamics of learning and recall at excitatory recurrent synapses and cholinergic modulation in hippocampal region CA3, *J. Neurosci.*, 15, 5249, 1995.
42. Freeman, W. J., *Mass Action in the Nervous System*, Academic Press, New York, 1975.
43. Hasselmo, M. E. and Linster, C., Modeling the piriform cortex, *Cortical Models: Cerebral Cortex*, 12, Jones, E. G. and Ulinski, P. S., Eds., Plenum Press, New York, 1998.
44. Guan, Z.-H. and Chen, C., On delayed implusive Hopfield neural networks. *Neural Networks*, 12, 273, 1999.
45. Wigstrom, H., Gustafsson, B., Huang, Y. Y., and Abraham, W. C., Hippocampal long-term potentiation is induced by pairing single afferent volleys with intracellularly injected depolarizing current pulses, *Acta Phsyiol. Scand.*, 126, 317, 1986.
46. Holmes, W. R. and Levy, W. B., Insights into associative long-term potentiation from computational models of NMDA receptor-mediated calcium influx and intracellular calcium concentration changes, *J. Neurophysiol.*, 63, 1148, 1990.
47. Grossberg, S., Some networks that can learn, remember and reproduce any number of complicated space time patterns, *Stu. Appl. Math*, 49, 135, 1972.
48. Levy, W. B., Colbert, C. M., and Desmond, N. L., Elemental adaptive processes of neurons and synapses: A statistical/computational perspective, in *Neuroscience and Connectionist Theory*, Gluck, M. A. and Rumelhart, D. E., Eds., Erlbaum, Hillsdale, NJ, 1990, 187.
49. Dudek, S. and Bear, M. F., Bidirectional long-term modification of synaptic effectiveness in the adult and immature hippocampus, *J. Neurosci.*, 13, 2910, 1993.
50. Levy, W. B., A sequence predicting CA3 is a flexible associator that learns and uses context to solve hippocampal-like tasks, *Hippocampus*, 6, 579, 1996.
51. Wallenstein, G. V., Eichenbaum, H. B., and Hasselmo, M. E., The hippocampus as an associator of discontiguous events, *Trends Neurosci.*, 21, 317, 1998.
52. LeMasson, G., Marder, E., and Abbott, L. F., Activity-dependent regulation of conductances in model neurons, *Science*, 259, 1915, 1993.
53. Fox, S. E., Wolfson, S., and Ranck, J. B. J., Hippocampal theta rhythm and the firing of neurons in walking and urethane anesthetized rats, *Exp. Brain Res.*, 62, 495, 1986.
54. Chappell, M. and Humphreys, M. S., An auto-associative neural network for sparse representations: Analysis and application to models of recognition and cued recall, *Psych. Rev.*, 101, 103, 1994.
55. Hasselmo, M. E., Rolls, E. T., and Baylis, G. C., The role of expression and identity in the face-selective responses of neurons in the temporal visual cortex of the monkey, *Behav. Brain Res.*, 32, 203, 1989.
56. McCormick, D. A., Connors, B. W., Lighthall, J. W., and Prince, D. A., Comparative electrophysiology of pyramidal and sparsely spiny stellate neurons of the neocortex, *J. Neurophysiol.*, 54, 782, 1985.
57. Abbott, L., Realistic synaptic inputs for model neural networks, *Network*, 2, 245, 1991.
58. Ermentrout, B., Phase-plane analysis of neural activity, in *The Handbook of Brain Theory and Neural Networks*, Arbib, M. A., Ed., MIT Press, Cambridge, 1998.

59. Wilson, H. R. and Cowan, J. D., Excitatory and inhibitory interactions in localized populations of model neurons, *Biophys. J.*, 12, 1, 1972.
60. Wilson, H. R. and Cowan, J. D., A mathematical theory of the functional dynamics of cortical and thalamic nervous tissue, *Kybernetik*, 13, 55, 1973.
61. Tsodyks, M. V., Skaggs, W. E., Sejnowski, T. J., and McNaughton, B. L., Paradoxical effects of external modulation of inhibitory interneurons, *J. Neurosci.*, 17, 4382, 1997.
62. Pinto, D. J., Brumberg, J. C., Simons, D. J., and Ermentrout, G. B., A quantitative population model of whisker barrels: re-examining the Wilson-Cowan equations, *J. Comput. Neurosci.*, 3, 247, 1996.
63. Hirsch, M. W. and Smale, S., *Differential Equations, Dynamical Systems, and Linear Algebra*, Academic Press, New York, 1974.
64. Patil, M. M. and Hasselmo, M. E., Modulation of inhibitory synaptic potentials in the piriform cortex, *J. Neurophysiol.*, 81, 2103, 1999.
65. Bland, B. H., Colom, L.V., Konopacki, J., and Roth, S. H., Intracellular records of carbachol-induced theta rhythm in hippocampal slices, *Brain Res.*, 447, 364, 1988.
66. Bressler, S. L., Spatial organization of EEGs from olfactory bulb and cortex, *Electroencephalogr. Clin. Neurophysiol.*, 57, 270, 1984.
67. Penttonen, M., Kamondi, A., Acsady, L., and Buzsaki, G., Gamma frequency oscillation in the hippocampus of the rat: intracellular analysis *in vivo*, *Eur. J. Neurosci.*, 10, 718, 1998.
68. Buzsaki, G., Hippocampal sharp waves: their origin and significance., *Brain Res.*, 398, 242, 1986.
69. Young, B. J., Otto, T., Fox, G. D., and Eichenbaum, H., Memory representation within the parahippocampal region, *J. Neurosci.*, 17, 5183, 1997.
70. Schoenbaum, G. and Eichenbaum, H., Information coding in the rodent prefrontal cortex. I. Single-neuron activity in orbitofrontal cortex compared with that in pyriform cortex, *J. Neurophysiol.*, 74, 733, 1995.
71. Hasselmo, M. E., Neuromodulation and cortical function: modeling the physiological basis of behavior, *Behav. Brain Res.*, 67, 1, 1995.
72. Ascher, P. and Nowak, L., The role of divalent cations in the N-methyl-D-aspartate responses of mouse central neurones in culture, *J. Physiol.* (London), 399, 247, 1988.
73. Zador, A., Koch, C., and Brown, T. H., Biophysical model of a Hebbian synapse, *Proc. Natl. Acad. Sci. U.S.A.*, 87, 6718, 1990.
74. Kapur, A., Lytton, W. W., Ketchum, K. L., and Haberly, L. B., Regulation of the NMDA component of EPSPs by different components of postsynaptic GABAergic inhibition: computer simulation analysis in piriform cortex, *J. Neurophysiol.*, 78, 2546, 1997.
75. Mayer, M. L. and Westbrook, G. L., Permeation and block of *N*-methyl-D-aspartic acid receptor channels by divalent cations in mouse cultured central neurones, *J. Physiol.*, 394, 501, 1987.
76. De Schutter, E. and Bower, J. M., Sensitivity of synaptic plasticity to the Ca^{2+} permeability of NMDA channels: a model of long-term potentiation in hippocampal neurons, *Neural Comput.*, 5, 681, 1993.

12 Modeling of Interactions Between Neural Networks and Musculoskeletal Systems

Örjan Ekeberg

CONTENTS

12.1 INTRODUCTION

All neuronal structures interact with their surroundings in one way or another. For example, single neurons send and receive synaptic signals, cortical networks communicate with other networks via a multitude of projections, spinal circuits send signals to muscles and receive sensory feedback, etc. In experimental setups, a neuron or a network may sometimes be studied in isolation but in general the

0-8493-2068-2/01/$0.00+$.50

environment must also be taken into account to establish the right conditions for normal operation. When experiments are done *in vitro*, this normally implies that the neuron or network is deprived of the natural input and output and, thus, works in isolation. *In vivo* experiments, on the other hand, may be able to keep most of the surroundings intact.

In the case of simulation studies the same consideration arises. Most models used in computational neuroscience to date correspond to an *in vitro* situation where the operation of the system under study is described and analyzed with no or very limited interaction with surrounding structures. However, there is now a growing need for what could be referred to as situated models, i.e., models that incorporate *in vivo*-like operating conditions. In some cases this may simply be a matter of extending the simulations to include surrounding neuronal structures, perhaps using much simpler neuron models for these peripheral parts. In other cases it may be necessary to incorporate completely new kinds of models.

For neural networks which interact closely with body mechanics via muscles and mechanosensors, models which describe the mechanical movement itself as well as actuators and sensors become imperative. This applies evidently to studies of classical motor functions like, for example, locomotion and posture control. Many sensory systems are also tightly coupled to motor acts and may therefore need a proper representation of the neuro-mechanical interaction. In this chapter we will describe some techniques to simulate the working environment for networks that interact with a mechanical environment. This relates in part to the methods used in the area of biomechanics. We should, however, keep in mind that the objective here is to provide a reasonable representation of the working environment, which means that we may allow ourselves to do rather drastic simplifications to the models of the mechanical subsystem.

In biomechanics it is common to distinguish between kinematic and dynamic descriptions of the motion. A *kinematic model* describes how the different parts of the body move in terms of positions, angles, velocities, etc. Such variables can often readily be measured, for example by using a video camera setup. The kinematic model is primarily useful for describing the movement itself, but is seldom enough for relating the motion to the neural activity. A *dynamic model* is more complete in that it also includes the underlying forces and torques. In most cases, a dynamic model is required to predict the motion that results from neuronal activation of muscles.

In this chapter, we will describe how a dynamic model of the body can be formulated so that it can be used to compute the motion as a result of our simulated motoneuronal activity. We will also discuss how the neuronal and mechanical models can be interconnected via model muscles and mechanoreceptors to form a complete feedback loop.

12.2 THEORETICAL FOUNDATION

In order to assemble a complete model for neuro-mechanical simulation the neuronal models described in earlier chapters may be used but they need to be complemented with models for body mechanics, muscles and, possibly, mechanoreceptors. In this section, we will go through these additional models one by one to provide the necessary pieces to complete this model jigsaw puzzle.

12.2.1 Body Mechanics

Before any mathematical model can be formulated for the body movements, we have to describe the body in a reasonably simplified way. Creatures, in general, come in an almost unlimited number of forms and it is not feasible to build a theoretical framework which covers every possible alternative. Instead, we will focus on a typical modeling situation, bearing in mind that in practice one may have to make several modifications to this framework to adapt it to the problem at hand.

Our primary assumption will be that the body can be regarded as a number of stiff sub-bodies, segments, interconnected via joints. These sub-bodies may typically correspond to limb segments, in the case of limb movements, or even separate bones of the skeleton. For bodies (or parts of the body) that are not naturally segmented in this way, it may, nevertheless, be possible to use the same technique by doing an artificial subdivision into sufficiently small segments. This method has been suggested for, for example, fish and snake bodies.

The segmented view of the body is not suitable for animals that utilize a hydroskeleton such as worms and snails. For these animals the pressure within body sections becomes essential and has to be simulated properly as it is generated by the contraction of several surrounding muscles. We will not go into the details of such models in this text but will focus on the more classical segmented systems.[1]

The neuronal system influences the movements of the linked segments via muscles; here considered in two forms: linear and angular. A linear muscle applies a pulling force between two *insertion points* located on different segments. In reality, most muscles apply force to the skeleton via a complex system of tendons, which means that this simplistic view of a linearly pulling muscle may not be adequate. In many cases it is more appropriate to use an angular model where the muscle directly generates torque around a joint.

In addition to the geometry of the segment model, it is necessary to decide on the physical parameters for each segment. This is comparatively simple since the only parameters needed are the mass and moment of inertia along with the location of the center of gravity. In the case of a full 3-D model, it is also necessary to know the direction of the principal axes of inertia which in practice is often evident from symmetries of the segments. It is generally practical to introduce a local coordinate system for each segment with the origin located at the center of gravity and the coordinate axes pointing in the principal directions of inertia. Things fixed to a segment, particularly joint and muscle insertion points, may then be expressed as constant positions in this local coordinate system.

The actual shape of a segment only becomes relevant in two situations: detection of contact with other objects and for graphical visualization of the resulting motion. Both these situations can normally be taken care of outside of the core simulation machinery.

Another aspect to consider when designing a mechanical model is whether it is necessary to incorporate the full 3-D motion or if a 2-D model is sufficient. In general, the mathematical description as well as the numerical solution becomes substantially simpler in 2-D. In some rare cases it may even be possible to reduce the model to a one-dimensional system. Examples of this include single-joint limb movements and pure horizontal or vertical eye movements.

12.2.1.1 Joints

The motion of completely independent segments would be trivial but not that interesting to simulate. What makes the mechanical model complex is the presence of joints which restrict how the segments may move in relation to each other and introduce *inner forces* which have to be deduced from the system as a whole. Joints can be of several different forms but they all have in common that they impose restrictions on how two joint segments may move in relation to each other. Such a restriction is called a *kinematic constraint* since it constrains the positions and velocities, i.e., the kinematic configuration, which may be reached during the motion.

In technical systems there are both sliding and rotational joints but in biomechanical systems the former are very rare. In a 2-D model, only one type of rotational joint is possible and it simply fixates one point in the first segment to another point in the second. The two segments are still able to rotate in relation to each other around the joint and their relative position can be described by a single variable, typically an angle.

In a 3-D model there are several possible variants of rotational joints and they may all be useful in biological models. The purest 3-D joint is the *spherical joint* which joins two points, as in the 2-D case, without any further restrictions. The human shoulder may be regarded as such a joint. Since the segments are free to turn in many ways around their common point, the relative position is not that trivial to describe. In fact, three angles are necessary to fully specify this kind of joint configuration and if not treated cleverly, this may be the cause of some rather tangled mathematics.

Another prevalent kind of 3-D joint is the *coaxial* or *hinge joint*. Here, the segments are only free to rotate around a common axis, like in the human knee. Again, it is sufficient to use one angle to describe the configuration since only one degree-of-freedom is left open. To characterize a hinge joint it is necessary to specify not only a point but also a direction within the local coordinate system of each segment. This direction forms the common axis of rotation.

A third, less common, type of rotational joint is the *universal joint* where two axes are involved, one from each segment. The motion is constrained such that the angle between the two axes is constant, often at right angle. The universal joint has two degrees-of-freedom and, thus, requires two angles to parameterize.

The constraints imposed by all the joints mentioned can be expressed as a set of algebraic equalities. This is important since it allows a uniform mathematical treatment which is independent of the precise configuration. All rotational joints specify that one point in the first segment coincides with another point in the second. The global coordinates for that point can therefore be expressed in two separate ways, one from the state variables describing the location of the first segment and another correspondingly from the second. As these coordinates have to be identical this constitutes an equality that has to hold during the entire motion. For hinge and universal joints, additional equalities emerge from the requirement on the angle between the two axes. The constant angle may be expressed in terms of a constant scalar product between unit vectors in the two directions, which again constitutes a simple algebraic equality.

12.2.1.2 External Forces

Besides direct muscle forces and inner forces emerging at the joints, segments are normally subject to a multitude of external forces originating from the mechanical surroundings. First of all, most movements take place within the gravity field of the earth. Gravity is easily incorporated into the mathematical model by adding a vertical force at the center of gravity for each segment. For water-living animals, buoyancy counteracts gravity and should be subtracted from the gravity vector. Gravity is proportional to the segment mass while buoyancy is proportional to its volume. Note, however, that the center of gravity and the center of buoyancy do not necessarily coincide. Thus, the two forces may have to be applied at different points.

Direct contact with other objects, such as foot-to-ground contacts during walking or a hand touching an object, is trickier to simulate properly. There are special problems associated with the simulation of contact forces. First, the kinematic constraints depend on whether or not the body is in contact with the object (which may be the ground). Second, the impacts may give rise to shock waves, which are not handled correctly by most numerical integrators.

With the numerical simulation methods we are suggesting here, the kinematic constraints are represented explicitly in a way which makes it possible to add and remove individual constraints while the simulation is running. One way of modeling contact is to add a new joint at the point of contact and include the corresponding constraints dynamically. A special problem is then to determine when this contact point should be removed again. One option is to calculate the contact force and check its direction.

The problem of shock waves arises when the impact forces at touchdown are very high giving rise to short duration, high amplitude forces throughout the body. Such a shock wave almost instantaneously changes the velocities of all segments. In less severe cases, when the body or object is compliant, it may be sufficient to use very small time steps right after the impact. In general, however, it may become necessary to utilize a separate method that explicitly calculates the velocity changes throughout the body. Friction between the body and ground/object is another phenomenon that may have to be taken into consideration but handling this properly becomes rather complex and is left out of our current framework.

For animals moving in water it may also be necessary to take the viscous and inertial forces of the water into account. A similar need arises when studying rapid movements in air, for example insect wing movements. The Navier–Stokes equations provide the mathematical tool for describing this properly. They state the relationship between the local flow field and the pressure (see Reference 2 for an introduction).

Numerical simulation of these equations is, however, no simple matter and generally requires substantial simplifications to be tractable. For larger animals the main problem is that a very thin boundary layer is formed around the body where the flow field changes rapidly. To adequately represent this numerically, extremely dense spatial grids are required and that can easily make the computational burden enormous. State-of-the-art methods for Navier–Stokes simulations rely on the use of non-uniform grids, which move or change shape during the motion.

It may be possible to use simplified models of the forces from the surrounding water. One technique is to ignore all dynamic effects of the flow and simply use the corresponding static drag force, which is proportional to the square of the velocity of each segment. Another technique is to use an "added-mass" model where the mass of each segment is increased to accommodate the water that has to move with it. Which approximation is adequate will vary from case to case and we will therefore not go deeper into this issue.

12.2.2 Muscle Models

The force produced by a muscle is primarily a function of its length, velocity and level of activation. In the most simplistic models the muscle force is assumed to be a linear function of all these variables. In reality the relationship is more complex but linear models are often sufficient for the purposes we are addressing here. The classical model of Hill[3] describes these three linear mechanisms as operating independently in the production of the force. The muscle is regarded as being composed out of three parallel components: a contractile (force generating) component, an elastic component and a viscous component. A serial elastic component may also be included (Figure 12.1).

Measured data on muscle force production are normally presented in terms of force-length and force-velocity diagrams. The parameters for a linear muscle model can then be estimated directly from the slope of the curves in the appropriate part of the diagrams. One parameter that is hard to measure experimentally is the scaling factor between motoneuron activity and force production. In most cases the motoneuron pool is simulated with a reduced number of neurons, which means that there is also a need to scale the activation itself properly. This is typically done by tuning this free parameter so that the resulting force becomes correct for some known level of activation of the motoneurons. Such a known calibration point may be maximal activation, mean activation during a rhythmic task, etc.

In the simplified muscle models we are considering here, the dynamics of the muscle activation is ignored and thus, the instantaneous activation level is used to set the force generated. This simplification, however, becomes a complication when the activation comes from spiking model motoneurons. The spike rate somehow has to be converted into a continuous muscle activation value. This may be achieved by incorporating a single-compartment passive "motor unit" with a proper time constant, which receives the motoneuron output and gives a smooth enough signal to control the instantaneous force.

Pairs of muscles often work in an antagonistic fashion around one joint to be able to generate torques in both directions. This is, for example, the case with flexor-

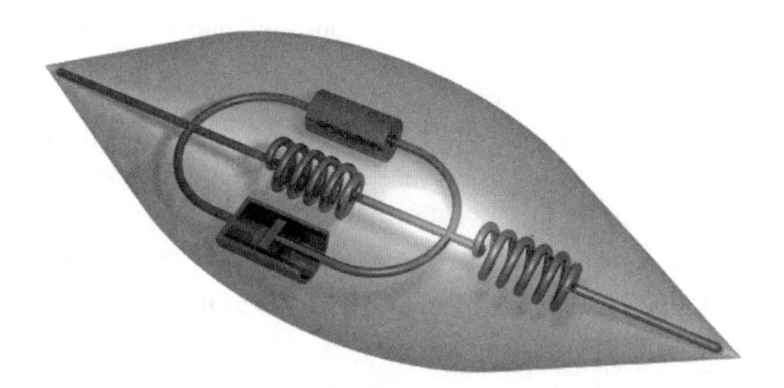

FIGURE 12.1 Schematic showing the parallel contractile, elastic, and viscous components of a model muscle. A serial elastic component is also included. See References 20 and 21 for more details. (Graphics by Tomas Ekeberg.)

extensor pairs of muscles in limb movement where one muscle flexes the joint while another extends it. Such pairs of muscles may be lumped into one force or torque generator in the mechanical model, thus reducing the complexity of the simulation.

12.2.3 Mechanoreceptors

Mechanoreceptors sense features of the mechanical state and translate this into neuronal activity which is fed back into the neuronal network by synaptic connections. In most circumstances it is relatively simple to provide a reasonable model for these receptors. As always, if a detailed account of the receptor properties is needed it may be necessary to use a sophisticated model also for this stage and this may become arbitrarily complex.

The first simplification that should be considered is whether the receptors can be regarded as pure input-output devices or if it is necessary to take some internal state of the receptors into account. In many cases it is sufficient to regard the mechanoreceptors as a simple transformation from mechanical state to neuronal activity levels. This greatly simplifies the simulation since the receptors themselves need not enter the numerical simulation machinery but can be handled separately as a function applied at each time step.

However, receptor activation is often not linear. Saturation effects and other non-linearities are commonplace and can often not be ignored. With a pure input-output model, non-linearities are trivial to incorporate. Measured calibration curves can, for example, be used directly by entering the measured values in a table and interpolate during the simulation to get fairly accurate activation values.

In practice, the trickiest part of incorporating mechanoreceptors may in fact be to get the appropriate mechanical state values in the first place. Most state variables that may influence these receptors are explicitly represented in the mechanical part of the simulation and pose no problem. This is normally the case for such variables as positions, angles and velocities. Other parts of the mechanical state, such as forces, torques, pressures, etc. may not be explicit in the simulation method and then have to be deduced from the variables available.

The output of the receptors needs to be converted into a form suitable as input to our neuronal simulation models. One simple technique is to enter the receptor activation as a membrane current in a modeled sensory neuron. If necessary, it may also be possible to formulate a model where the true activation method of the receptor is mimicked more closely. One example of such a technique is to use a membrane leakage, which depends on the stretch or force applied.

12.3 SIMULATING THE MECHANICAL MODEL

12.3.1 CHOICE OF STATE VARIABLES

First thing to do when putting the mechanical model in mathematical clothes is to choose a kinematic parameterization, i.e., a set of state variables which uniquely defines the positions and velocities of all the segments. There are many ways of doing this parameterization and the choice is important because it affects how complex the mechanical model will be to formulate and simulate.

One seemingly natural choice is to use a set of state variables that only describe motions that are actually possible, that is, in accordance with the kinematic constraints imposed by the joints. This is the case if we, for example, use joint angles to describe the relative position between neighboring segments. In this type of description, each degree of freedom corresponds to one state variable such as an angle or a position. The drawback with this kind of natural parameterization is that the corresponding equations of motion typically become very complex unless the system is extremely simple.

A perhaps less obvious alternative is to use a redundant state representation in combination with a set of algebraic constraints. The redundant representation then has more degrees-of-freedom than the mechanical system itself but the constraints restrict the solutions to a subspace of the correct dimensionality. For example, the position of each segment can be represented with six values (in 3-D) or three values (in 2-D) independently of any joints. The kinematic constraints from the joints are then incorporated by utilizing a solution method, which ensures that the state does not diverge from the allowed configuration.

The advantage with the redundant parameterization technique is that the state variables for each segment can be made independent to make the dynamic equations very simple, for example linear. The drawback is that the differential equations describing the dynamics have to be complemented with a set of algebraic equations imposing the kinematic constraints. The numerical simulation technique used thus has to be able to handle such differential-algebraic equation systems. In practice, the advantages often make the redundant systems the best choice.

The use of angles to represent the orientation of the segments may seem straight forward, but it inevitably introduces trigonometric dependencies into the equations. In 2-D, this is not so much of a problem but in 3-D things may become quite complex. An alternative is then to utilize pure Cartesian coordinates to completely do away with angles.[4] The orientation of each segment is then represented via a set of orthogonal unit vectors, i.e., the base vectors of the local coordinate system. These vectors are stored using Cartesian coordinates in the global coordinate system.

The Cartesian representation is highly redundant. For example, 12 values are used to represent the position and orientation of a segment in 3-D, a system with only 6 degrees of freedom. Extra constraints are therefore required to restrict the solutions to allowed values. These extra constraints emerge from the restriction that base vectors have unit length and are mutually orthogonal. It is straightforward to formulate this as a new set of algebraic constraints that are added to our set of kinematic constraints. Numerically, the same simulation machinery can handle both sets of constraints. One disadvantage with the Cartesian representation is that the number of state variables increases. In practice, however, the simpler equations make up for the burden to compute the extra state variables.

12.3.2 Equations of Motion for a Planar System

We will now look more in detail into how a planar (2-D) mechanical model can be formulated mathematically and simulated numerically. By necessity, this will force us to use a mathematical formalism, which may be hard to follow for a more biologically oriented reader. In particular, we will make use of partial differential equations and compact matrix notation.

As argued earlier, redundant representations are normally preferable because they can make the dynamic equations simpler. In the 2-D case the pure Cartesian representation may not pay off and we generally save that technique for 3-D problems where angles cause more problems. Here, we will use a set of state variables where the position of each segment is described in terms of the location of its center of gravity (x,y) and one additional angle φ determining its orientation (Figure 12.2). That this is a redundant representation should be obvious since we can select values which position the individual segments independently, possibly violating any kinematic constraints enforced by the joints.

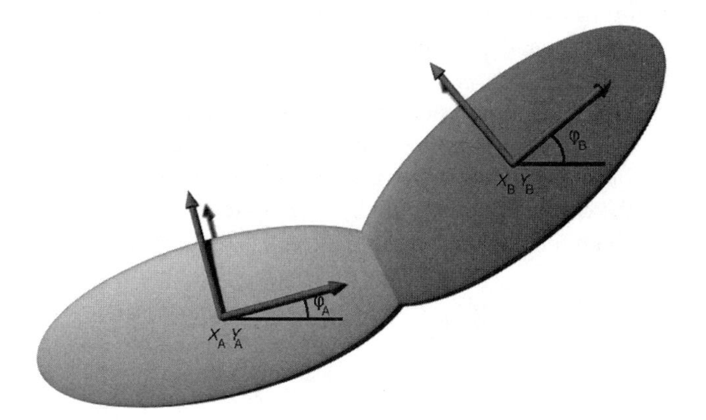

FIGURE 12.2 For a planar system, the position of each segment can be described by three state variables which also defines the location of a local coordinate system. The joint between the two segments is at a fixed position in both local systems. (Graphics by Tomas Ekeberg.)

The dynamics of each segment is given by Newton's equations:

$$m\ddot{x} = f_x$$

$$m\ddot{y} = f_y$$

$$I\ddot{\varphi} = \phi$$

where m is the mass and I the moment of inertia. The right hand sides are the forces f and torques ϕ, which act on the segment. To simplify the mathematical formulation, these equations are often written in matrix form:

$$M\ddot{\bar{q}} = \bar{f}$$

where

$$M = \begin{bmatrix} m & 0 & 0 \\ 0 & m & 0 \\ 0 & 0 & I \end{bmatrix},\ \bar{q} = \begin{bmatrix} x \\ y \\ \varphi \end{bmatrix},\ \bar{f} = \begin{bmatrix} f_x \\ f_y \\ \phi \end{bmatrix}$$

For a system of interconnected segments, the dynamic equations for each segment are simply combined into a larger system of equations of the same form, except that we have to include the inner forces, $\bar{\Gamma}$, acting at the joints.

$$M\ddot{\bar{q}} = \bar{f} + \bar{\Gamma}$$

Incorporation of joints in 2-D is straightforward since there is basically only one type of joint: rotation around a common point. As we have argued above, the kinematic constraint equations should be derived from the fact that this common point may be mathematically expressed in terms of the location of each of the two segments involved in the joint. Let us call the two segments, A and B. The common point is of course fixed within the local coordinate systems of both A and B. To make the equations simpler, we will use polar coordinates $\langle \alpha_A, \beta_A \rangle$, and $\langle \alpha_B, \beta_B \rangle$ for the location of the common point in each of these coordinate systems. Since this is the same point, both the x and y coordinates have to be equal:

$$x_A + \alpha_A \cos(\varphi_A + \beta_A) = x_B + \alpha_B \cos(\varphi_B + \beta_B)$$

$$y_A + \alpha_A \sin(\varphi_A + \beta_A) = y_B + \alpha_B \sin(\varphi_B + \beta_B)$$

This constitutes our algebraic constraint equations corresponding to the joint between segments A and B. In general, by constructing similar equations for all joints and moving all terms to the left hand side, these kinematic constraints can be collected in one compact vector equation:

$$\bar{g}(\bar{q}) = \bar{0}$$

The inner forces $\bar{\Gamma}$ emerge at the joints and are related to the constraint equation we have just formulated. In fact, the vector $\bar{\Gamma}$ is always of the form $G^T\bar{\lambda}$ where G is the Jacobian matrix corresponding to the kinematic constraint equation and $\bar{\lambda}$ is a vector which scales the different constraints appropriately. A Jacobian matrix can be regarded as the generalization of a derivative for vector-valued functions. It is the matrix comprised of all the partial derivatives of each component with respect to each component of the argument vector:

$$G = \begin{bmatrix} \frac{\delta g_1}{\delta q_1} & \frac{\delta g_1}{\delta q_2} & \cdots \\ \frac{\delta g_2}{\delta q_1} & \frac{\delta g_2}{\delta q_2} & \\ \vdots & & \ddots \end{bmatrix}$$

The components of the vector $\bar{\lambda}$ are uniquely defined since they must be chosen so that the constraint equation is fulfilled. Thus, the entire mechanical system can be written as the following differential-algebraic equation system:

$$M\ddot{\bar{q}} = \bar{f}(\bar{q}, \dot{\bar{q}}, t) + G^T(\bar{q})\bar{\lambda}$$

$$\bar{g}(\bar{q}) = \bar{0}$$

12.3.2.1 Numerical Solution Techniques

Since we are faced with a differential-algebraic equation system, we can not simply use an ordinary differential equation solver to simulate the system we have just described. The problem is that the factors $\bar{\lambda}$ are unknown and have to be computed at each time step so that the solution adheres to the constraint equation:

$$\bar{g}(\bar{q}) = \bar{0}$$

Once we have a method to compute the values of $\bar{\lambda}$, any standard differential equation solver, such as the Runge–Kutta method, can be applied to the remaining pure differential equation:

$$M\ddot{\bar{q}} = \bar{f} + G^T\bar{\lambda}$$

Fortunately, the values of $\bar{\lambda}$ can be calculated from the equations we already have. First, we note that the constraints have to be fulfilled at all times. Therefore, we may take derivatives with respect to time of the constraint equation and still have a valid equation. The first derivative yields:

$$G\dot{\bar{q}} = \bar{0}$$

The second time derivative gives us:

$$G\ddot{\bar{q}} + \dot{G}\dot{\bar{q}} = \bar{0}$$

The trick is now to eliminate $\ddot{\bar{q}}$ by substituting $\ddot{\bar{q}}$ from the dynamics equation, i.e.,

$$\ddot{\bar{q}} = M^{-1}\left(\bar{f} + G^T\bar{\lambda}\right),$$

which results in:

$$GM^{-1}\left(\bar{f} + G^T\bar{\lambda}\right) + \dot{G}\dot{\bar{q}} = \bar{0}$$

or

$$\left(GM^{-1}G^T\right)\bar{\lambda} = -GM^{-1}\bar{f} - \dot{G}\dot{\bar{q}}.$$

This may look complicated but in fact this is a simple linear equation system where our unknowns $\bar{\lambda}$ can easily be solved using standard techniques, e.g., Gaussian elimination. The only potential problem with this equation is that the inverse of the mass matrix M is needed. The way M has been constructed, however, guarantees that it is a diagonal matrix, which is trivial to invert.

Once $\bar{\lambda}$ is known, it is straightforward to integrate the dynamic equation

$$M\ddot{\bar{q}} = \bar{f} + G^T\bar{\lambda}$$

over time using conventional techniques.

12.3.2.2 Potential Problems with the Numerical Solution

There is one serious problem associated with the solution method just described. If the solution for some reason diverges from the kinematic constraints, there is no guarantee that the solution method will bring the values back toward kinematically correct solutions again. In fact, numerical errors will always enter the solution and gradually accumulate until the simulation eventually breaks down due to instabilities.

The underlying reason behind this problem is that we have not been utilizing the kinematic constraints directly but only via their second time derivatives. This means that the accelerations will follow the constraints while the positions and velocities may drift away so that, for example, two segments are no longer coinciding at a common joint. In fact, the drift of the velocities is far more problematic and generally needs to be compensated for unless the simulation time is very short.

Fortunately, there is a rather simple cure to this problem suggested by Alishenas.[5,6] The idea is to bring the positions and velocities back to the subspace defined by the constraints by projecting them to the closest point in this subspace. It is generally sufficient to do this for the velocities so we will restrict our description to this method here.

Let's assume that we, after some time of simulation, have the velocities $\dot{\bar{q}}$, which are not perfectly coherent with the constraints. We may then compute a vector $\bar{x}$ by solving

$$\left(GM^{-1}G^T\right)\bar{x} = G\dot{\bar{q}}.$$

Note that the matrix $(GM^{-1}G^T)$ has already been used in the ordinary computations and need not be computed again. The vector $\bar{x}$ is now used to calculate a new set of velocities $\dot{\bar{q}}^*$:

$$\dot{\bar{q}}^* = \dot{\bar{q}} - M^{-1}G^T\bar{x}.$$

These new velocity values are guaranteed to meet the constraints while being as close to $\dot{\bar{q}}$ as possible. They may therefore replace $\dot{\bar{q}}$ in the dynamic simulation. This velocity stabilization is comparatively inexpensive in terms of computation. Furthermore, it is not necessary to apply this projection at every time step since it takes a while for the errors to accumulate. Application at every tenth step is typically sufficient to ensure a stable simulation.

12.3.3 Equations of Motion for a 3-D System

The simulation technique we have just described for a planar mechanical system can in principle also be used in the 3-D case but many things in the mathematical formulation become far more complex. In particular, the use of angles for segment orientation results in complex trigonometric dependencies. We will here sketch an alternative method, which in general is better for 3-D systems.

As argued before, we can use a pure Cartesian parameterization of the mechanical state. The location of each segment is then described by twelve variables: three for the location of the center of gravity and three coordinates for each of the three directions corresponding to the principal axes of inertia (Figure 12.3).

$$\bar{q} = \left[x \; y \; z \; \bar{e}_x^{(x)} \; \bar{e}_y^{(x)} \; \bar{e}_z^{(x)} \; \bar{e}_x^{(y)} \; \bar{e}_y^{(y)} \; \bar{e}_z^{(y)} \; \bar{e}_x^{(z)} \; \bar{e}_y^{(z)} \; \bar{e}_z^{(z)}\right]^T$$

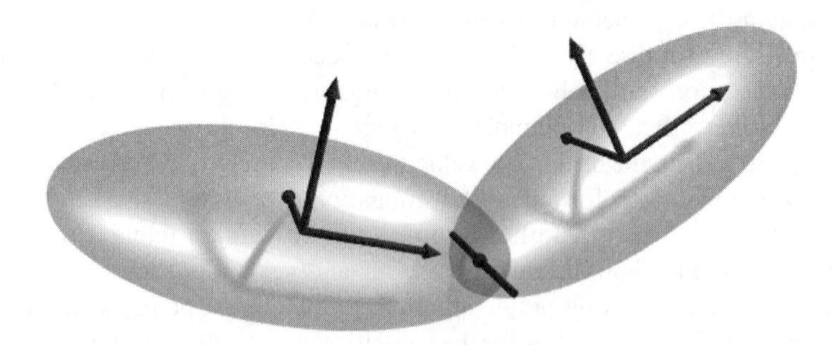

FIGURE 12.3 The location of each segment is described by the position of the center of gravity and the orientation of the base vectors defining the local coordinate system. Here, a hinge joint is constraining the relative movements of the two segments by stipulating one common point and one common axis. (Graphics by Tomas Ekeberg.)

Here, we have introduced the unit vectors $\bar{e}^{(x)}$, $\bar{e}^{(y)}$, and $\bar{e}^{(z)}$, which form the basis for the local coordinate system.

The equation of motion still has the form

$$M\ddot{\bar{q}} = \bar{f}$$

Provided that the unit vectors $\bar{e}^{(x)}$, $\bar{e}^{(y)}$, and $\bar{e}^{(z)}$ are selected to point in the principal directions of inertia, the mass matrix becomes a simple diagonal matrix:

$$M = diag\left(m\ m\ m\ I_x\ I_x\ I_x\ I_y\ I_y\ I_y\ I_z\ I_z\ I_z\right)$$

The force vector $\bar{f}$ now has the following interpretation:

$$\bar{f} = \left[f_x\ f_y\ f_z\ \phi_x^{(x)}\ \phi_y^{(x)}\ \phi_z^{(x)}\ \phi_x^{(y)}\ \phi_y^{(y)}\ \phi_z^{(y)}\ \phi_x^{(z)}\ \phi_y^{(z)}\ \phi_z^{(z)}\right]^T$$

where the different ϕ's represent external torques around various axes.

This representation has twelve variables describing the location of each segment and needs extra constraints to limit the degrees of freedom to six, which is what is expected for a fixed object in 3-D space. The constraints are obtained from the requirement that the vectors $\bar{e}^{(x)}, \bar{e}^{(y)}$, and $\bar{e}^{(z)}$ should be orthogonal unit vectors:

$$\begin{aligned}
\bar{e}^{(x)} \cdot \bar{e}^{(x)} &= 1, & \bar{e}^{(x)} \cdot \bar{e}^{(y)} &= 0, \\
\bar{e}^{(y)} \cdot \bar{e}^{(y)} &= 1, & \bar{e}^{(y)} \cdot \bar{e}^{(z)} &= 0, \\
\bar{e}^{(z)} \cdot \bar{e}^{(z)} &= 1, & \bar{e}^{(z)} \cdot \bar{e}^{(x)} &= 0,
\end{aligned}$$

These extra constraints are added to the kinematic constraints derived from the joints to form a $\bar{g}$-function which includes all the constraints. Thus, the constraints are condensed down to the same form as before:

$$\bar{g}(\bar{q}) = \bar{0}$$

12.3.3.1 Numerical Solution

Simulations of complex mechanical systems are traditionally caught in a computational dilemma. Counteracting forces that vary rapidly with position or velocity often cause problems because even small numerical errors give rise to large reacting accelerations. This constitutes what is known as a *stiff system* and makes it necessary to take extremely small time steps to ensure numerical stability unless *implicit* numerical integration techniques are employed. Implicit methods, however, generally require a full non-linear equation system to be solved at each step. For a complex mechanical system this may be an overwhelming task.

A traditionally separate problem is that the constrained dynamic motion is described by a differential-algebraic system, which needs to be resolved. In the 2-D method described above we solved this by computing the unknown factors $\bar{\lambda}$. A simpler way to handle the constraints would be to regard them as very stiff counteracting forces, which would ensure that the motion proceeds according to the constraints. The mathematics then becomes simpler but we definitely end up with a stiff system, even if the underlying mechanical system is not stiff by itself. Thus, this really necessitates the use of implicit integration.

An elegant way to solve this dilemma has been suggested by de Jalón and coworkers.[7,8] They noted that by using the Cartesian representation, all the constraints emerging from joints of any kind are simple second order equations. The Jacobian matrix, being the partial derivatives of the constraints, is therefore always a linear function of the positions. This makes the implicit integration techniques tractable and, thus, makes it possible to utilize the straightforward technique of stiff counteracting forces to handle the constraints.

The traditional way of integrating the equations of motion is to first solve the accelerations and then use them to calculate the velocities and positions. Problems with drift and instabilities can, however, be avoided by instead using the positions as primary variables. To achieve this, velocities and accelerations are replaced by appropriate difference approximations:

$$\dot{\bar{q}} = \frac{2}{h}\bar{q} - \hat{\dot{q}} \qquad \text{where } \hat{\dot{q}} = \dot{\bar{q}} + \frac{2}{h}\bar{q}$$

$$\ddot{\bar{q}} = \frac{4}{h^2}\bar{q} - \hat{\ddot{q}} \qquad \text{where } \hat{\ddot{q}} = \ddot{\bar{q}} + \frac{4}{h}\dot{\bar{q}} + \frac{4}{h^2}\bar{q}$$

Here, h is the time step and $\hat{\dot{q}}$ and $\hat{\ddot{q}}$ are expressions calculated from the state variables of the previous time step. Note that the positions and velocities of the step only enter as constants. The only remaining unknowns are the positions at the next time step, which have to be solved from the equations of motion.

The implicit solution method typically involves two stages. First a *predictor* step is taken which gives a reasonable guess for the positions in the next time frame. This can typically be a simple extrapolation using the known positions and velocities from the previous step. Next, a *corrector* is applied which improves the new position iteratively by means of Newtons method.[9] Normally, only a single iteration is necessary.

12.4 SOFTWARE

There are a number of commercial software packages available for mechanical multi-body simulations. The most famous product of this sort is the ADAMS simulator (*http://www.adams.com*). The primary market for these products is the mechanical industry and there is generally little support for biomechanical models per se. There are a few products on the market aimed especially for biomechanical modeling. One example is SIMM "Software for Musculoskeletal Modeling" (*http://www.musculographics.com*). Typically, these packages are not designed to support interaction with a neuronal model, normally running in another simulator. One would, however, expect future versions of these commercial packages to support our needs as the interest from both neuroscience and biomechanics is growing. At present, however, we are forced to either write our own simulators or arrange a clever interface to the existing packages.

12.5 EXAMPLES

12.5.1 Lamprey Swimming

The lamprey has been used as a model animal for vertebrate locomotion in a number of studies.[10] The spinal rhythm generating neuronal network has been characterized in detail and constitutes one of the best known vertebrate circuits. A number of simulation studies have also been done of this system from single neuron models[11,12] up to extensive network simulations.[13,14] Since the spinal circuit communicates directly with the mechanics via muscles and mechanoreceptors a natural extension of the simulations was to incorporate also the mechanical subsystem.[15,16]

The swimming motion of the lamprey is basically horizontal: lateral undulations of the body travel from head to tail with increasing amplitude and propel the body forward. In the neuro-mechanical simulation studies a planar mechanical model was therefore used, where the motoneurons activate longitudinal muscles located on each side of the body. The mechanical model is based on a chain of interconnected segments with muscles pulling on each side in an antagonistic fashion. Stretch receptors sensing the curvature of the body synapse into the rhythm generating circuit. Linear, stateless models were used for both muscles and mechanoreceptors.

The mechanical part of the simulation was handled with the technique we have just described for planar systems. One extra complication was that the forces from the surrounding water could not be neglected in order to mimic the true motion pattern. Since the lamprey is several decimeters long while the boundary layer of quickly varying water velocities is only a fraction of a millimeter, using the Navier–Stokes equations was considered too risky. Instead, the static drag force was used as an approximation, which is also much simpler to compute from the local velocities of the body segments.

The neuronal system itself was modeled with simple non-spiking cells to be able to simulate the activity along the entire spinal cord. The parameters for these simplified neurons where in fact derived from earlier more detailed simulation studies comprising multi-compartment neurons with appropriate channel dynamics.

The integrated neuro-mechanical model of the lamprey could be used to address a number of scientific questions. The role of the sensory feedback could be studied by tampering with the feedback and studying the effects. The effect of different kinds of control input from the brainstem could be mimicked to explain how various motor behaviors come about. Later, a 3-D model of lamprey swimming has been constructed to make neuro-mechanical studies of pitch and roll turns possible.[17]

12.5.2 Mammalian Walking

The neuronal control of four-legged walking is a complex task, which involves a number of simultaneously operating mechanisms. Only fractions of the neuronal circuitry is known in any detail while a considerable amount of knowledge is available on the systems level. Neuro-mechanical simulations can here play a role as a way to put known neuronal mechanisms in a context to see whether they are sufficient to explain the observed behavior and perhaps indicate where to search for missing pieces of information.

In a series of simulation studies of mammalian walking, manually designed neuronal circuits were used to implement mechanisms known from experimental data.[18,19] These networks are critically dependent on sensory feedback for their operation, which made it necessary to incorporate the mechanical system into the simulations. While the neuronal models as such are general enough to describe many different species of walking mammals, it was necessary to fix the parameters of the mechanical system to describe an animal with realistic proportions. Since most experimental data on walking is available from the cat, model parameters, such as segment weights and sizes, where set to resemble a cat.

The body was regarded as a system of nine segments, two for each leg and one for the trunk. Knee joints where represented by hinge joints, thereby restricting the knee motion to a back-forward swing. The hip was a two degree-of-freedom universal joint, which allowed both back-forward and sideways (abduction-adduction) motions. Each leg was actuated by three angular muscles, each representing the net effect of several real muscles working around a common joint. The muscle parameters where linearly controlled directly from the simulated motoneuron output. Mechanoreceptors were incorporated to sense joint angles and ground contact. This mechanical information was conveyed via transfer functions to neuronal activation levels in special sensory neurons, which could then synapse onto the rest of the network.

A Cartesian representation was used for the mechanical system to avoid the more complex mathematics of a state description built on angles. The numerical integration technique described by de Jalón was utilized with additions to handle ground contact. The mechanical simulation and the neuronal controller were run as two separate communicating programs on a workstation and the resulting motion could be observed in close-to-realtime in a graphics window. One advantage with a system running that fast is that it becomes possible to interact with the simulation, e.g. by perturbing the motion and observing the reaction.

This system has been used to address questions regarding, e.g., coordination of muscle activation for a single leg at different walking speeds, leg coordination and posture stabilization.

REFERENCES

1. Alschers, C., Numerische Metoden zur Lösung differentiell-algebraischer Gleichungen mit einer Anvendung auf das Hydroskelett. Diplomarbeit, Fakultät für Matematik, Universität Bielefeld, Germany, 1995.
2. Massey, B. S., *Mechanics of Fluids*, 5th ed., van Nostrand Reinhold, UK, 1983.
3. Hill, A., The heat of shortening and the dynamic constants of muscle, *Proc, R. Soc. London, Ser. B*, 126, 136, 1938.
4. de Jalón, J. G., Unda J. and Avello, A., Natural coordinates for the computer analysis of multibody systems, *Comput. Meth. Appl. Mech. Engin.*, 56, 309, 1986.
5. Alishenas, T., Zur numerishen Behandlung, Stabilisierung durch Projection und Modellierung mechanischer Systeme mit Nebenbedingungen und Invarianten. Ph.D. thesis, Royal Institute of Technology, Stockholm, 1992.
6. Ólafsson, Ö. and Alishenas, T., *A Comparative Study of the Numerical Integration and Velocity Stabilization of Two Mechanical Test Problems*, Tech. Rep, TRITA-NA-9210, Dept. Numerical Analysis and Computing Science, Royal Institute of Technology, Stockholm, Sweden, 1992.
7. Bayo, E., de Jalón, J. G., Avello, A., and Cuadrado, J., An efficient computational method for real time multibody dynamic simulations in fully cartesian coordinates, *Comput. Meth. Appl. Mech. Engin.*, 92, 377, 1991.
8. de Jalón, J. G. and Bayo, E., *Kinematic and Dynamic Simulations of Multibody Systems*, Springer-Verlag, Heidelberg, 1993.
9. Press, W. H., Teukolsky, S. A., Vetterling, W. T., and Flannery, B. P., *Numerical Recipies in Fortran 77: The Art of Scientific Computing*, 2nd ed., Cambridge University Press, Cambridge, 1992.
10. Grillner, S., Deliagina, T., Ekeberg, Ö., Manira, A. E., Hill, R. H., Lansner, A., Orlovsky, G., and Wallén, P., Neural networks that coordinate locomotion and body orientation in lamprey, *Trends Neurosci.*, 18, 270, 1995.
11. Ekeberg, Ö., Wallén, P., Lansner, A., Tråvén, H., Brodin, L., and Grillner, S., A computer based model for realistic simulations of neural networks. I: The single neuron and synaptic interaction, *Biol. Cybern.*, 65, 81, 1991.
12. Brodin, L., Tråvén, H., Lansner, A., Wallén, P., Ekeberg, Ö., and Grillner, S., Computer simulations of N-methyl-D-aspartate (NMDA) receptor induced membrane properties in a neuron model, *J. Neurophy.*, 66, 473, 1993.

13. Wallén, P., Ekeberg, Ö., Lansner, A., Brodin, L., Tråvén, H., and Grillner, S., A computer-based model for realistic simulations of neural networks. II: The segmental network generating locomotor rhythmicity in the lamprey, *J. Neurophy.*, 68, 1939, 1992.
14. Wadden, T., Hellgren-Kotaleski, J., Lansner, A., and Grillner, S., Intersegmental co-ordination in the lamprey — simulations using a continuous network model, *Biol. Cybern.*, 76, 1, 1997.
15. Ekeberg, Ö., A combined neuronal and mechanical model of fish swimming, *Biol. Cybern.*, 69, 363, 1993.
16. Ekeberg, Ö. and Grillner, S., Simulations of neuromuscular control in lamprey swimming, *Philos. Tran. Royal Soc. London, Ser. B*, 354, 895, 1999.
17. Ekeberg, Ö., Lansner, A., and Grillner, S., The neural control of fish swimming studied through numerical simulations, *Adap. Behav.*, 3, 363, 1995.
18. Wadden, T. and Ekeberg, Ö., A neuro-mechanical model of legged locomotion: single leg control, *Biol. Cybern.*, 79, 161, 1997.
19. Wadden, T., *Neural Control of Locomotion in Biological and Robotic Systems*, Ph.D. thesis, Royal Institute of Technology, Stockholm, 1998.
20. Zajac, F. E., Muscle and tendon: properties, models, scaling, and application to biomechanics and motor control, *Crit. Rev. Biomed. Eng.*, 17, 359, 1989.
21. Winter, D. A., *Biomechanics and Motor Control of Human Movement*, John Wiley & Sons, New York, 1990.

Index

D

G

H

I

J

K

L

N

S

T

U

V

W

X

Z

LIMITED WARRANTY

CRC Press LLC warrants the physical disk(s) enclosed herein to be free of defects in materials and workmanship for a period of thirty days from the date of purchase. If within the warranty period CRC Press LLC receives written notification of defects in materials or workmanship, and such notification is determined by CRC Press LLC to be correct, CRC Press LLC will replace the defective disk(s).

The entire and exclusive liability and remedy for breach of this Limited Warranty shall be limited to replacement of defective disk(s) and shall not include or extend to any claim for or right to cover any other damages, including but not limited to, loss of profit, data, or use of the software, or special, incidental, or consequential damages or other similar claims, even if CRC Press LLC has been specifically advised of the possibility of such damages. In no event will the liability of CRC Press LLC for any damages to you or any other person ever exceed the lower suggested list price or actual price paid for the software, regardless of any form of the claim.

CRC Press LLC specifically disclaims all other warranties, express or implied, including but not limited to, any implied warranty of merchantability or fitness for a particular purpose. Specifically, CRC Press LLC makes no representation or warranty that the software is fit for any particular purpose and any implied warranty of merchantability is limited to the thirty-day duration of the Limited Warranty covering the physical disk(s) only (and not the software) and is otherwise expressly and specifically disclaimed.

Since some states do not allow the exclusion of incidental or consequential damages, or the limitation on how long an implied warranty lasts, some of the above may not apply to you.